Sensors, Transducers, & LabVIEW

ISBN 0-13-081155-6

VIRTUAL INSTRUMENTATION SERIES

Lisa K. Wells • Jeffrey Travis

- LabVIEW For Everyone

Mahesh L. Chugani • Abhay R. Samant • Michael Cerra

- LabVIEW Signal Processing

Rahman Jamal • Herbert Pichlik

- LabVIEW Applications

Barry Paton

- Sensors, Transducers & LabVIEW

Prentice Hall PTR, Upper Saddle River, New Jersey 07458
http://www.phptr.com

Library of Congress Cataloging-in-Publication Data
Paton, Barry E.
Sensors, transducers, and LabView / Barry E. Paton.
p. cm.
Includes index.
ISBN 0-13-081155-6
1. Transducers—Computer simulation. 2. Detectors—Computer simulation. 3. LabVIEW. I. Title.
TK7872,T6P34 1998
681′.2—dc21 98-19173
CIP

Acquisitions editor: Bernard M. Goodwin
Cover designer: Bruce Kenselaar
Cover designer director: Jerry Votta
Manufacturing manager: Alexis R. Heydt
Marketing manager: Kaylie Smith
Compositor Production services: Pine Tree Composition, Inc.

Prentice-Hall, Inc.
A Simon & Schuster Company
Upper Saddle River, New Jersey 07458

Prentice Hall books are widely used by corporations and government agencies for training, marketing, and resale.

The publisher offers discounts on this book when ordered in bulk quantities. For more information contact:

Corporate Sales Department
Phone: 800-382-3419
Fax: 201-236-7141
E-mail: corpsales@prenhall.com
Or write:
Prentice Hall PTR
Corp. Sales Dept.
One Lake Street
Upper Saddle River, New Jersey 07458

Printed in the United States of America

10 9 8 7 6 5 4 3 2 1

ISBN: 0-13-081155-6

Prentice-Hall International (UK) Limited, *London*
Prentice-Hall of Australia Pty. Limited, *Sydney*
Prentice-Hall Canada Inc., *Toronto*
Prentice-Hall Hispanoamericana, S.A., *Mexico*
Prentice-Hall of India Private Limited, *New Delhi*
Prentice-Hall of Japan, Inc., *Tokyo*
Simon & Schuster Asia Pte. Ltd., *Singapore*
Editora Prentice-Hall do Brasil, Ltda., *Rio de Janiero*

To my girls—
Janet, Tara, Cassie
and my son, Mark,
another digital wizard
for all of their support and understanding

Contents

2 Invisible Fields 19

3 Random Numbers 39

▼12

Beer's Law: Determination of the Concentration of Impurities in a Liquid 209

▼13

Hunt for Red October 227

Foreword

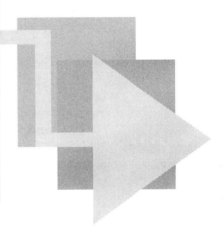

In 1975, the Intel Corporation announced the single chip microcomputer. This was the first time that the four parts of a computer; the central processing unit, the memory, the input/output drivers and the clock circuitry had been fabricated on a single piece of silicon. A functioning computer could now be built with just this chip, an external crystal, a resistor, a capacitor and a 5 volt power supply. The I/O drivers were able to sink sufficient current so that a small light emitting diode could be used for output and a single wire shorted to ground could form the simplest input. Many scientists, engineers and instrument designers across the country could see the new vision of computer controlled instruments, sensors and transducers. Within two years, at Dalhousie University, a new grass roots course in microcomputer instrumentation was born. It was called "Microcomputers in the Real World," a title chosen to emphasis the hands-on nature of computer instrumentation. The primary goal of this course was to introduce engineers and scientists to advantages of computer control in experimental science. It soon became one of the most popular junior level courses.

The course encompassed a variety of technologies and skills including the study of

sensors and transducers
computer interfacing
data acquisition and logging
data analysis
and data presentation and reporting.

In the early years only a few of these goals could be achieved. Our earliest efforts used an Intel 8080 processor and homemade analog-to-digital boards. Just to write the code to acquire a single channel of pressure information over a period of time was a great achievement. Later as memory became cheaper and better software tools were developed, more and more of our goals could be accomplished within a half semester course.

In 1988, national Instruments introduced LabVIEW 2.0. It was used in my research lab for data acquisition, analysis, and presentation. It was clear from its inclusion in the research program that LabVIEW had all the potential to achieve our undergraduate requirements for computer interfacing. However, it was still too expensive for the undergraduate labs. We had to wait till 1996 when National Instruments developed LabVIEW Student Edition and the excellent resource guide by Lisa Wells. Suddenly, it was now possible to achieve all of our goals in a single half semester course consisting of three hours lecture and three hours lab per week. Several other developments had also occurred over the intervening years. First sensors and transducers had matured. Now reliable and low cost sensors and transducers were available. The operational amplifier now made light work of interfacing sensors to the computer, and the computer revolution had placed high performance computers on our desktop at a cost no one could resist. In 1996, LabVIEW Student Edition was introduced on trial into "Microcomputers in the Real World" as a companion to the usual curriculum. By the end of the trial, it was clear that LabView would be the chosen software for future years. Productivity for students using Lab-

View far outstripped their fellow student's using the older software. In the following year, LabVIEW SE edition was used exclusively. All our long cherished goals were now achievable by all students in the half semester course. Currently we are using LabView 5.0 in the undergraduate labs. Come visit the sensor lab at "http://sensor.phys.dal.ca".

Barry Paton

Preface

Good Stuff to Know Before You Get Started

Sensors, Transducers & LabVIEW presents a new approach to learning microcomputer interfacing. It is designed for people who want to make or see things happen. My premise is that the best way to learn is with *'hands-on-experience'*. From beginning to end, the reader is challenged with interesting exercises and problems drawn from the real world. Not too long ago, we challenged students with interesting problems to contemplate, to model and to calculate. Today, we not only contemplate, model and calculate but can also measure, interface and report.

Areas of study are chosen from a wide range of application fields including physics, chemistry, mathematics, engineering, and medical sciences. Each chapter comes with a library of LabVIEW programs called VIs. *You will find a list of over 170 programs at the back of the book and on the enclosed CD.* These programs are interlaced with the text material so that interaction with LabVIEW is both natural and experience expanding. Besides the VI's used in the text, various *Exercises* and *LabVIEW Challenges* are sprinkled throughout. Exercises are short programming challenges to extend the understanding of a specific point in the discussion and should be completed before continuing in the text. LabVIEW Challenges are more comprehensive exercises requiring further reading, development or thought.

Sensors, Transducers and LabVIEW is a book of solutions. It is packed full of interesting devices, circuits and problems to challenge any person interested in making measurements in the real world.

What is LabVIEW

LabVIEW is simply the most elegant programming language for data acquisition, analysis, simulation or computer control of instruments, techniques or processes. LabVIEW is an acronym for **Lab**oratory **V**irtual **I**nstrument **E**ngineering **W**orkbench and was created by National Instruments as an intuitive and user friendly interface for writing computer programs. LabVIEW is an object oriented language and its style, syntax and data flow is different from conventional linear programming languages. For example, the linear language instruction X=X+1 implies that X is to be replaced by the old value of X incremented by one. In LabVIEW this instruction requires a shift register structure that clearly distinguishes between the new value of X and the old value of X. LabVIEW uses data flow concepts to execute subroutines called sub-VIs. In LabView, a subroutine can not execute until the inputs are satisfied and available. Only then does the processing continue within the subroutine. Consider the following program

```
LOOP X = X + A SIN (θ)
     GOTO LOOP
```

In linear programming, each line is executed one after the other. Each time the loop is executed the program calculates a new value of X. In LabVIEW data flow demands a different structure. Load from Sensors, Transducers and LabVIEW library a program called **Data Flow.vi.** Select **Show Diagram** from the

Windows menu. Click on the Icon with a light bulb. Now run the program by pressing the ⇨ key and watch the action. Small colored balls represent packets of data. Note how the data flow is stopped at a program node until all the data packets arrive

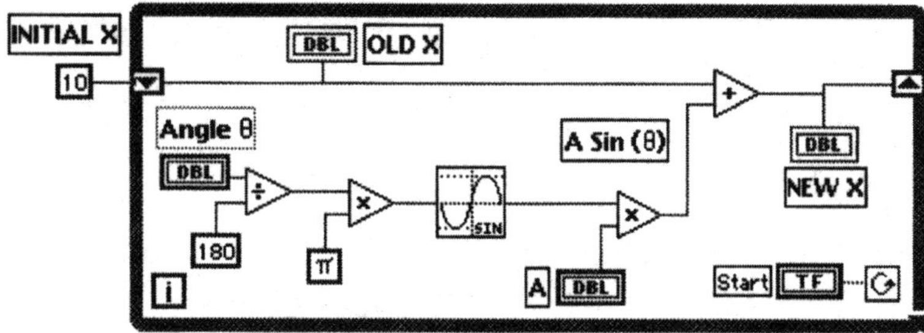

LabVIEW programming is different from conventional linear programming and requires a change in the designer's mind set. However once you experience the advantages of data flow programming, you too will be a fan.

Organization

In the first six chapters of Sensors, Transducers and LabVIEW, this new form of object oriented programming is presented with numerous applications to learn and hone your LabVIEW skills. Areas covered include the LabVIEW environment, modular programming, sub-VI's, programming structures, string operations, array operations and plotting routines. Sensors, Transducers & LabVIEW is about 'doing' things in the real world. Starting in Chapter 4, the input/output drivers for parallel and serial ports are introduced so programs can be transformed into actions. Along the way, numerous sensors (temperature, pressure, and magnetic field) are introduced together with transducers (stepping motors, LED displays, printers and plotters) as a means of solving real world interfacing problems. By the end of the first six chapters, all the basic LabVIEW structures and programming styles have been presented so that advanced concepts, sensors, and challenges may be tackled.

The advanced Chapters 7–15 are presented as LabVIEW projects covering topics such as the electronic barometer, remote video over the WEB, faster scanning, the electronic compass, and so on. All the common interfaces Centronics, RS232C, IEEE488, DAQ cards and TCP/IP are covered within the projects.

Each chapter follows the same format. The first section introduces the subject or the problem to be studied. The second section covers background material, hardware or software information necessary to understand the remainder of the chapter. As an example, in Chapter 2, the physics of Hall effect sensors and the design for a Hall sensor interface are discussed. This section is optional. For a reader who is already sensor or transducer savvy, it can be skipped. Section three simulates the sensor and its interface. The following sections build on measurement skills developed in the early part of the chapter to produce a system solution. Modular programming techniques are used throughout. With bricks and mortar in the early sections, walls are built, then roofs, windows and finally the entire house. LabVIEW is a wonderful language to learn the lesson that mighty programs arise from small modules.

Readers with real sensors and interface cards can use them to link the projects with the real world. For readers without sensors, the simulated sensor VIs given in the library can be used. Every effort has been made to make the simulated sensors look and feel like real sensors. In some cases actual measured data sets are built into the sensor simulation.

Objectives of the Book

This book is written for students, scientists, researchers, and practicing engineers who are interested in learning how to use LabVIEW in a wide range of sensor and transducers applications. This book is written for the novice LabVIEW user who learns best from examples, for the practicing engineers who wants a refresher course on practical LabVIEW applications and for the expert who needs a sensor driver in a hurry. The text is a casebook of interesting sensors, circuits, drivers and applications. Over 150 sample programs can be found in the program library on the CD.

This book expects you to have a basic knowledge of your computer's operating system. You should know how to access menus, use the mouse, open and save files. Although previous LabVIEW experience is not necessary, it is an advantage for the sensor projects in the later chapters. The CD provides LabVIEW 5.0 Evaluation Software with a LabVIEW tutorial, highly recommended for the novice user.

After reading this book and working through the many exercises and challenges, you should be able to

- Understand the workings of a wide range of sensors and transducers
- Design LabVIEW drivers for sensors and transducers
- Understand the differences between serial, parallel, GPIB and TCP/IP interface ports
- Build LabVIEW interfaces for serial, parallel, GPIB and TCP/IP ports
- Create application with a DAQ card
- Appreciate the elegance and simplicity of a graphical programming language
- Understand the power of modular (subVI) programming
- Build sensor and transducer applications simulated or in the real world using the ports and real sensors
- Create system applications

How to Use Sensors, Transducers and LabVIEW

For the scientist, researcher, practicing engineers or technologists

This is definitely a book you will want to read, with your computer by your side. "Reading" is eye opening, "doing" is mind expanding but "using" is real learning. As you read the text, ideas and questions will pop into mind. Stop reading and experiment with LabVIEW. Prove to yourself that events do occur as presented. *Exercises* along the way elucidate some subtle point. Now is the time to try out these ideas. *LabVIEW Challenges,* much like homework, can be done after the material in the chapter has been understood. Some challenges are simple extrapolations of ideas presented in the chapter, others require exploration, further reading and thought. Completed challenges will deepen your understanding of sensors, transducers and especially LabVIEW. You will soon be creating your own challenges.

I wonder if I could add an altimeter to my bicycle?
What would be the altimeter output?
How would I get the signal into my laptop computer?
How could I calculate how much work is done riding up old Smoky?

For the students

You cannot get enough of LabVIEW. The length of the text and CD allows only some of the many properties of LabVIEW to be covered. Read the LabView manuals, study how other applications are coded, but above all explore. Deep in many LabVIEW menus and sub-menus, you will find many interesting functions that will make for efficient programming. When you find one of these little gems, try it out with a demo program to be sure it does exactly what you expect. Icons and even the HELP textboxes allow only limited information and can be confusing. Remember in designing a LabVIEW application to solve a problem, there are many paths to the answer. In the absence of constraints such as the fastest or the most efficient program, all solutions are correct. You can learn a great deal by studying other student's solutions. Adding sensors and transducers to your project makes it come alive. The LabVIEW motto is "It's OK to have fun." My motto is "If it is not fun, it is not Physics". My students and myself have had a lot of "fun" using LabVIEW to instrument and control many applications. You can too.

For the Instructor

The basic concepts of LabVIEW programming and simple input/output devices are presented in the first six chapters. Each chapter requires about two 1 hour lectures. LabVIEW diagrams, sensor or transducer specifications can be embedded into an electronic slide-show or on overhead transparencies. Have LabVIEW running in the classroom so that the static LabVIEW images can be run in real time. I like to have both applications running concurrently so that I can switch back and forth during my lecture. Many of the text ideas are suitable for both virtual and real demonstrations. It's one thing to see a demonstration such as stoplights executing some complex 'advanced flashing green' operation on the computer screen, but the student eye opener is when real lights or colored light emitting diodes start flashing the designed operation.

For the hobbyist

Once you have mastered LabVIEW concepts and built a few instrument drivers, the whole world is your stage. Many electronic parts catalogs and electronic magazines describe all kinds of inexpensive sensors and transducers. Utilizing dedicated microcontrollers such as the SMI102 introduced in Chapter 4 provide a very low cost interface for these sensors. Although limited in input/output capability they are ideal for hobby projects. Let your imagination wander, and LabVIEW will provide an easy path to bring computer control to your favorite application.

Sensors, Transducers & LabVIEW CD

The CD included with this book contains demonstration, example, and solution VIs for each chapter of the text. The most effective way to learn a new programming language is to use it. Have your computer close at hand as you read the text. See an idea, try out that idea and check it out with the examples in the library. If you do not have LabVIEW currently installed on your computer, then you can install LabVIEW 5.0 Evaluation Software included on the CD. You will have unlimited edit time but the runtime is limited to 5 minutes.

The program libraries are simply the quickest method to get you up and running. All program VIs discussed in the text are included in the VI library. You will find a complete listing of the 172 programs near the end of the book and on the CD. These VIs are a template, or a starting point for program development. The teaching strategy is, first to understand, then to imitate, then to improve the VIs. Your challenge is to make them even better. All sensors and transducers are simulated in the text so that no additional hardware is required, other than your computer, and the Sensors, Transducers & LabVIEW CD. However, if you do have real sensors or real transducers and access to the RS232, GPIB or TCP/IP ports you can see *virtual instrumentation in action.* Replace the simulated sensor or transducer VIs with real devices and drivers and let your project 'sing or dance'.

Also included on the CD is over 50 megabytes of Quicktime and Mpeg movies of selected sensors and transducers in action. For best results use Quicktime version 3.0 or later. Enjoy!

■ What You Need to Get Started

Place the Sensors, Transducers & LabVIEW CDROM into your computer (PC or MAC). Install LabVIEW 5.0 Evaluation Software on your computer hard drive. You will need

40 Megabytes of hard disk space
16–32 Megabytes RAM
fast CPU (Pentium, 68040 or PowerMac)

If you have LabVIEW already installed, you will only need one of the four Sensor, Transducers & LabVIEW libraries.

Chose PC Version 4 or Version 5, MAC Version 4 or Version 5
See the ReadMe file on the CD for further comments

Note: It is not necessary to put the libraries onto your hard drive, but response will be faster if you do. Double click on your favorite program and enjoy the benefits of an easy-to-use graphical programming environment.

Conventions used in the text

Convention	Definition
Bold	Bold text denotes VIs, functions menus and menu items For example: **Tools**
>	Angle bracket to the right is used to indicate sub-menus For example: **Functions>Numeric>Trigonometric>Sine**
<>	Text between left and right angle brackets indicate a key For example: <CR> indicates the Return or Enter key
[]	Text between square bracket is used to indicate a button For example: [Start Your Engines] Text between square brackets is also used to indicate a terminal block label or a subVI icon. Examples: [Init] a link labeled Init*ialize.* [AV3] a sub-VI called AV3.
italic	Italic denotes emphasis, key term or concept or a VI Examples: *Friend/Foe.llb.*

About the Author

Dr. Barry Paton is Professor of Physics at Dalhousie University in Halifax, Nova Scotia. His research interests include optics, optical diagnostics, fiber optics, and quantum optics. He has been active in fiber optic sensing and communication networks for underwater instrument pods and ROV's (Remotely Operated Vehicles). He has developed optical diagnostic instruments for the medical field. Recent research activities have focused on optical switching techniques for the 'next' generation of high speed computers. In May 1996, he was involved in an experiment to grow high quality semiconductor crystals aboard the space shuttle Endeavour (STS-77) using telepresence. Dr. Paton is a strong supporter of Applied Science and Engineering and has been a consultant for many Maritime companies. Dr. Paton is an avid outdoor fan enjoying windsurfing, sailing, skiing, and hiking.

Over the past twenty years, Dr. Paton has developed a unique style of teaching microcomputer instrumentation. Students relate best to real world experiences, and this course challenges students to make things happen in the real world. LabVIEW was instrumental in his research laboratory for in excess of eight years and has become the choice programming tool in the undergraduate laboratory. Dr. Paton has also developed a well received 12 week LabVIEW course for Continuing Education. For National Instruments Educational series, Dr. Paton authored *Fundamentals of Digital Electronics,* a LabVIEW approach to studying and simulating digital electronic circuits. It is available from National Instruments, Part Number 321948A.

Lisa Wells, author of two books, *LabVIEW Student Edition User's Guide* and *LabVIEW for Everyone* was excited by Dr. Paton's approach to teaching LabVIEW. A suggestion was made that this unique approach be shared with the printed world. *Sensors, Transducers & LabVIEW: An Application Approach to Virtual Instrumentation* is the result.

Find out more about Dr. Paton's academic activities by visiting his WEB site at sensor.phys.dal.ca or by sending e-mail to barry.paton@dal.ca.

Acknowledgments

The author wishes to thank several generations of students who were able to turn his challenges and ideas into working microcomputer circuits and systems. Without their devotion to detail and an urge to succeed, the challenges would still be just good ideas. Each success made it necessary to find new sensors, new transducers, or new techniques to challenge each new class. These challenges keep me at the forefront of technology.

Lasting thanks go to two summer students, Tim Salzman and Rene Coulomb who turned my ideas and the prototype VIs into the Sensor, Transducers & LabVIEW library.

Very special thanks goes to Lisa Wells whose boundless energy and kind words keep me on course during the critical writing phase.

Much appreciation goes to the folks at National Instruments, Dan Philips and Ravi Marawar for ongoing support, comments and providing LabVIEW 5.0 Evaluation Software found on the CD.

Finally, I would like to thank Bernard Goodwin, Diane Spina from Prentice Hall and Patty Donovan at Pine Tree Composition for turning a set of coffee stained notes and battered Zip disks into a shiny new book.

Overview

As scientists, engineers, technologists, or just the curious, we design experiments to better understand the world around us. Sensors are used to probe and measure physical parameters that describe the world. Transducers stimulate the world in a known way so our sensors can read the response. To automate measurements and responses, a microcomputer drives and controls the sensors and transducers. LabVIEW is the most powerful measurement and control language available to execute the control algorithms and to present the results is a user-friendly format. This chapter introduces the LabVIEW workspace and demonstrates how analog and digital signals are measured and manipulated, all in a friendly and transparent graphical language.

Goals

- Enter the LabVIEW environment
- See the graphical language at work
- Build a digital-to-analog converter
- Take the LabVIEW challenge: Vector Calculator

Key Terms

- LabVIEW workspace
- LabVIEW Tools, Controls, and Functions
- LabVIEW Palettes everywhere
- Analog Amplifier
- Digital Amplifier
- Digital-to-Analog Converter

LabVIEW Environment

1

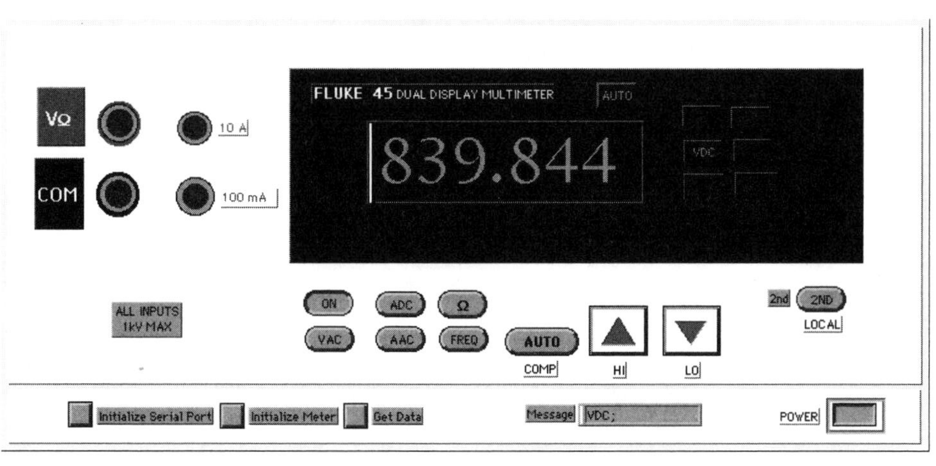

LabVIEW Virtual Instrument for a Popular Digital Voltmeter FLUKE 45

Welcome to the world of LabVIEW and virtual instrumentation. A virtual instrument consists of a computer simulation of a traditional hardware instrument. A window on the computer screen called the front panel replaces and in many cases looks just like the front panel of the real instrument.

When you click the buttons, rotate the knobs, or just turn on the power switch, all the functionality of the real device is simulated on the screen. You can quickly get a feel for the capability and the operation of an instrument. In some cases, the simulation can be used as a training exercise, to gain familiarity, and learn the operation of the instrument. In other cases, groups of instruments are combined to demonstrate measurement techniques, and the simulation becomes a valuable educational tool. Indeed, training on a LabVIEW simulation may be of lower cost than using a hardware simulator.

The difference between simulation and virtual instrumentation is that behind the scenes, through various communication lines (RS232, IEEE488, or TCP/IP), information on the computer screen is connected to a real instrument by software. When you click on a simulated button on the video monitor, the real switch operates. Herein lies the power of LabVIEW. The virtual instrument cannot only simulate the instrument, technique, or process, but also control the real instrument, real technique, or real process from the same front panel. As a teaching tool, the instrument can be simulated and used in a way that it cannot be harmed or harm any other device. Once the skills in operation are learned, the simulation can be connected to a real instrument to provide real-time computer control over that instrument.

The Front Panel

Launching LabVIEW yields two panels or windows labeled with the default titles "Untitled 1" and "Untitled 1 Diagram." The top window is the front panel of the virtual instrument.

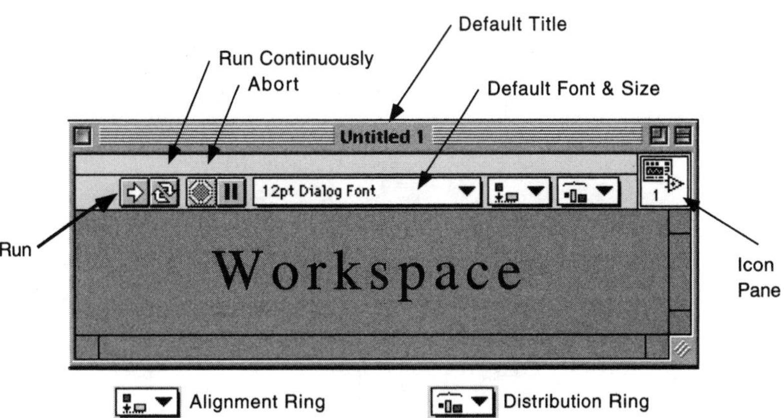

The central gray area is the "Workspace" where the virtual instrument's front panel will be assembled. Scroll bars at the side and across the bottom will be available to center the instrument once components are added to the front panel. Above the Workspace is a toolbar, with different tools to run, stop, or pause the program; set the text font, size, and style; and align components in the workspace; and a LabVIEW default icon pane that ultimately will be the graphical representation of the program. Additional tools and a library of components are necessary to help us build the front panel.

The Front Panel Palettes

A palette is defined as a group of objects that are placed together in a separate window. The palette can be "picked up" by clicking on it and moved to a new location by dragging the palette with the mouse button down. When the mouse button is released, the palette stays at the new location. Two particularly useful palettes are the **Tools** palette and the **Controls** palette. You can show these by clicking on the Windows menu and selecting **Show Tools Palette** and **Show Controls Palette.**

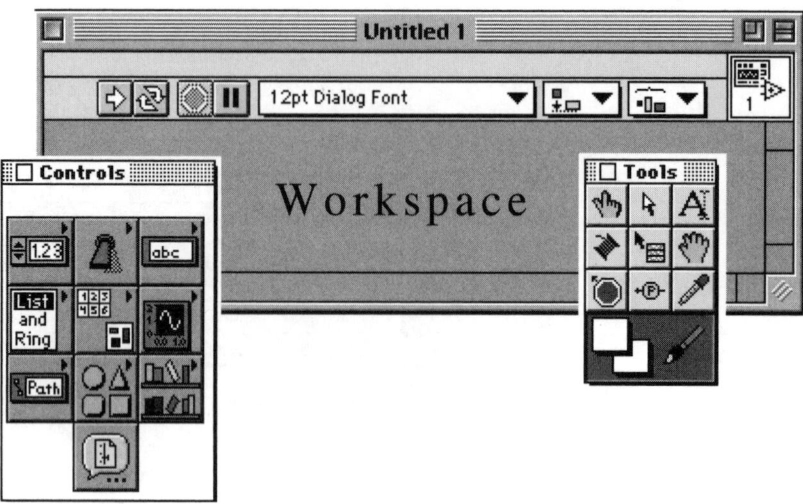

Each palette has a variety of objects or icons that add additional capabilities. Let's look at the **Tools** palette first. Each icon, when selected, changes the cursor to a new function when used in the Workspace. Outside the workspace, the cursor is depicted as an arrow and executes the usual mouse-type operations, opening up a menu, selecting and deselecting an item, highlighting an object, and dragging an object across the window. Inside the workspace, the cursor has multiple functions. Which function it executes is determined by the selection of the tool icon. Here is a short list:

The **Operating** tool lets your finger do the walking through the front panel, changing controls and indicators even when the program is running.

The **Positioning** tool is used primarily for editing the front and diagram panels and allows you to select, move, and resize objects.

The **Text** tool allows you to make and edit free labels that can be used for instructions or titles. The font, style, size, and color of all text boxes can be changed by first selecting the text with the text tool and then selecting the new style.

The **Wiring** tool on the diagram panel connects terminal boxes and functions together by assigning a data path between objects. On the front panel, the wiring tool is used to assign controls and indicators to sub-VI inputs and outputs for converting a program into a module.

The **Pop-up** tool opens up an object's menu to reveal properties than can be changed.

The **Scroll** tool allows the panels to be moved around and viewed differently.

The **Probe** tool creates probes on wires so you can view the data as it passes through the wire.

The **Breakpoint** tool sets points on the VI where the program will pause. It is used for debugging complex programs.

The **Color Copy** tool allows the user to select a color from an existing object. Useful for matching colors between objects so that the same shade is selected.

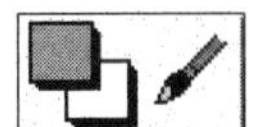

The **Color** tool allows you to select the foreground or background colors of an object from a multitude of shades and hues.

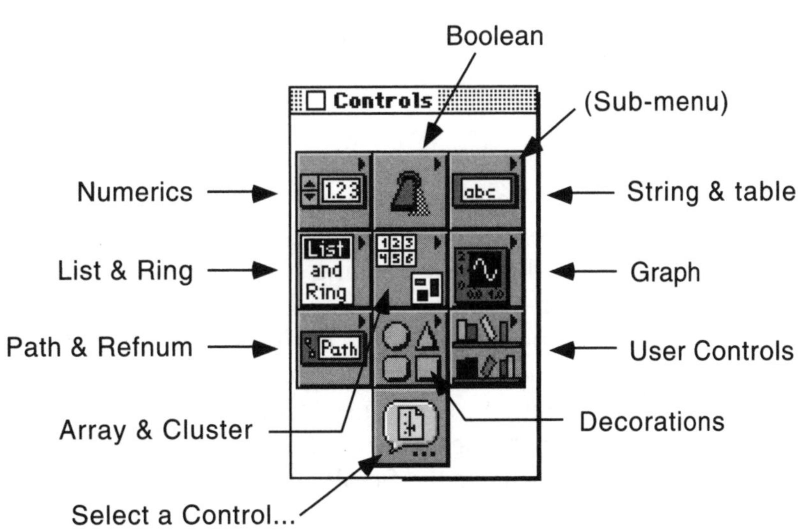

The **Controls** palette contains a wealth of LabVIEW controls and indicators. These are conveniently divided into data types (numerics, Booleans, strings & tables, arrays & clusters, paths & refnums, graphs, and user controls). The remaining button, called **Decorations,** is used to provide simple shapes that can be moved, resized, and colored to add shape and sparkle to your panels. Each palette that contains a black arrow in the top right corner indicates that there are more menus in a subpalette. By clicking on that control button and holding down the mouse button, the associated subpalette appears. For example, observe what happens when the graph button is clicked. Five graph types appear on the subpalette.

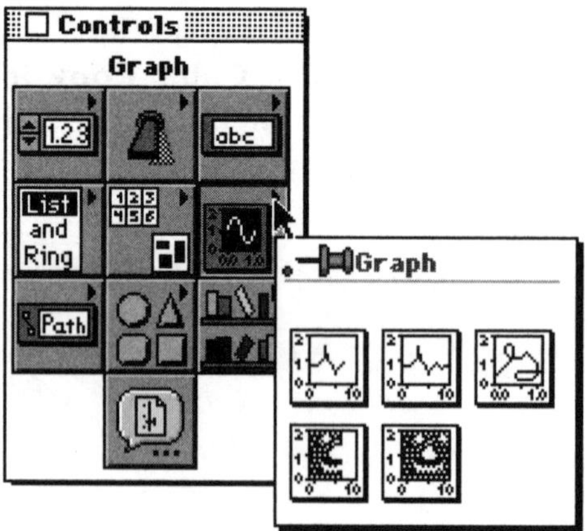

To place a control or indicator on the front panel, it is first necessary to find in a subpalette the appropriate object, then to select it by clicking on that object. The cursor then changes to the **Scroll** tool so you can now drag the object into the workspace. When it is in the correct location, just click and drop the object, which "magically" appears. In the example below, a gauge indicator is found on the numeric menu. By clicking on that object, a box appears around it and the **Positioning** cursor becomes the **Scroll** cursor.

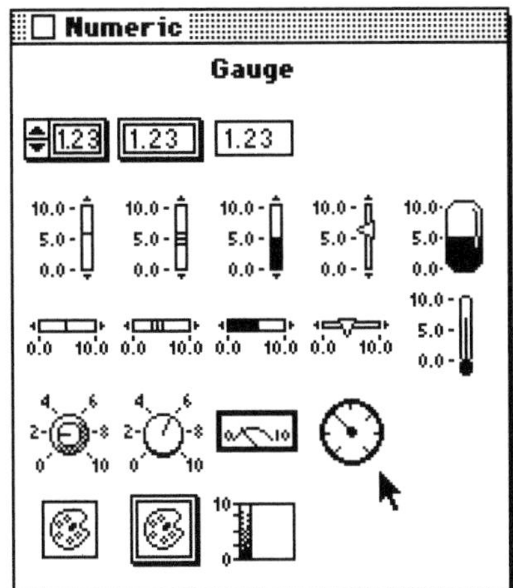

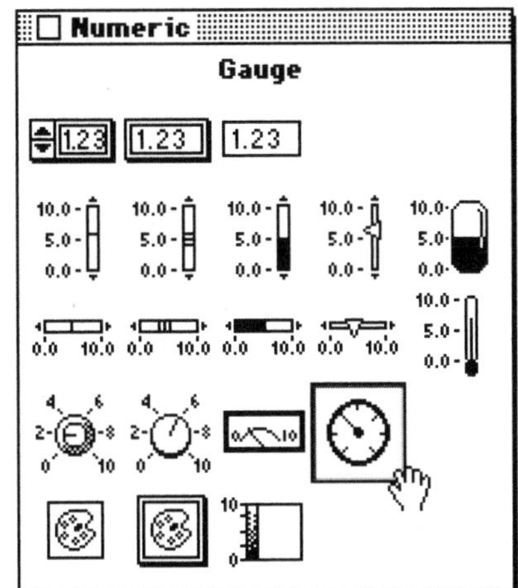

Now just drag the object into the workspace,

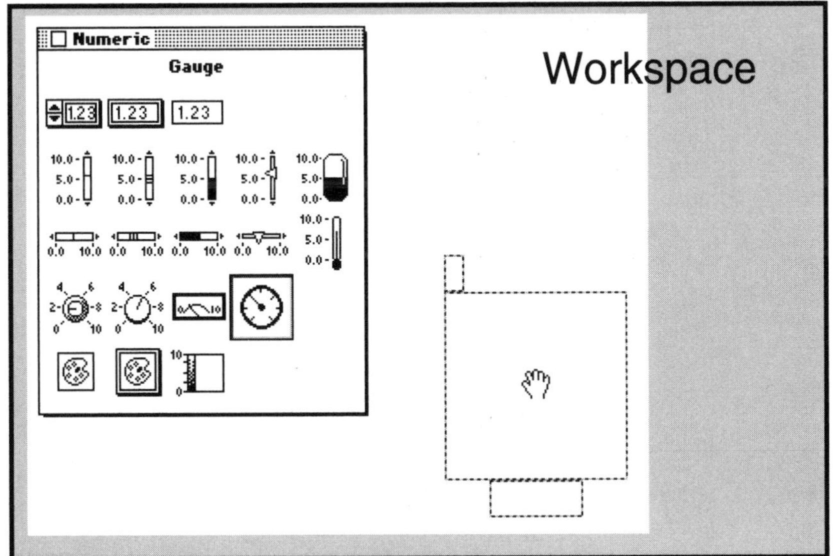

and presto, the gauge appears. The black box indicates that a label is requested. You can now type a name for this object. When the cursor is clicked in the workspace, the label is set. It will be carried onto the diagram panel to provide a name for this numeric output.

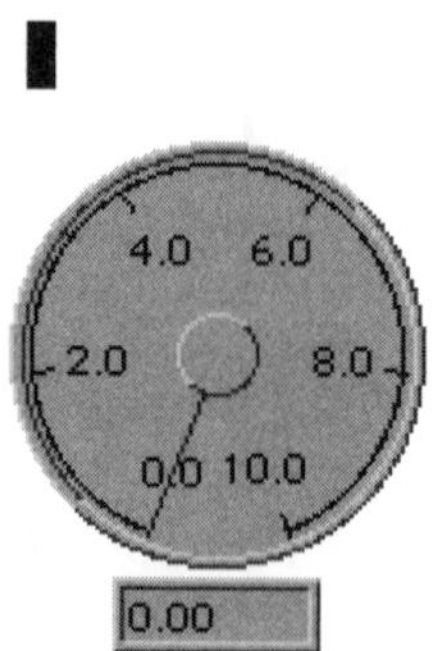

The controls palette is only active and visible when the front panel is the active window. When the diagram panel is selected, a new **Functions** palette appears. If it is not visible, it can be selected from the **Windows>Show Functions** palette. Like the control palette, it contains a wealth of LabVIEW functions conveniently divided into areas of common operations such a numerics, Booleans, strings, arrays, clusters, and so on.

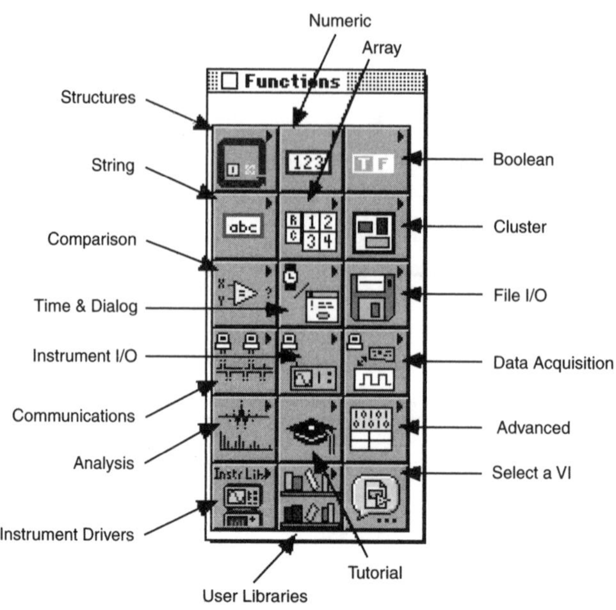

LabVIEW has an extensive list of mathematical functions and operations, and in some cases several subpalettes must be navigated to get to the desired function. In the following case, the tangent operation is found on the Numeric subpalette. This palette contains further subpalettes for Conversions, Trigonometric, Logarithmic, Complex, and Constants. In the trigonometric subpalette the tangent operation is found. Selection and placement of a function is similar to that of controls and indicators.

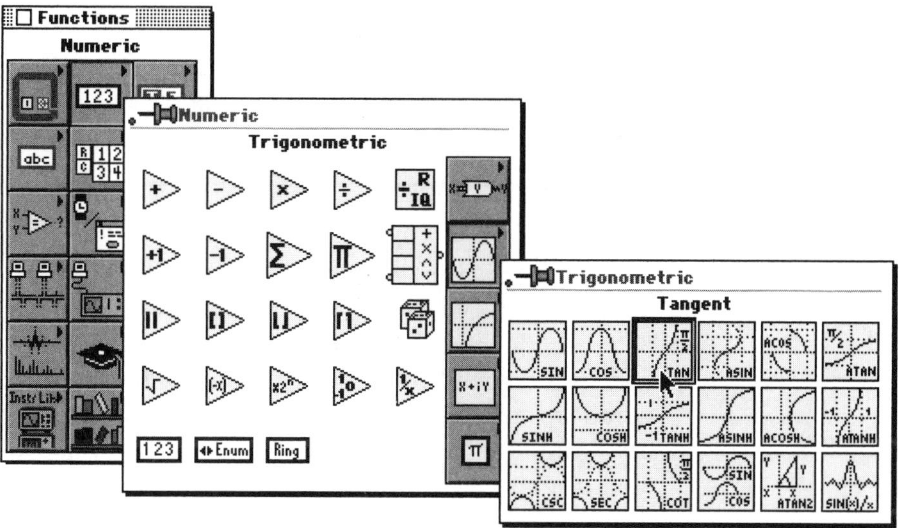

■ Panels, Panels Everywhere!

The original concept for LabVIEW was in the form of an electronic spreadsheet for instrument control applications. In a spreadsheet, data is input and output on a matrix of boxes on the front panel. Behind every box on the front panel is an arithmetic equation. When the spreadsheet was run, the inputs were converted to outputs by the hidden formulas and the results showed up on the front panel. How do you build a spreadsheet for instrumentation?

The front panel of LabVIEW contains all the input and output devices: the switches, the relays, the digital voltmeters, the displays, and so on. Look at a simple instrument where an input level is to be amplified with a gain from 0–10 and displayed on a meter. Each of the input/output functions has a graphical representation. A thumbwheel switch is used to set the input level, a potentiometer to set the gain, and a meter to display the result. Oh yes, do not forget the ON/OFF switch to power up the amplifier.

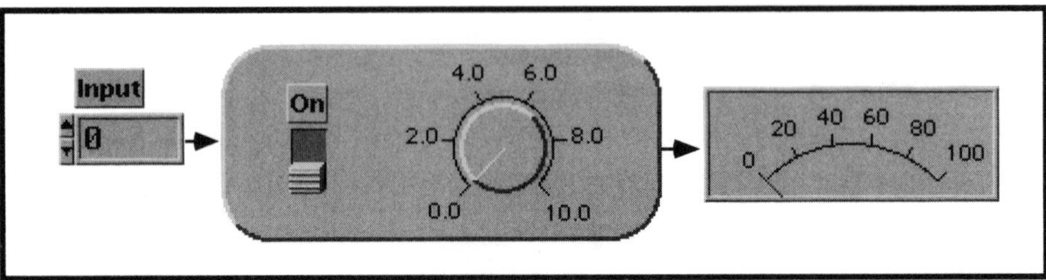

Behind the front panel input/output devices is the diagram panel or window, where the actual circuit is to be built. The link between the two panels is comprised of the terminal blocks. Each input or output device will have a terminal box that can be wired up to other components (functions). The lettering inside the terminal box and its color tells what type of data is generated or expected. Suitable labels aid in the program layout and documentation.

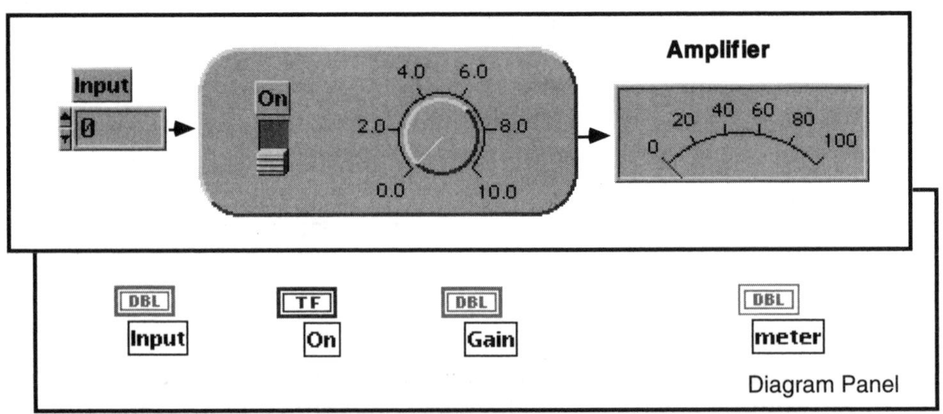

The input, gain, and meter values are all numeric values and are indicated by an orange box with the letters "DBL" inside the terminal box. DBL stands for double precision floating point variable. Note the different shaped boxes for input and output. Inputs have heavy outlines around the terminal, while outputs have a single line around the terminal. The ON/OFF switch is a different data type, a Boolean. It only has two possible states ON or TRUE and OFF or FALSE. Its input terminal is green in color, heavy outlined with the letters "TF" inside the box.

Analog Amplifier

To convert these inputs and outputs into an instrument, the inputs and outputs must be connected to the mathematical functions that simulate the required operation. In this instrument, the relationship is Output = Input × Gain. In LabVIEW, graphical symbols or icons are used to represent mathematical and other functions. One wires up the inputs to the input side of the function and the output to the display. To do this, we will use the **Positioning** tool to move the input and output terminals into a more natural programming arrangement.

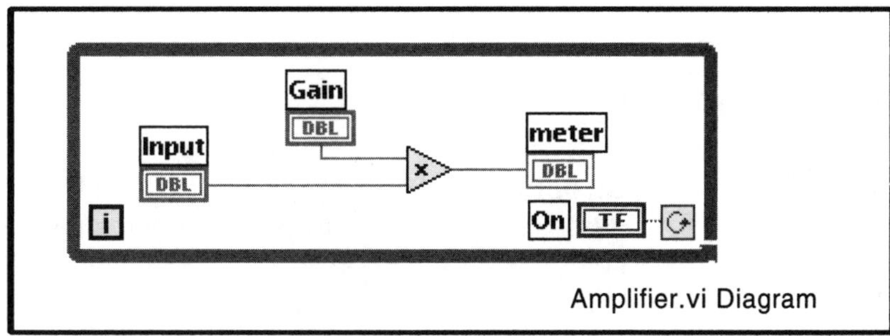

Amplifier.vi Diagram

The window behind the front panel is often called the *block diagram,* since the program design resembles an electronic schematic diagram. Here is where we have built the circuit to simulate the amplifier. Note that the inputs are usually on the left and the outputs on the right. The wiring tool is used to connect input/output terminals to the functions. Just click on the input/output, drag to the appropriate location, and click. Presto, in a data path is made.

What about that power switch? In LabVIEW, programs are often placed into a while loop, shown above as the heavy gray box. It has a control terminal [⟳] and even an index [i]. The [while . . . loop] structure executes everything inside the box, then reads the control terminal [⟳]. If the Boolean input is true, the loop repeats and executes all functions inside the box. If the Boolean input is false, the process stops. In this example, the while . . . loop simulates the ON/OFF power switch on the front panel.

■ LabVIEW Challenge: Offset and Gain

The most common form of calibration for a linear sensor is subtraction of a constant offset from the sensor output followed by multiplication of the difference signal by a scaling constant. *Design a LabVIEW program to execute the calibration equation*

*Output = (Input – Offset) * Gain*

Use Amplifier.vi as a starting point and modify the program to execute the calibration equation. Save your program as Amplifier2.vi.

Digital Amplifier

In our second example, we will look at a digital amplifier, whose input is only amplified when a switch is thrown. The bit switch is depicted on the front panel as a vertical throw switch. It outputs a Boolean data type, true (ON) or false (OFF). These two states are converted into a numeric (1) or (0) respectively by a Boolean-to-numeric transformation function. A bit indicator (LED display) displays the state of the switch when the program is run. The output of the switch (0) or (1) is multiplied by a constant weighting factor and the resultant value displayed in the output box labeled "Value."

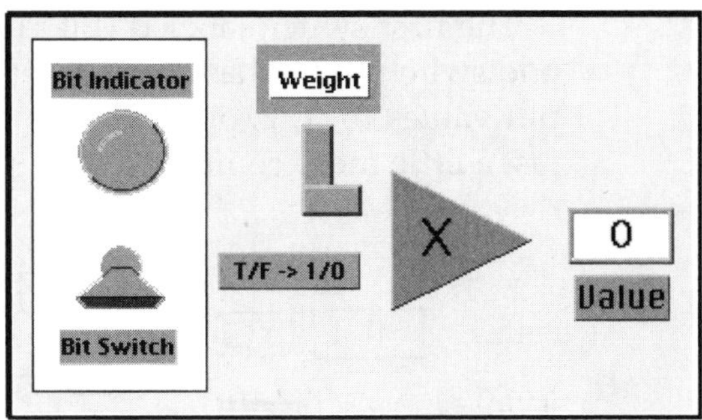

For example, if the weight is 64, then the output can only have two values: 0 if switch is off and 64 if the switch is on. In this example, there are two inputs (Bit Switch and Weight) and two outputs (Bit Indicator and Value). These four input/output links show up on the block diagram as four terminal boxes.

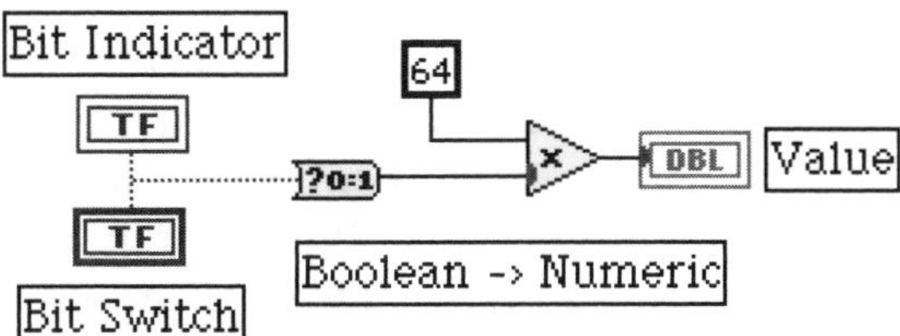

Note the special icon [?0:1] that converts the Boolean output of the switch into a numeric value. This is essential, since the multiply function can only operate on numerics.

When a second binary switch is added in parallel with the first one, the outputs of the two switches are added together to form a sum.

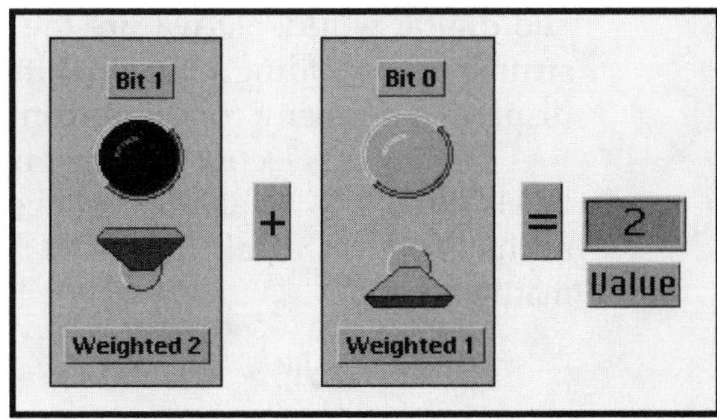

The first switch labeled [Bit 1] has a weight of 2 and the second switch [Bit 0] has a weight of 1. There are four possible output values (0, 1, 2, or 3). On the block diagram, the program is just a little more complicated.

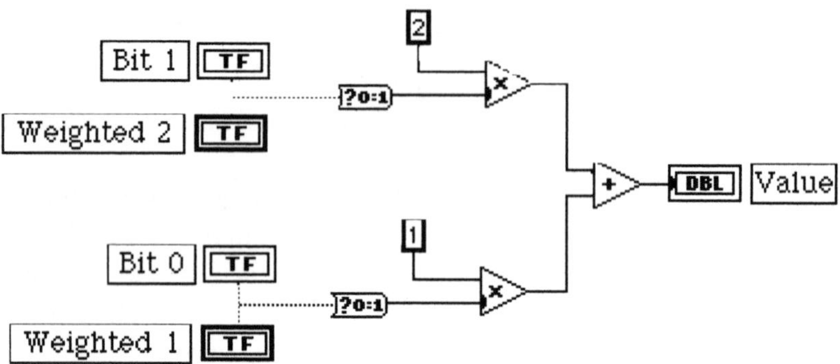

A two-input ADD function has been added to create the output sum. By careful arrangement and alignment of the icons with the alignment tools, the functionality of the front panel can be preserved in the program design depicted on the block diagram. This program, called DAC2.vi, is in essence a 2-bit digital-to-analog converter.

Digital to Analog Converters

The digital-to-analog converter, better known as the DAC, is a major interface circuit that forms the bridge between the analog and digital worlds. DACs are the core of many circuits and instruments including digital voltmeters, plotters, oscilloscope displays, and many computer-controlled devices.

A DAC is an electronic component that converts digital logic levels (bits) into an analog voltage. The output of a DAC is the arithmetic sum of all the input bits weighted in a particular manner:

$$DAC = \sum_{i=0} w_i b_i$$

where w_i is a weighting factor, b_i is the bit value (1 or 0) and i is the index of the bit number. In the case of a binary weighting scheme, $w_i = 2^i$, the complete expression for an 8-bit DAC is given by the expression

$$\mathrm{DAC} = 128\, b_7 + 64\, b_6 + 32\, b_5 + 16\, b_4 + 8\, b_3 + 4\, b_2 + 2\, b_1 + 1\, b_0$$

Any number between 0 and 255 can be represented by an 8-bit binary number.

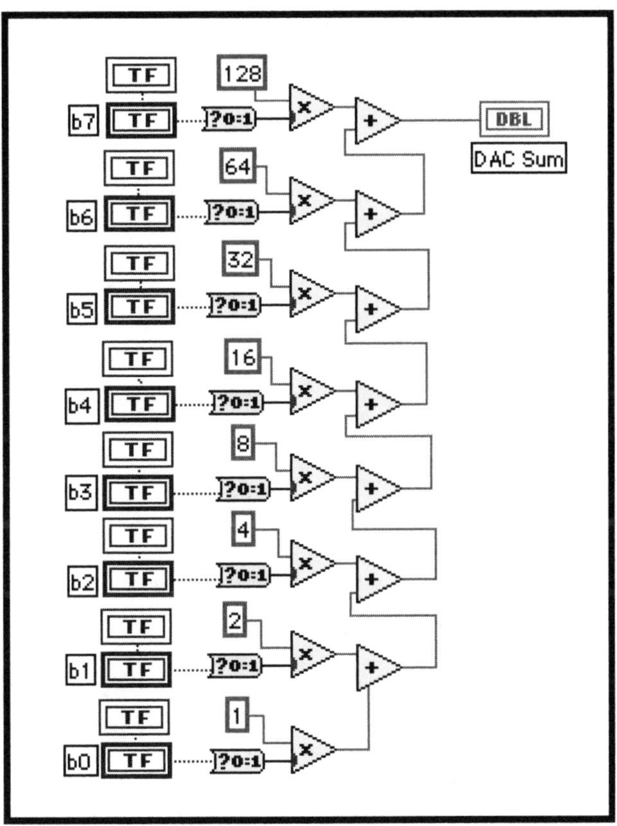

An 8-bit DAC can be simulated by summing the outputs of eight digital amplifiers, weighted with 128, 64, 32, 6, 8, 4, 2, and 1. The block diagram for such a simulation (DAC8.vi) is shown above.

Note how each bit circuitry is similar to all other bits differing only by a different weighting factor. In a program with this much redundancy, wise use of cut-and-paste operations and the LabVIEW alignment tools make the circuit easy to build, easy to to read, and easy to debug. In the future chapters, modules and arrays will be introduced as a means of writing programs that are even more readable, concise, and manageable.

On the front panel, eight Boolean switches are used to set the input bits b_0 through b_7. Eight LED indicators are also used to display the binary value of the input bits when the simulation is run. The analog output is displayed in a numeric display. The front panel for DAC8.vi is shown below.

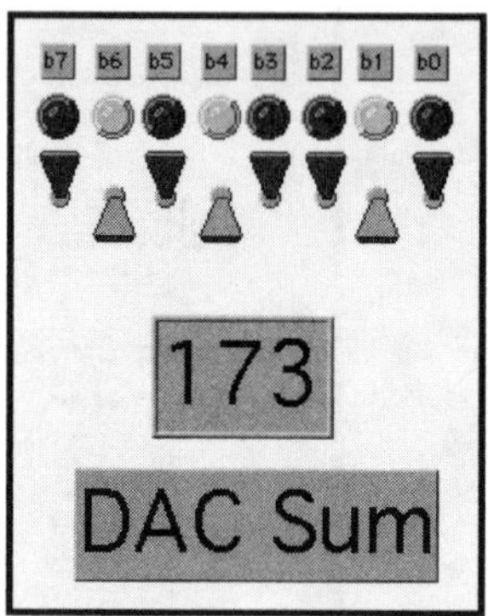

Load from the chapter 1 library DAC.vi. By running this program continuously, observe the relationship between the binary codes and their numeric equivalent value. Check out the weighting factors for each bit by toggling each switch on and off.

Here are some useful binary patterns. What are their DAC sums?

00000000 ____________

10101010 ____________

11001100 ____________

11110000 ____________

10000000 ____________

01010101 ____________

00110011 ____________

00001111 ____________

These binary patterns generate useful and interesting waveforms when the bits are output sequentially.

In the next chapter, you will see how a program can be converted into a sub-program or in LabVIEW, a sub-VI. DAC.vi is also a sub-VI and its simulation will be used many times in future chapters to convert a binary pattern into a numeric number.

LabVIEW Challenge: Vector Calculator

Many scientific variables (voltage, current, etc.) can be represented as vectors. Vector manipulation and arithmetic provides a concise method to represent and solve many types of problems. *Build a vector calculator that adds two vectors* **A** *and* **B.**

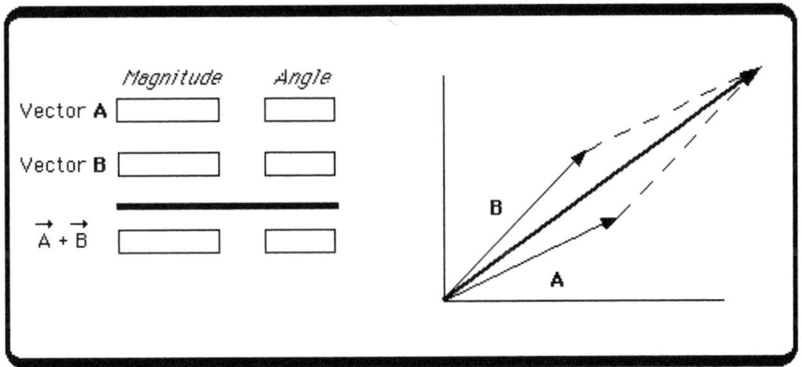

Recall each vector can be represented by its rectangular components

$$\mathbf{A} = A_x\mathbf{i} + A_y\mathbf{j} \text{ and } \mathbf{B} = B_x\mathbf{i} + B_y\mathbf{j}.$$

The resultant vector R is

$$\mathbf{R} = \mathbf{A+B} = (A_x + B_x)\,\mathbf{i} + (A_y + B_y)\,\mathbf{j}$$

To make the challenge more interesting, assume the vectors are represented by polar coordinates, where the vector is described by its length or magnitude $|\mathbf{A}|$ and the angle θ it makes with the horizontal axis.

Overview

The world around us is filled with magnetic fields. This chapter introduces the Hall effect sensor as a means of measuring this invisible field. A Gaussmeter is designed and built using the 'real' properties of a typical Hall sensor. The Gaussmeter is then turned into a callable subroutine called a sub-VI so it can be used in a virtual experiment to observe the angular dependence of a Hall probe. Modularity in LabVIEW and the ability to run modules as either a VI or a sub-VI is one of the most useful and important concepts. It allows small sections of the code to be tested and debugged independently while maintaining a compact graphical format in the main program.

Goals

- Understand Magnetic field sensors
- Learn LabVIEW controls and functions
- Build a LabVIEW simulation of a Gaussmeter
- See the power of modules, sub-VIs
- Conduct an experiment with the virtual Gaussmeter

Key Terms

- Hall Effect
- Gaussmeter
- Virtual Experiment
- LabVIEW controls and functions
- Modular programming
- Connector panels
- LabVIEW icons

Invisible Fields

2

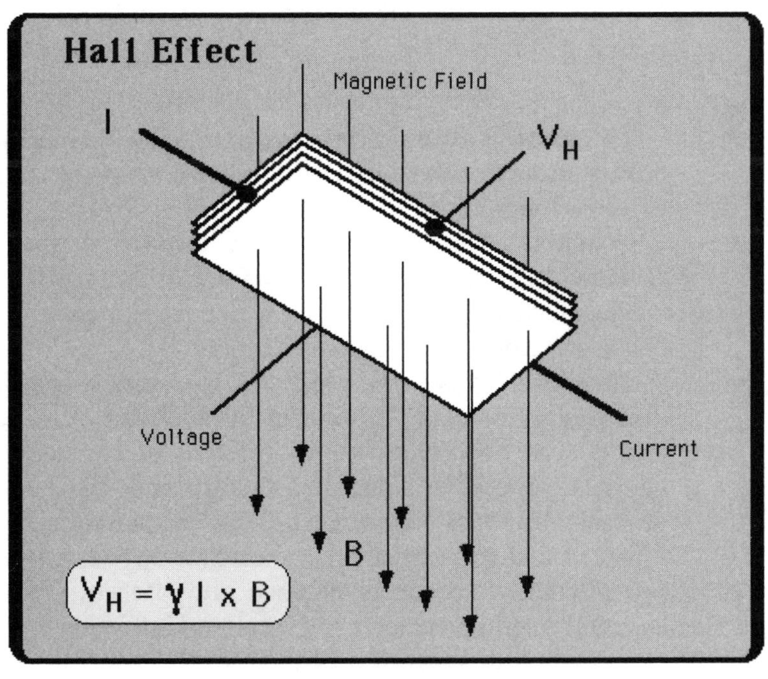

Principle of the Hall Effect

In our modern society, we are surrounded by invisible fields: radio waves, microwaves, radar waves, and magnetic fields—fields that we cannot see, feel, or smell. Today, there exist many sensors that can measure these fields and convert the magnitude of the field into a voltage. With some simple interfacing, LabVIEW can be used to characterize these fields. In this chapter, we take a look at one of these sensors, the Hall Effect sensor, used to measure the invisible magnetic field.

The Hall Effect

In 1879, Edwin Hall discovered that when a current-carrying conductor is placed in a magnetic field, a voltage is generated in a direction perpendicular to both the current and the magnetic field. This observation, known as the Hall Effect, arises from the deflection of the moving electronic charge inside a conductor or semiconductor in the presence of a magnetic field. From Maxwell's equation, the response of an electron to a magnetic field is given by Lorentz's Law

$$\mathbf{F} = q\,\mathbf{v} \times \mathbf{B}.$$

The force, **F** experienced by a charge q is given by the vector cross product of the average velocity of the electron, **v**, and the external magnetic field, **B**.

Consider a semiconductor bar of thickness w placed in a magnetic field **B** with a current **I** flowing through it. The force acting on the electron ($-q$) is in a direction perpendicular to the motion of the electron **v** and the external field.

This force causes the electrons to move to one side of the semiconductor bar, resulting in the buildup of an excess negative charge on one side and a deficit of electrons or positive charge on the other side. This separation of charge over the distance w generates an electric field across the semiconductor and a force on the electrons in a direction opposite to the Lorentz force. These two forces balance and the equilibrium voltage across the semiconductor is the called Hall Voltage, V_H. It is easy to show that this condition generates the transducer equation

$$V_H = \gamma\,\mathbf{I} \times \mathbf{B}$$

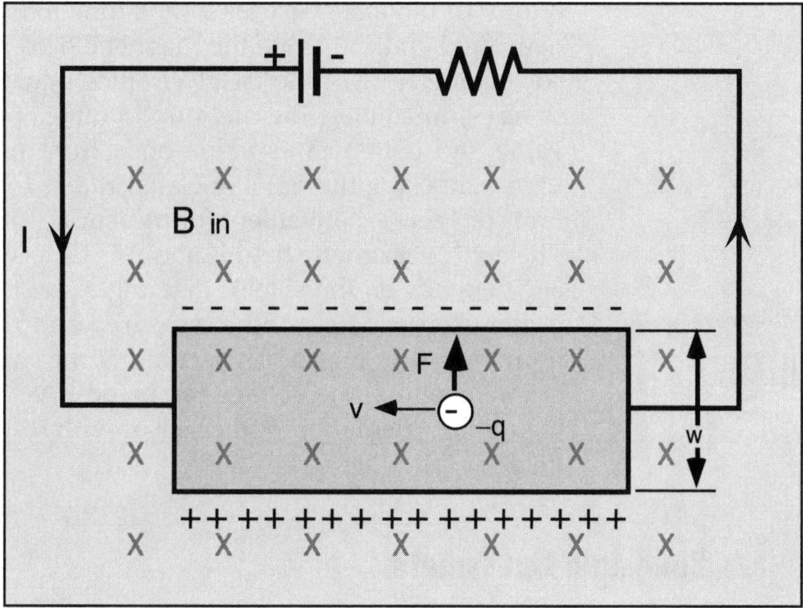

The constant γ is called the Hall coefficient and is related to the properties of the semiconductor and the sensor geometry. The vector cross product comes from the Lorentz equation. A more useful form of the transducer equation is

$$V_H = \gamma \, \mathbf{I} \, \mathbf{B} \, \mathrm{Sin} \, \theta$$

where θ is the angle between the direction of current and the direction of the external magnetic field. The Hall voltage is a function of three variables: current, magnetic field, and angle. Holding any two constant allows the third to be determined from a measurement of the Hall voltage. Thus the Hall sensor can be used to build a Gaussmeter, a current meter, or an angle encoder.

The Gaussmeter

A Gaussmeter uses a Hall effect sensor to convert a magnetic field into a voltage. Inside the Gaussmeter, an electronic circuit provides a constant current to drive the Hall Effect sensor. The output of the meter is thus only a function of the external magnetic field and the angle the magnetic field makes with the sensor geometry. Additional electronics add the calibration constant (γ) to convert the measured voltage into a magnetic field value and display this value on a front panel in appropriate units. Choosing the units of Gauss rather than the conventional units of Telsa is convenient for measuring small magnetic fields. The earth's magnetic field is about ½ Gauss. Since the Hall voltage depends on the sine of that angle, the Hall sensor is placed in the magnetic field to be measured and rotated until a maximum signal is found. That value is the magnetic field at that point. Note that the voltage can be positive or negative depending on the orientation of the sensor with respect to the magnetic field.

■ Building a Gaussmeter

To simulate a basic Gaussmeter, we will need a front panel meter that has both positive and negative ranges. Launch LabVIEW, which opens up a new workspace called "Untitled 1" and its associated diagram window, called "Untitled 1 Diagram." On the front panel window, we will build the Gaussmeter. If the **Controls** palette is not showing, select and click on **Windows>Show Controls Palette.**

From the **Controls** palette select **Numeric>Meter.** Drag the outline of the meter into your workspace. When the mouse is clicked, the meter icon appears together with a black box on the upper left-hand corner of the icon. This is the label box and its black color tells you that you can type a name into the icon label box. Type "Hall Voltage" on the keyboard, which shows up in the label box. Clicking the mouse terminates the entry. We will now change the limits of the display from its default value of (0, 10.0). to (–10.0, 10.0). With the operating tool selected, move the hand cursor over the lower limit of the Hall Voltage icon. The cursor will change into an **Insert** indicator. Select this limit by clicking and swiping the cursor over the number 0. The number will now be displayed as white numbers on a black background. Enter the number –10 and click outside the icon space. Voila! We have the front panel meter of the Gaussmeter.

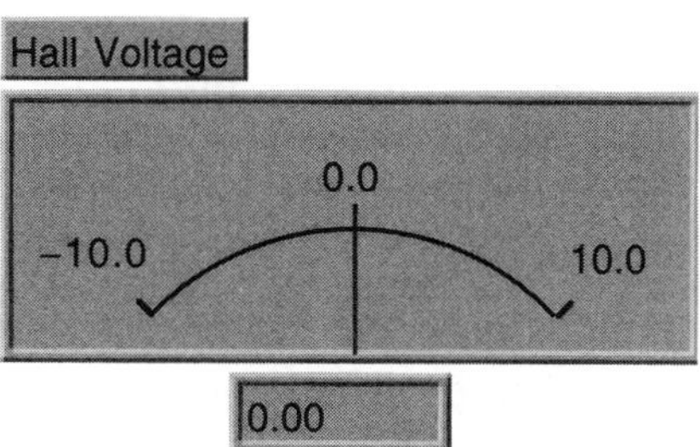

The constant current source in the Gaussmeter is simulated using a numeric control with a range from 0–15 ma. Select a vertical slide control from the **Controls>Numeric** palette. Enter its label as "Current (ma)."

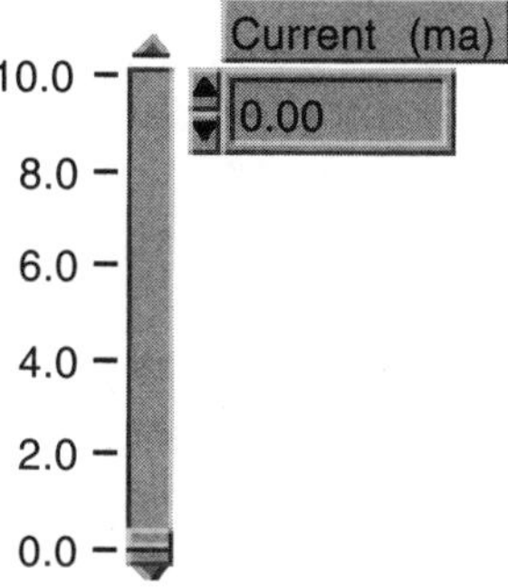

Place the indicator in the lower left part of your Gaussmeter front panel. Change the range of the display from its default upper limit of 10.0 to 15 using the same technique as with the Hall Voltage icon.

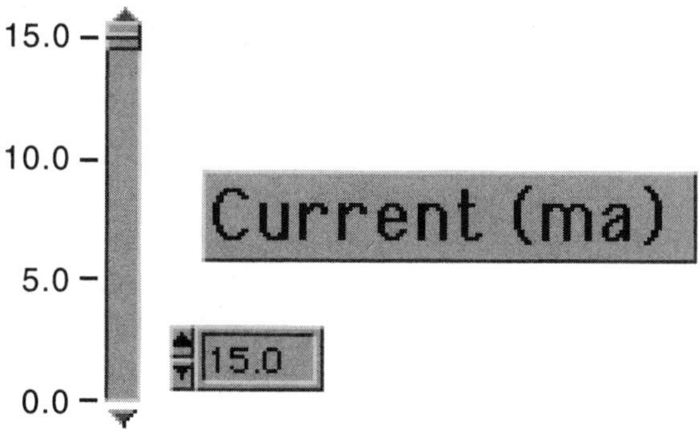

To make the label more readable, you can use another font or a different size, style, or color. All these options are available under the **Text Settings** in the menu bar. Use the **Text** tool to select the text label by placing the cusor over the text box. Click and swipe over the text you wish to change. Type any text changes then modify the selected text by clicking on the **Text Settings** in the menu bar. Use the **Positioning** tool to place the label at a convenient point.

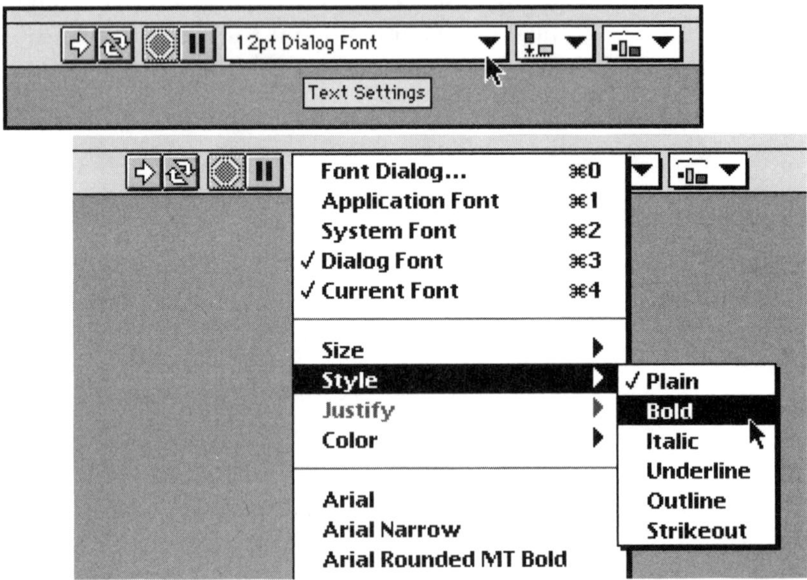

To simulate the magnetic field, a numeric horizontal fill slide control is used. Selected it from the **Controls**>**Numeric** palette and label it "Magnetic Field (Gauss)."

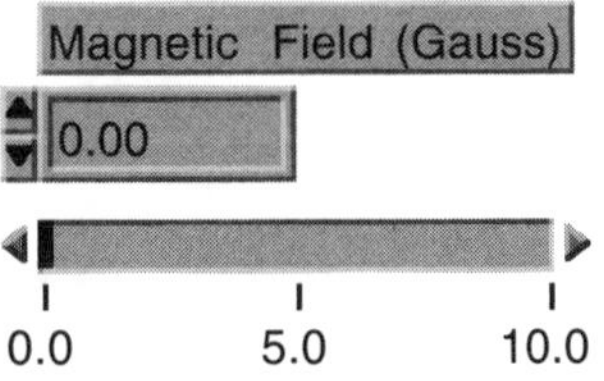

Change the limits of the slide control from (0.0, 10.0) to (0, 400). In this case, only the horizontal slide indicator is used, not its associated digital display. Pop up on the slide wire to display the **Properties** menu. Select **Show>Digital Display** and the indicator will be hidden. As this is a toggle function, repeating this action will bring back the digital display. Use the text tool to enhance the label.

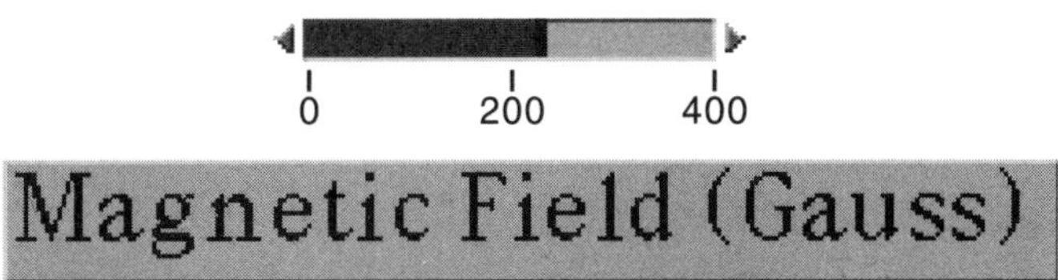

To simulate rotating the Hall sensor in a fixed magnetic field, a rotary knob is selected from the **Controls>Numeric** palette. Label this control as "Angle (degrees)."

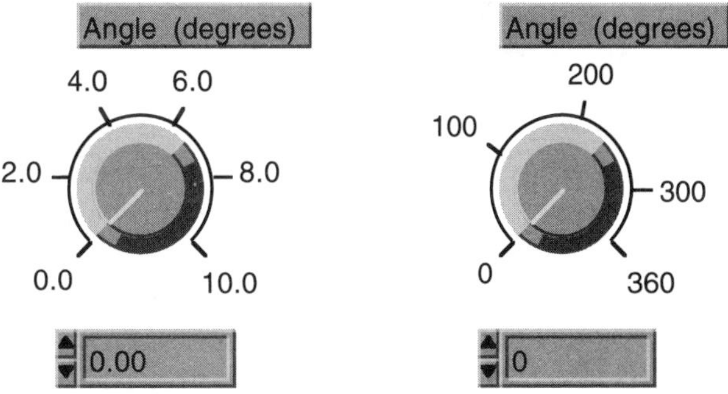

Use the operating tool to set the range from 0 to 360 degrees. In practice, the angle is only measured to 1 degree, hence the precision is changed from 360.0 to 360 using a format option. Pop up on the knob control to bring up the Properties menu. Select **Format & Precision ...** and set the **Digits of Precision** to 0. To better simulate rotation of the sensor, the limits of the rotary display are overlapped—that is, 0 and 360 are the same value.

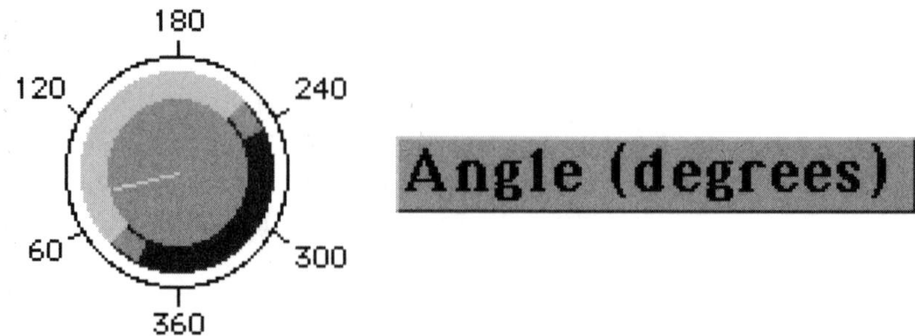

Select the knob control by placing the **Positioning** tool over the rotary display and clicking. Move the cursor to the right-hand corner and the cursor becomes a rotation tool. Click and drag the 360 value onto the 0 value. The smallest interval label will now read 50. Using the **Operating** tool, change this label to 60, a more convenient interval.

Together these controls (current, magnetic field, and angle) and the indicator (Hall Voltage) form the front panel of the Gaussmeter.

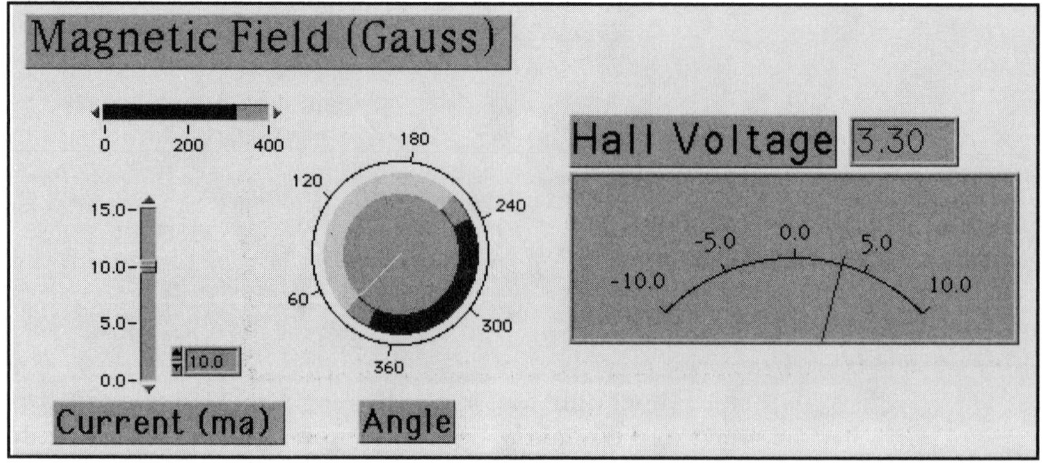

To complete the program, it is necessay to code the Hall Effect transducer equation onto the block diagram.

The mathematical functions or LabVIEW icons needed are multiplication, division, and the sine operator. Select the Gaussmeter diagram panel. If the functions palette is not visible, select **Windows>Show Functions** palette. From the **Functions>Numeric** palette select Multiply and Divide icons. The Sine function is found in **Functions>Numeric>Trigonometric** subpalette. The physicial input is measured in degrees, while the Sine function uses radians. Conversion of degrees to radians is accomplished using the constants Pi and 180 together with another multiply and divide function. The numeric constant terminal box is found in the **Functions>Numeric** palette. The calibration constant γ for this sensor is $1/600$ or 0.00167 and is added as the last operation in a muliplication function.

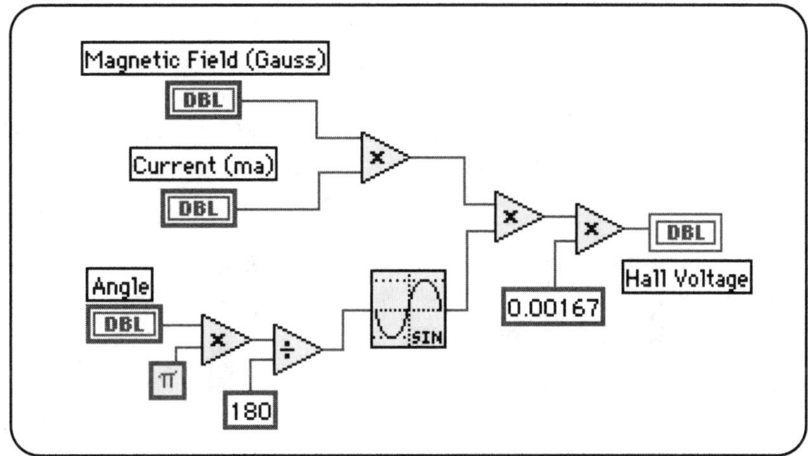

Position the icons as shown in the above diagram and wire programming links between the terminal boxes and functions using the wiring tool.

Upon returning to the front panel, the program can now be run continuously. Observe the Hall Voltage output as the current is changed, the magnetic field modified or the Hall sensor rotated. In a typical Gaussmeter application, the current is fixed and the calibration constant programmed into the electronic circuit so that the display is calibrated and reads directly in Gauss. The sensor is placed into a magnetic field and is rotated until a maximum signal is found and that value is the magnitude of the external magnetic field. Save your program as Gaussmeter.vi.

Converting a Program.vi into a Callable sub-VI

In a larger application, the magnetic field may be just one of several measurements and it is necessary to link the Gaussmeter to a control program. To convert a LabVIEW program into a subprogram (sub-VI), it is necessary to ensure that each (sub)program has a unique name. This was set when our program was saved as Gaussmeter.vi. The (.vi) extension is important as LabVIEW recognizes only subprograms that have this extension. In the graphical language of LabVIEW, each control, indicator, function, or subprogram is displayed on the diagram panel as a unique icon. Creating an icon for Gaussmeter.vi is accomplished by popping up on the icon panel (upper right-hand corner) of the front panel window and selecting **Edit Icon . . . .** A new window opens up to reveal the icon editor. Tools on the left allow you to draw points, lines, or boxes in the icon workspace. Text can be added and colors may be selected. On the right side, a black-and-white or colored icon can be selected.

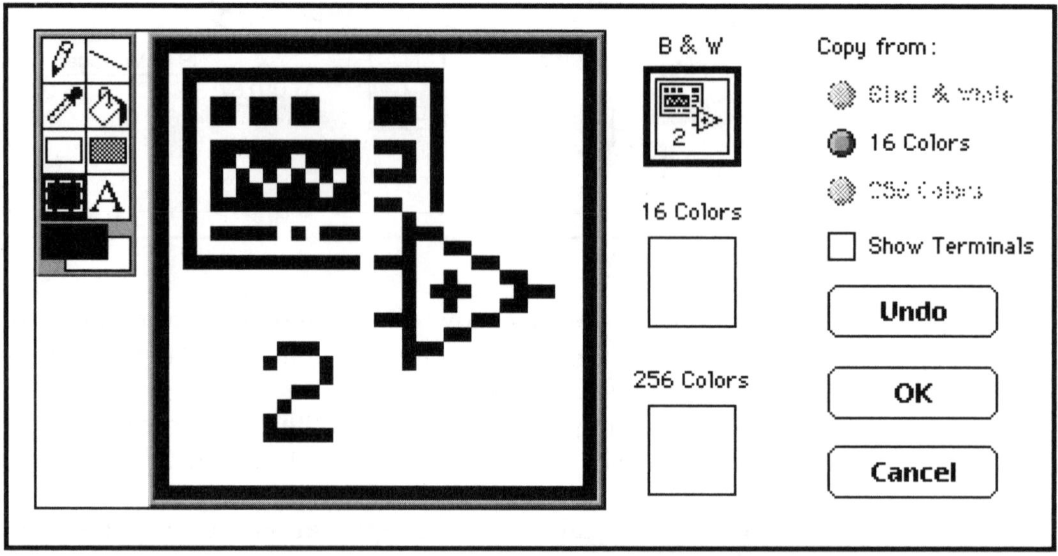

To design a new icon, first select the color type (B&W, 16 colors, or 256 colors). Use the tools on the left to create your icon. Colored backgrounds can be added by selecting a color and clicking the paint bucket at appropriate locations.

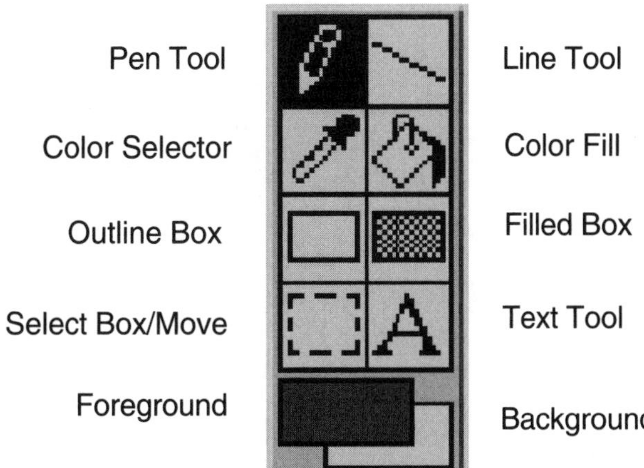

When completed, click on [OK] and your icon will replace the default black-and-white LabVIEW icon. To link Gaussmeter.vi to a control program and thus allow parameters to be passed from the sub-VI to the main VI, the required inputs and/or outputs must be assigned as a link. This is accomplished using the **Show Connector** option, part of the pop-up menu for the program box.

On the front panel, pop up on the new VI icon and select **Show Connector** from this menu. The icon box will turn white, subdivided into as many boxes as there are inputs and outputs in the VI. In this case, there are three inputs (current, field, and angle) and one output (Hall voltage). Popping up a second time on the VI icon provides a new menu that allows you to select the link pattern and some other handy editing features. The **Patterns** menu allows the designer to select the number of links for the VI. Select from the **Patterns** menu two boxes. One will be used for the input [Angle] and the other for the output [Hall Voltage]. To activate these links, it is necessary to click the cursor, which becomes the **Wiring** tool over one of the boxes. The selected box will turn black. Move the cursor to the input rotary knob and select it by clicking on it. It will now be highlighted by a dashed outline around the control. Move outside the selected area and click a third time. The black box in the icon area will now turn dark gray indicating the connection has been made. Repeat this operation for the meter display. The result of the second operation is shown below for assignment of the Hall Voltage output link.

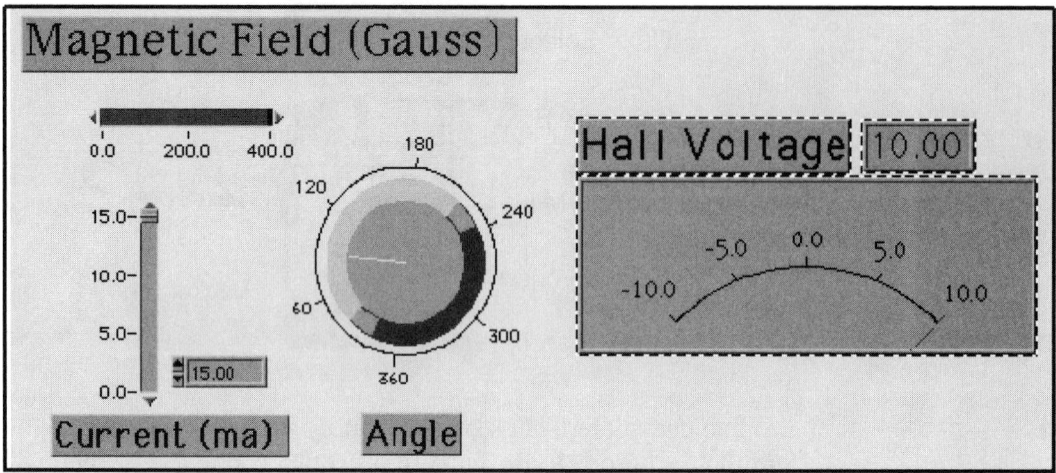

Return to the VI icon and pop-up to select **Show** icon. Gausmeter.vi is now both a program VI and a sub-VI. Set the input current to 15 ma and the magnetic field to 400 Gauss. From the **Operate** menu select **Make the Current Values Default** and save your program (sub-VI) as Gaussmeter2.vi.

LabVIEW Challenge: Building a DAC sub-VI

Convert your program DAC.vi from Chapter 1 into a sub-VI. You will need a pattern selection with nine boxes, eight inputs, and one output.

The Angular Dependence of a Hall Probe

The transducer equation predicts that the Hall voltage follows a sine curve if the sensor is rotated about an axis perpendicular to an external magnetic field. Consider an experiment to demonstrate this dependence. First use the Hall sensor simulator Gaussmeter2.vi to chart the Hall voltage. Open up a new program window and name it Hall Chart.vi. On the front panel, all

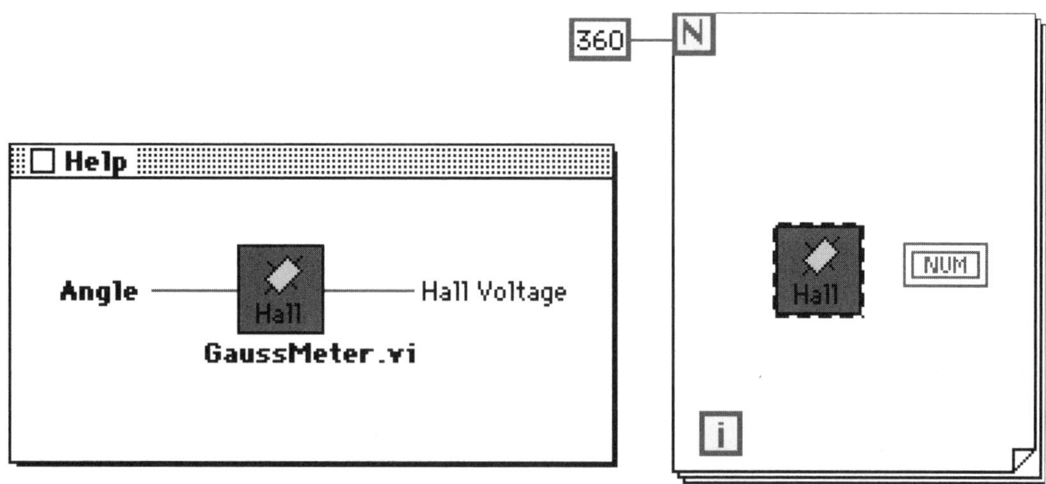

that will be needed is a **Waveform Chart** found under the **Controls>Graph** palette. On the diagram window, place a For . . . Loop from the **Functions>Structures** palette. From the **Functions>Numeric** palette, a numeric constant of 360 is also needed. Inside the For . . . Loop sub-diagram add your Gaussmeter2.vi from the last section. Select **Functions>Select a VI . . . .** A file menu pops-up and allows you to select the Gaussmeter2.vi. To see how the wiring is to be connected, select the **Help** function (^H) when the VI is selected. In this case, help shows one input, Angle, and one output, Hall Voltage.

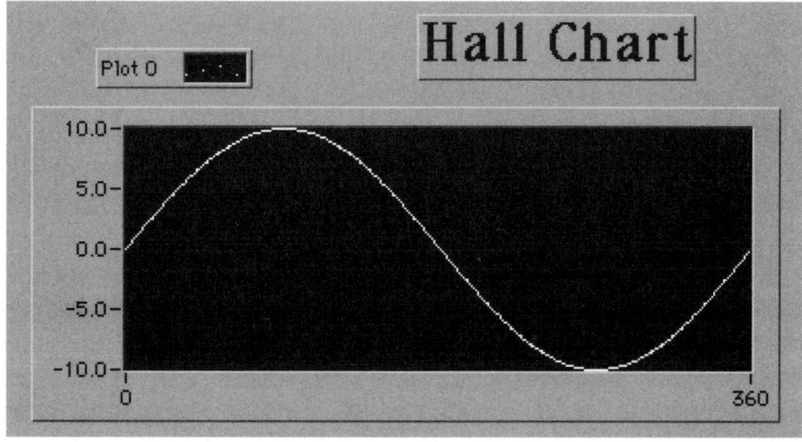

The index of the For . . . Loop is used to simulate rotating the Hall Probe through an angle [i] degrees. Run the program to observe the angular dependance of the Hall Voltage.

One can compare the simulated Hall experiment with a real experiment (shown below) that uses a Honeywell commercial Hall sensor.

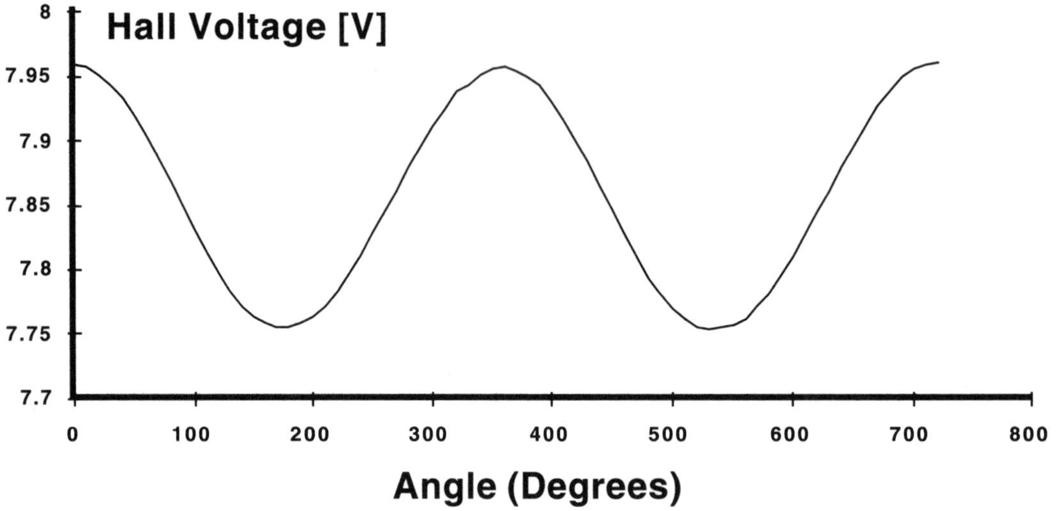

Both the real experiment data set and the simulated data sets follow the sine variation expected from the transducer equation. However, in the real experiment, the Hall signal is offset by a constant voltage and always positive. This is a result of the sensor design which uses a single power supply to excite the sensor. Subtraction of the sensor offset yields a curve almost exactly as predicted by the transducer equation and shown in the simulation.

Honeywell Hall Effect Sensor

The Honeywell magnetic field sensor PK 8764-4 is representative of commercially available Hall effect sensors. The Hall sensor and electronics are housed inside an integrated circuit chip. The IC is mounted on a ceramic base. Thick film resistors on the ceramic plate are laser trimmed to provide temperature compensation and calibration.

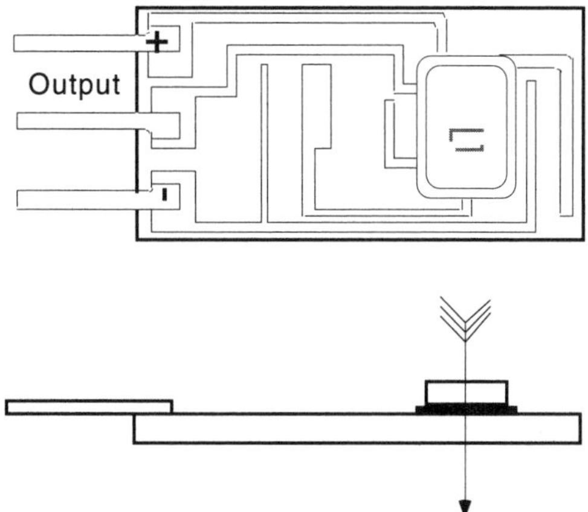

The sensor package occupies a small cross-section area of 15.2 by 5.3 mm and a thickness of less than 2 mm. Two of the leads labelled + and – are connected to an external power source of 8 to 16 volts. Internal circuitry provides a constant current source for the Hall element. The transducer equation is simplified to become

$$V_H(\text{volts}) = \text{Offset} + \gamma\, B\, \text{Sin}\, \theta$$

where B is measured in Gauss, γ is a calibration contant, and the offset voltage is ½ of the supply voltage. This device has a

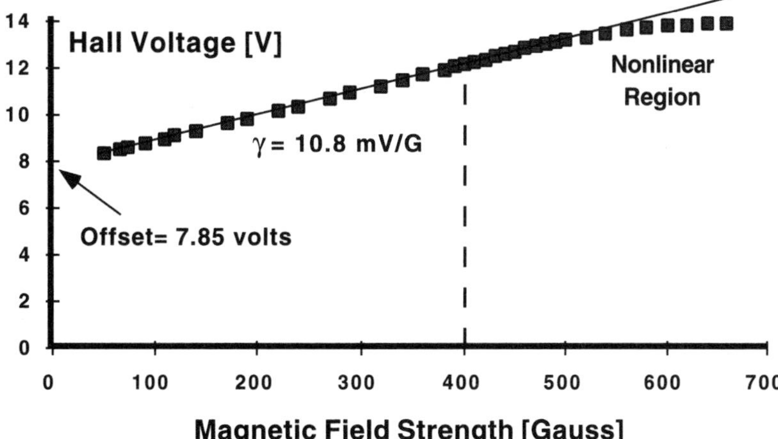

range of (+/−) 400 Gauss with a sensitivity γ of about 10 mV/Gauss. The device cannot be damaged by exposing it to a magnetic field larger then the 400 Gauss calibrated range. Outside the calibrated linear region, the sensor displays a non-linear response. A calibration curve for a Honeywell sensor is shown on the bottom of the previous page.

Hall Effect Simulator Version 2

In the real-world environment, magnet motors, fluorescent lights, ground loops, and so on, all add noise to the sensor signal. As seen in the Honeywell sensor response curve, the signal is offset and noisy. A better simulation for a Hall Probe is to add noise and an offset to the transducer equation. Gauss2.vi simulates such a sensor. As in the previous simulation, each time this VI is called, the measured Hall voltage is returned at the angle input. A chart of the Hall Voltage versus orientation angle for the simulator Version 2.0 follows.

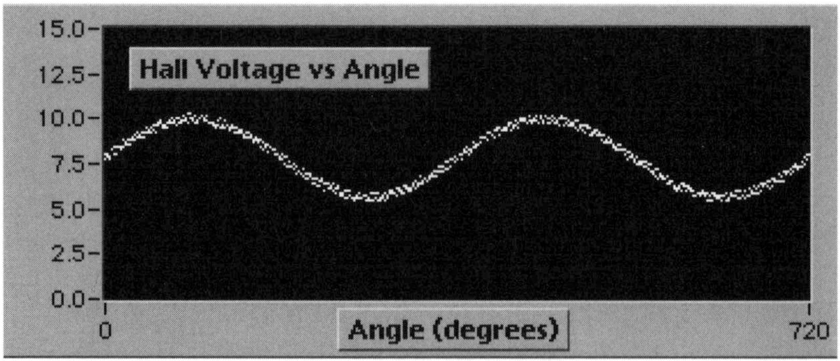

LabVIEW Challenges: A Real Hall Effect Sensor

Using the Gauss2.vi as the sensor input, design a series of programs to answer the following questions.

1. What is the offset voltage in the Hall signal?
2. What is the magnitude of the external magnetic field value?

3. Does the Hall Voltage follow a sine wave when the sensor is rotated in a magnetic field?
4. From the chart output, estimate the magnitude of the noise signal.
5. With an estimate of the noise, can a better estimate for the external magnetic field be determined?

Hints

By rotating the sensor in a magnetic field, a chart of the maximum and minimum signals can be plotted. The average value of the maximum and minimum levels gives an estimate of the offset. This would be exact except for the noise that affects the average. After the offset is removed, the amplitude of the Hall signal gives an estimate of the external field. You will need the transducer equation and the calibration constant. One property of a sine wave is that the negative part of the cycle should be equal but opposite in sign to the positive part of the cycle. If you expand the signal out along the x-axis (angle) and zoom in on the maximum or minimum signal levels, an estimate of the noise is given from the local peak signals. With the average noise level, a better estimate of the offset and maximum sensor signals can be found.

In latter chapters, averaging of the signal levels using array functions will be introduced as a means of filtering the noise signal.

LabVIEW Experiment: Magnetic Field of a Muffin Fan Motor

An environmental agency wants to know the maximum magnetic field that is present at the back of a muffin fan motor used to cool a microcomputer. A Hall sensor was slowly moved across the back side of the motor and the magnetic field was measured with a Honeywell type sensor and its output displayed on an oscilloscope. The amplitude of the Hall voltage as a function of distance from the axis of the motor can be observed by running a program called Motor.vi.

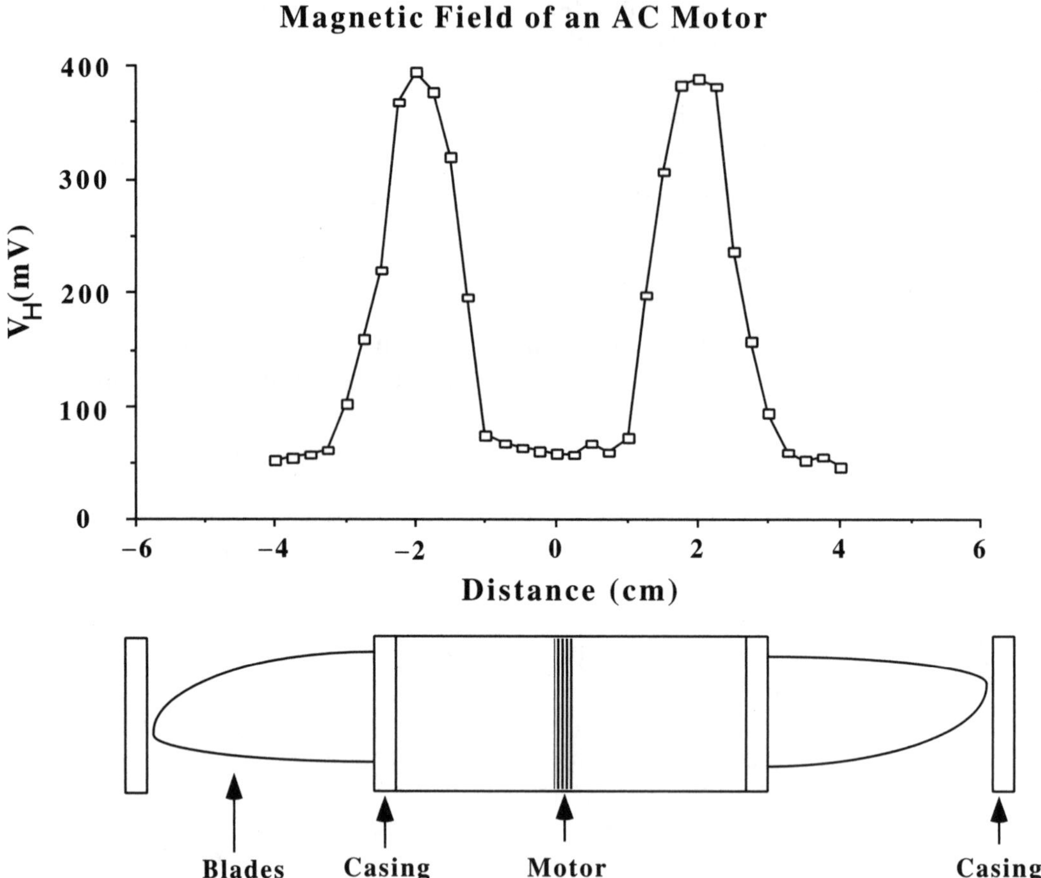

If the tranducer equation is $V_H = 7.5\ (\text{mV/gauss}) \times \mathbf{B}$, what is the maximum magnetic field generated by the fan motor?

Overview

Random numbers are studied as a means to learning about LabVIEW's four program structures, the For . . . Loop, the Case statement, the While . . . Loop, and the Sequence structure. A virtual experiment in coin flipping demonstrates the randomness of the LabVIEW random number function. With a While . . . Loop and Case structures, virtual dice are designed and thrown. Shift registers with multiple elements allow an important analog and digital instrument, the pseudo-random number generator, to be built and tested. In LabVIEW, programs follows the data flow and events or functions cannot proceed without input data. In the real world many events are sequential and LabVIEW uses a special sequence the "film strip" structure to execute linear programming.

Goals

- Study the four LabVIEW structures
- For . . . Loop, Case, While . . . Loop, and Sequence
- How random is random?
- Design shift registers and ring counters
- Build a pseudo-random number generator
- Measure the execution time of a sub-VI

Key Terms

- For . . . Loop
- Case(s)
- While . . . Loop
- Shift Register(s)
- Ring Counters
- Pseudo random numbers
- Sequence structure
- Execution time

Random Numbers

3

Catch a Falling Star!

For centuries mathematicians, scientists, engineers, and gamblers have been fascinated with random numbers and random number generators. In this chapter, a study of random numbers is used to investigate the programming structures of LabVIEW: For . . . Loop, While . . . Loop, Cases, and Sequences.

For . . . Loop

In the last chapter, the For . . . Loop was used to look at a finite number of iterations of a calculation. The For . . . Loop executes all the code contained within its boundary a specific number of times. The count terminal **[N]** is wired to a constant, a control, or a variable outside the loop. The iteration terminal **[i]** contains the number of completed loops: 0 for the first loop, 1 for the next loop and so on all the way up to *N*–1.

In LabVIEW, a random number generator (dice icon) is found under the **Functions>Numeric** submenu. Each time this function is called, a random number will be generated from 0 to .999999. To observe the operation of the random number function, build a VI consisting of a For . . . Loop, a count numeric control, the random number icon, a digital display, and a Waveform Chart.

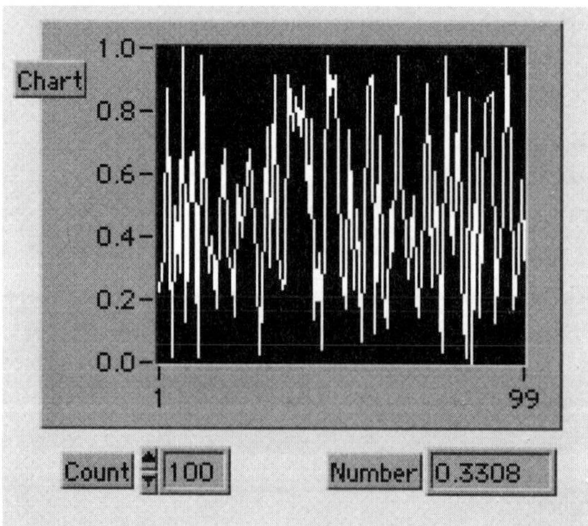

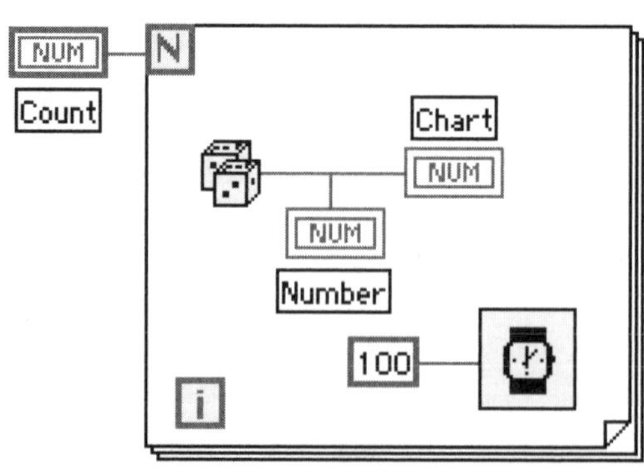

You might wish to add a Wait function from the **Functions>Time & Dialog** submenu into the loop to slow down the display. The number of milliseconds between loops, a numeric constant is wired to the function **Wait (ms).**

Suppose you wish to simulate the flipping of a coin N times. A random number generator will need to generate only two states: a head or a tail labeled 1 or 0, respectively. A **Round to the Nearest Integer** function is selected from the **Functions>Numeric** submenu. It rounds all values from .5–.9999 to 1 and all values from 0–.4999 to 0. To make the simulation more interesting, a LED display is used to represent the coin flipping. A LED indicator requires a Boolean data type, hence a **Comparison > Equal?** function is used to convert the numeric value (0 or 1) into a Boolean data type (True or False).

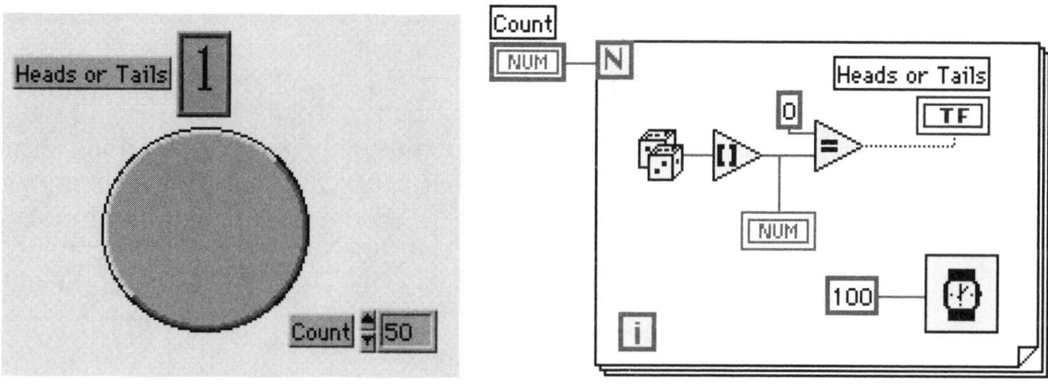

Case Statement

LabVIEW uses the Case statement as a means of redirecting the program flow when there are two or more choices of path. For two choices, the cases are labeled "**True**" or "**False.**" If the condition True is met, execute all inside the true subdiagram; if the other condition False is selected, execute all inside the false subdiagram. The Case statement executes all the code contained within the subdiagram before continuing. The conditional switch [**?**] is a Boolean input terminal located on the side of the Case structure, which selects the case.

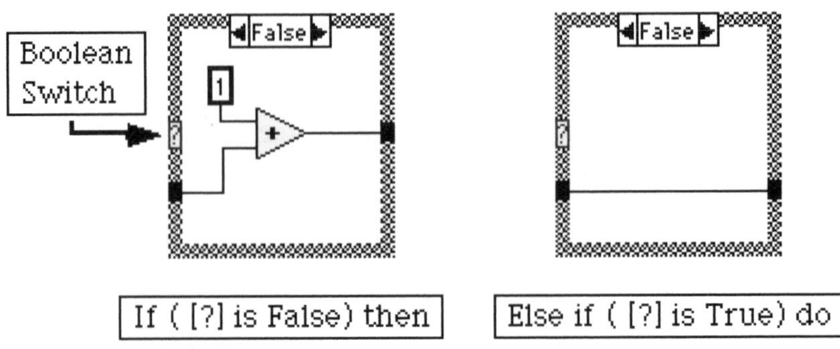

How Random Is Random?

A true random number generator will yield an equal number of heads or tails in the limit where the number of flips becomes large. Let's test this hypothesis on the LabVIEW random number generator. When a head is flipped, add one to a heads counter. When a tail is flipped, do nothing. The ratio of the number of heads over the number of flips should approach 0.5 in the limit of many coin tosses.

Modify your coin flipping program to count the number of heads or tails. A Case statement is used to increment the number of heads count. When a head is flipped, the <|**False**|> subdiagram is active and one is added to the heads count. When a tail is flipped, the <|**True**|> subdiagram is active and the count is wired straight through from one tunnel ■ to the other tunnel ■. To accumulate the number of heads, the previous sum must be recalled. In LabVIEW, memory of a previous result is accomplished with a shift register. Pop up on the boundary of the For . . . Loop and select **Add Shift Register.** Two new terminals appear on the loop structure. The terminal on the left is the previous result, and the terminal on the right is the new value. The shift register is initialized by wiring a constant outside the shift register to the left terminal. Save your program as HeadsSum.vi.

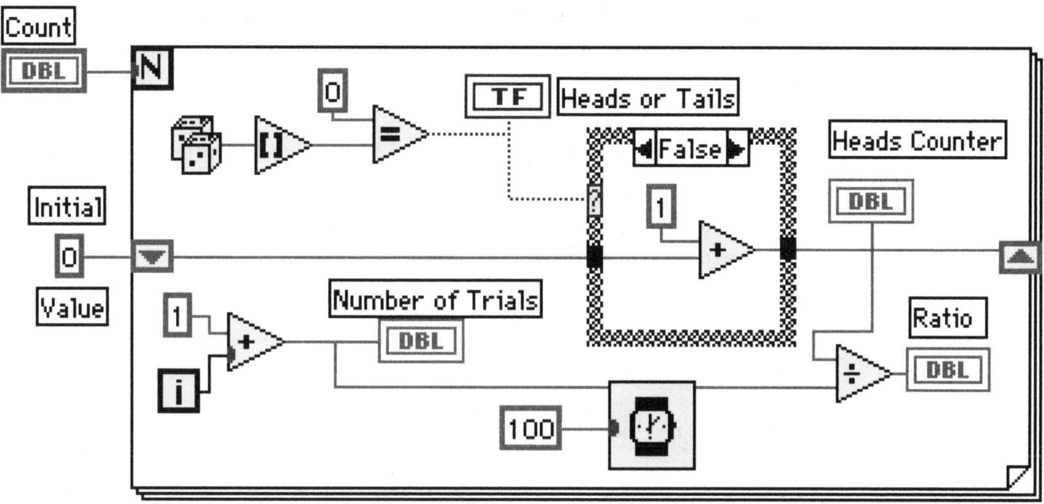

The loop index **i** is offset by one and is used to count the number of flips. [Ratio] is just the heads count divided by the number of flips. By trying different numbers of coin tosses, one can observe that the ratio approaches 0.5 in the limit of many tosses. It is also interesting to look at an experiment or a trial where the coin is tossed n times. Repeating the trial a few times and averaging the results shows the randomness in LabVIEW's random number generator. Here is a sample set of experiments.

Number of Flips	Trial 1	Trial 2	Trial 3	Trial 4	Average
1	1	0	0	0	0.25
8	6	4	2	5	0.53
64	35	37	32	28	0.516
512	252	245	239	267	0.490
4096	2086	1991	2052	2069	0.500

While . . . Loop

The While . . . Loop executes all the programming code contained within its subdiagram repeatedly, as long as the Boolean loop condition is True. When the condition become False, the

While . . . Loop completes the current subdiagram program and exits. The index counter [i] counts the number of times the loop is executed. Note that when the loop condition is not wired to a Boolean, the While . . . Loop will execute once and only once. This is an important property and is used in many sensor simulations for sequential evaluation of a transducer equation.

In LabVIEW, it is good programming practice to place all your program code inside a main While . . . Loop and wire the loop condition terminal to a front panel switch.

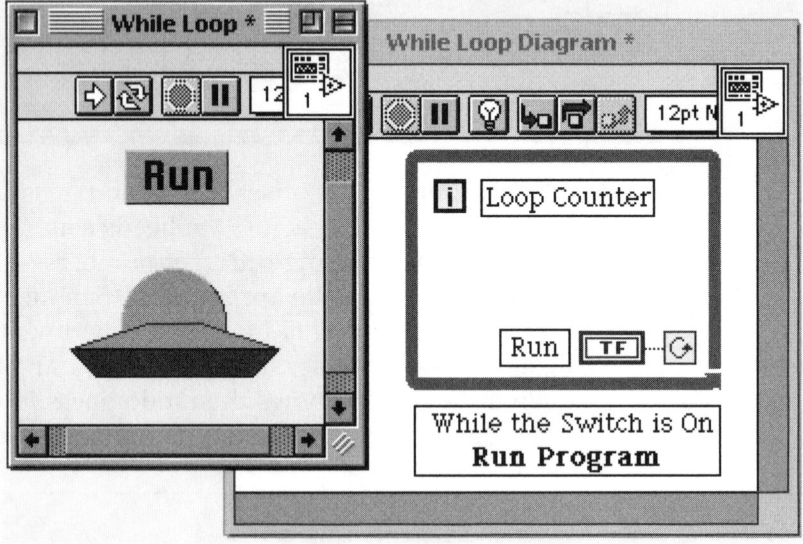

It is a bit like ensuring that each instrument has it own on/off switch. Programs interrupted by the main loop condition do so cleanly, leaving program variables in a known state.

■ LabVIEW Challenge: Replacing Structures

Replace the For . . . Loop in the above program with a While . . . Loop. Pop up on the For . . . Loop and select **Replace>Structure>While Loop** using the positioning tool. You will need to add a Boolean switch to the front panel and remove the number of Trials control. Note the difference between using the front panel switch or the run icon to start and stop the program.

Random-N Counter

Multiplying the LabVIEW random number generator output by a constant (*N* – 1) yields a random-N counter. For example, multiplying the LabVIEW random number by 5 and rounding the result to the nearest integer yields a random-6 integer counter. Here only six integers 0 through 5 are generated in a random sequence.

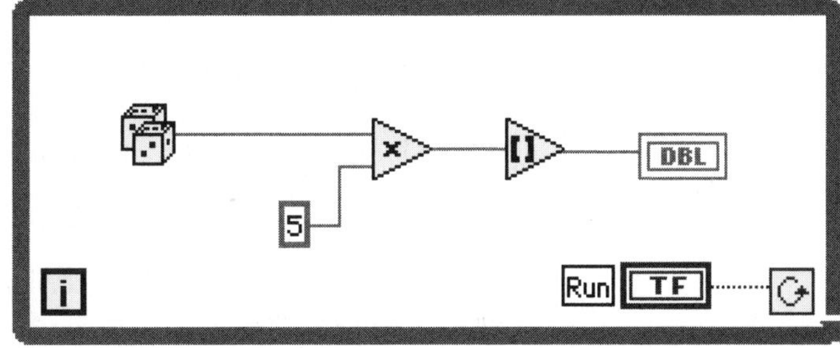

Random-N counters are particularly useful in situations where random numbers or events require random numbers within a fixed range (1 to N).

Digital Dice

A die consists of a six-sided cube with spots on its sides running from one spot to six spots. Seven spot locations in the form of an H are used to code the numbers from one to six. Look at a set of dice to remind you of the spot locations used for each number. In this section, a multiple Case statement is used to create a virtual die on the front panel.

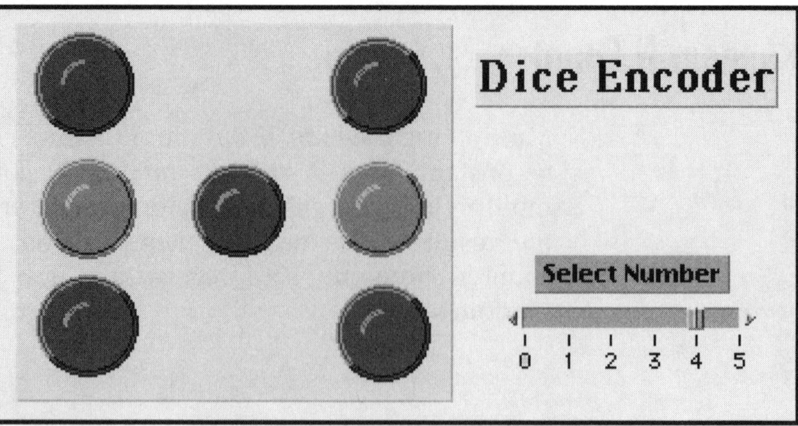

Seven LED displays are chosen and placed in an H formation. It is a good idea to use the resize tool to enlarge the LED displays. Select a LED and move the cursor to the bottom right corner. Click and drag the resize cursor to enlarge the LED. You may also want to color the LED so it is easy to distinguish between the on state the off state. Use copy and paste operations to create the other six LEDs. Note that the six states on the numeric control have been labeled as 0 to 5. Recall LabVIEW always counts from 0 not 1.

Create a new program called Encoder.vi using the front panel shown above. Use the positioning tool to select the integer range for the **Horizontal Pointer Slide** by popping up on the control and selecting **Format & Precision> [0] Digits of Precision.** The six different LED displays of the virtual die can be programmed by using multiple Case statements.

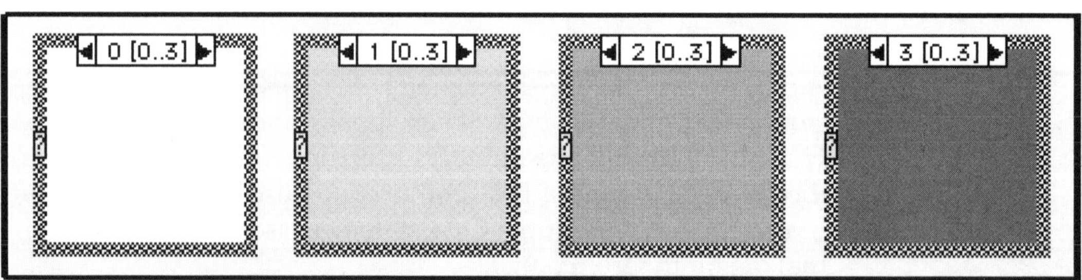

On the diagram page, pop up and select the Case icon from the **Functions>Structures** submenu. Click and drag to create a subdiagram to the workspace. On the boundary of the Case statement pop up a new menu and select **[Add Case After].** The <|True|>/<|False|> header will now be replaced by <|0 [0..2]|> /<|1 [0..2]|> or <|2 [0..2]|>. You have created a Case statement with three possible paths to follow. The choice of case is set by wiring the condition terminal [**?**] to a numeric integer that has the value of 0, 1, or 2. Repeat this process until you have six cases. The cases are stacked one on top of the other with only the selected case showing. Any cursor clicked in the header will move the stack up |> or down <| to reveal the case of interest.

Inside Case <|0|> place seven Boolean constants from the **Functions>Boolean** submenu. These are to be wired up to the seven LED indicator terminals. The <|0|> Case corresponds to the die (number one), hence only the center LED is to be on. With the operation tool, set the center LED to be true and all others to be False. The numeric slide selector labeled [Case] is wired to the **?** terminal.

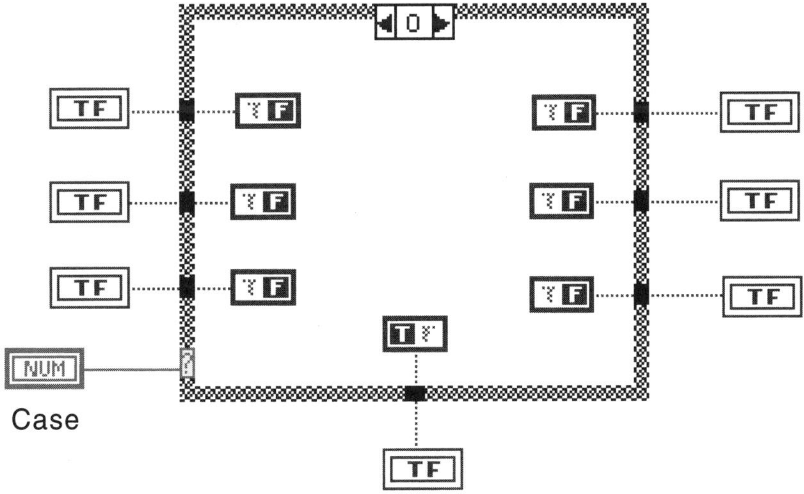

The remaining five cases are to be wired in a similar fashion. An easy way to reduce the number of cursor operations is to select the seven constants from the <|0|> case subdiagram and copy them into the clipboard. As each other case is selected, paste the seven constants into the new case workspace. You will still have to wire up the links to the appropriate tunnels.

Returning to the front panel, run the program continuously. With the operating tool, slide the [Case] control back and forth to see the different sides of the virtual die displayed. Moving between the states so fast that the eye cannot follow the changes is a method of rolling the die. At some time, the user stops the action and the resting place defines the roll.

LabVIEW Challenge: Let the Good Times Roll!

To roll the virtual die, it is necessary to replace the slider with a high-speed Random-6 counter. When the roll switch is asserted, the Random-6 counter outputs a random sequence of numbers from 0–5 at a speed much faster than the eye can see. All the LED displays blend together into an blur of numbers. When the roll switch is stopped, the Random-6 counter stops and the encoder displays the random number on the virtual die.

Note: Replace the slide switch on the Encoder front panel with a Boolean switch called "Roll." On the diagram page, place the Random-6 counter and the encoder structure inside a While . . . Loop. Wire the Roll switch terminal to the While . . . Loop terminal and let the "good times roll."

Shift Registers

LabVIEW's program execution follows the data flow. At each programming node—be it a function, a VI, or an operation—calculations can only proceed when all the data inputs are available. Then the program continues through the node until all the elements have completed their operations. Finally, outputs are passed to the next node(s). As a result, sequential programming such as the linear programming operation

$$X = X + 1$$

cannot be accomplished in the conventional manner. To execute this operation, (add 1 to the previous value of X), the program must have memory. LabVIEW provides memory in the form of the shift register.

On a new diagram panel open up a While . . . Loop and pop up on the node boundary. A new menu is presented that allows a shift register to be added to the While . . . Loop. Select **Add Shift Register** and two new terminals show up on the While . . . Loop boundary.

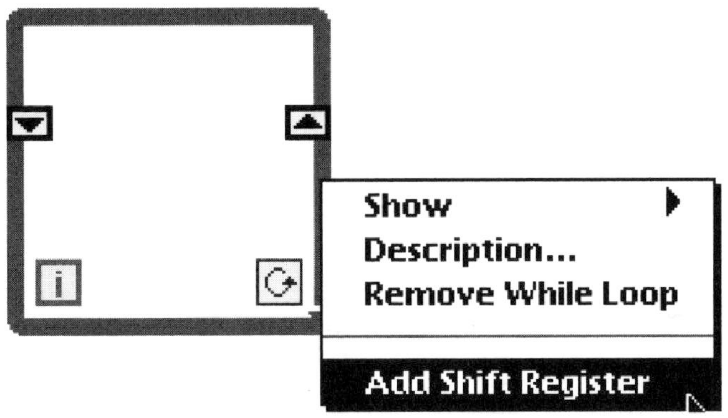

The terminal with the up arrow ▲ is the input and the terminal with the down arrow ▼ is the output of the shift register. After execution of the first iteration, the contents of the input terminal are passed to the output terminal. This value, the previous result, then becomes the value used in the next iteration.

For readers who may be familiar with digital electronics, the While . . . Loop with a shift register is equivalent to a D-latch. The ▲ terminal is the data input D, the ▼ terminal is the output Q, and the loop index [**i**] is the clock input, clk.

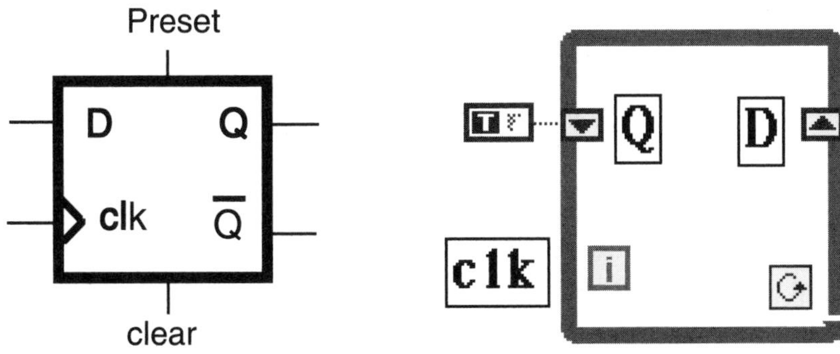

Multiple shift elements can be added by popping up on the boundary and adding another element by selecting **Add Element.** If seven more elements are added, the structure becomes an 8-bit shift register.

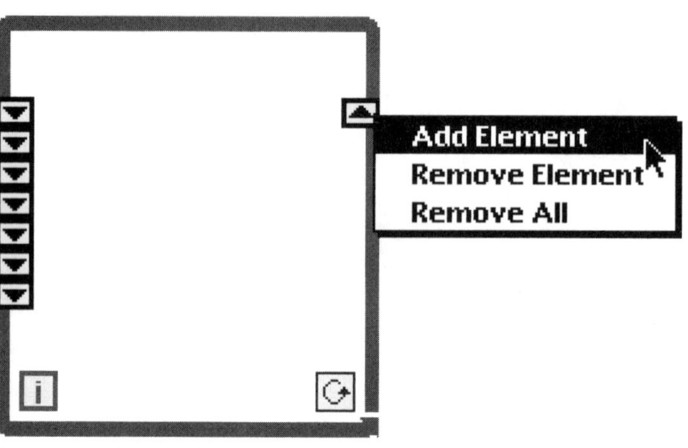

Eight-Bit Shift Register

To visualize the operation of an 8-bit shift register, open up a new VI and on its diagram panel place a While . . . Loop with a shift register of eight elements. The top ▼ is the least significant bit while the last element added is the most significant bit.

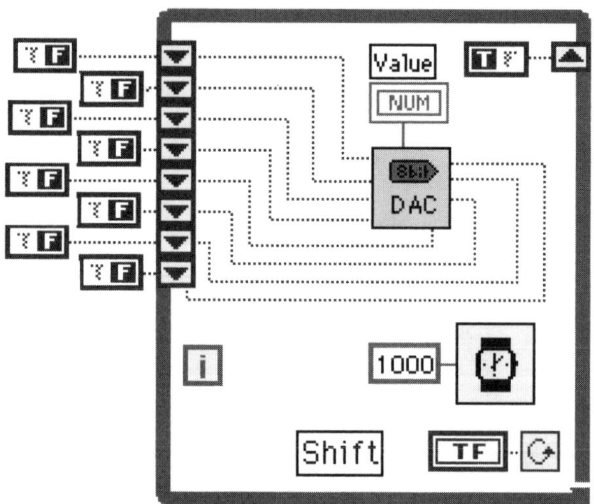

To initialize the shift register, eight Boolean constants are added outside the While . . . Loop and wired to the respective shift register elements. This forms the basis of a parallel load into the shift register. For a serial load, a ninth Boolean constant or variable is added inside the loop and wired to the shift register input. Wire up the eight shift register outputs to eight LED displays (optional) and add the DAC.vi designed in Chapter 1. The output of the DAC goes to a numeric display on the front panel. Try to predict the sequence of numbers that will appear on the LED and numeric displays before running the program. Save your program as Shift8.vi. *Note:* Remember the binary weighting scheme of the DAC.vi.

■ LabVIEW Challenge: Catch a Falling Star!

Design a program to simulate a star falling into the earth's atmosphere. The star is depicted by a yellow LED display and its terminal is wired on the diagram panel to the least significant bit of a 8-bit shift register. After some random time, the pattern in the register starts to shift at a rapid rate and the star begins its journey into the earth. Seven LED displays colored red and wired to other output states simulate the falling star. Another very large LED colored blue signifies Mother Earth, which will change color when the star strikes earth's surface. The chapter opener art is an example of one such design.

Ring Counters

When the output of the most significant bit, the bottom ▼, is wired to the input, the n-bit shift register becomes an n-bit ring counter. A ring counter will repeat the output shift register pattern after a finite number (N) of iterations. In the following case, the most significant bit is initialized true while all other bits are false. After eight iterations, the output pattern on the DAC output or on the LED displays will repeat. Again try to predict the sequence of outputs before running the program. In this arrangement, the 8-bit ring counter is also called a modulo-8 counter.

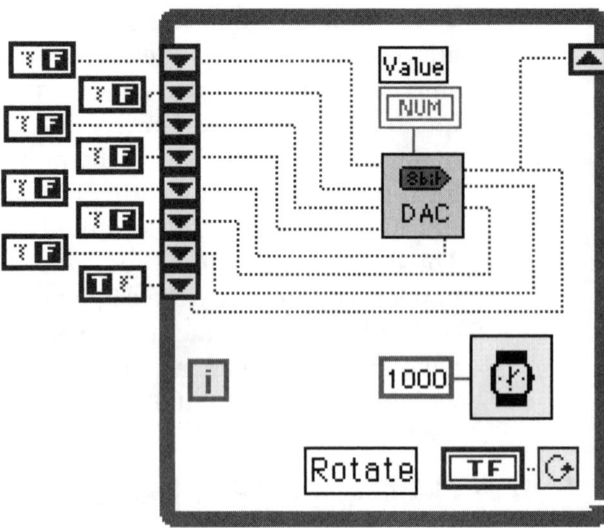

A modulo-n counter is a common component used in digital electronic circuits to repeat an operation n times.

■ LabVIEW Challenge: Switch Tail Ring Counter

A switch tail ring counter inverts the output of the most significant bit before feeding it into the input of the shift register. *Modify the above configuration to form an 8-bit switch tail ring counter.* This program now becomes a modulo-X counter. What is the value of X?

Pseudo-Random Number Generator

In general, a n-bit ring counter repeats the pattern after n loops; hence, the ring counter is also a modulo-n counter. With the addition of various logic conditions on the Boolean outputs, the input can be prepared so that the counter simulates any type of module n counter. In the special case where the Boolean outputs 3, 4, 5, and 7 are XORed together and this result is fed back into the input, the 8-bit ring counter becomes a pseudo-random number generator (PRNG). This device forms the basis of many bit error testing instruments. When the binary output of the PRNG is fed into the DAC.vi, a pseudo-random sequence of numbers in the range 0 to 255 is generated. Each number appears only once in the 255-step sequence, then the sequence repeats.

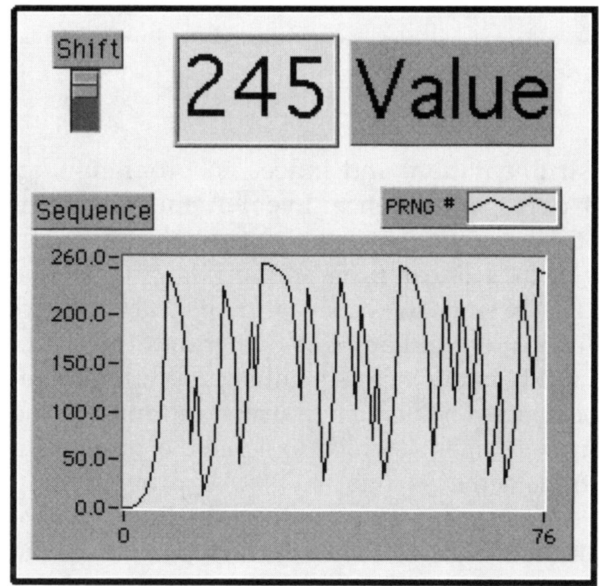

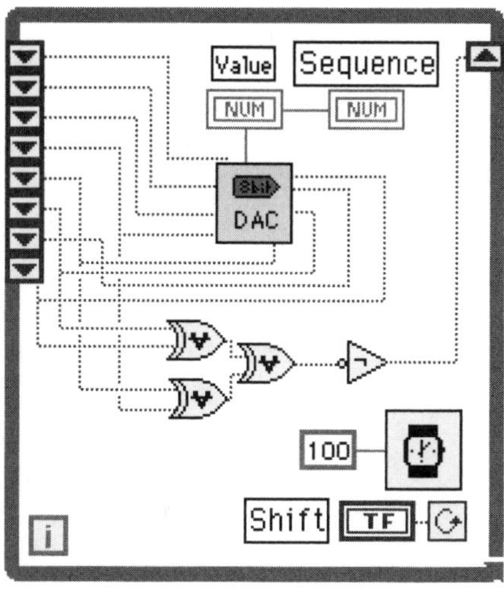

Using a Waveform Chart, observe the PRNG sequence of the generated numbers. Over a small number of iterations, the sequence appears random. However, the complete number sequence repeats every $(2^n - 1)$ loops, where n is the number of elements in the counter. For eight bits ($n = 8$), $2^n - 1 = 255$. Let the PRNG.vi program run until over 300 numbers have been generated. With the operating tool, the upper and lower limits of the x-axis in the chart display can be set so one can observe a smaller region of the data set. Observe that the sequence of numbers in the range 0 to 20, agrees exactly with the set of numbers in the range 255 to 275. With the pseudo-random number generator, over the short interval, the values are "random," but over the long interval the sequence repeats itself. Hence, future values can be predicted from a current value and the number of iterations.

■ LabVIEW Challenge: Predicting PRNG Values

Modify the PRNG program so that a finite number of iterations take place. If the initial value of the shift register was set to the binary pattern (10101010), what would the numeric value of the PRNG read after 9 loops? Save this program as PRNG2.vi.

Sequence Structure

Many times in instrumentation, and indeed in programming, some events must occur in sequence. Event A must be completed before event B, which must be completed before event C. This is an easy task in a linear programming languages like BASIC, C, or FORTRAN since the order of the instructions dictates the program execution. In LabVIEW, program flow is dictated by data availability and hence sequential program flow requires a special construct. The Sequence structure found in the **Functions>Structures** submenu, which looks like a frame of a piece of 35 mm film, achieves this result. All programming placed inside one frame executes before the next frame is executed. The frame number, <|**n**|> is found in the top center of the subdiagram. Popping up on the frame boundary provides a new menu to add or remove frames. Sequence locals from the same popup menu allow local variables to be passed between frames. As a example, let's calculate how long it takes to execute the 255 iterations of the PRNG2.vi.

Calculating the Execution Time for a Sub-VI

On a new diagram panel, select the **Sequence** structure from the **Functions>Structures** submenu. Popping up on the sequence boundary, add two frames **Add Frame After** and one variable **Add Sequence Local**. Into the first frame add the **Tick Count** function from the **Functions/Time & Dialog** submenu. This function returns the current value of an internal clock measured in milliseconds with a resolution of 55 ms for the PC and 17 ms for the MAC. By wiring the Tick Count output to the local variable, the initial value of the internal clock can be forwarded to another frame. In the middle frame place the PRNG2.v1 with the [# of loops] control set to 255. The last frame calls another Tick Count function, which gets the completion time. Subtraction of the initial clock value from the final value divided by 1000 yields the execution time in seconds for 255 iterations of the pseudo-random number generator.

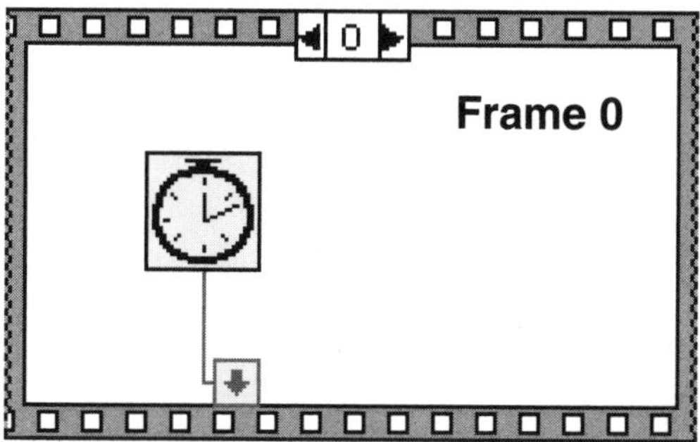

Start the Timer

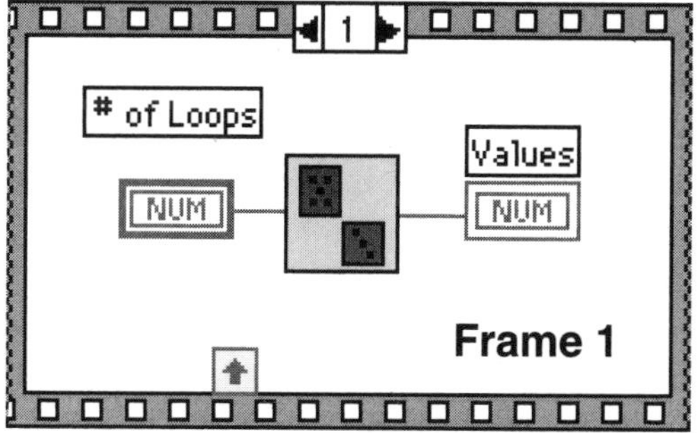

Call PRNG
Loop Times

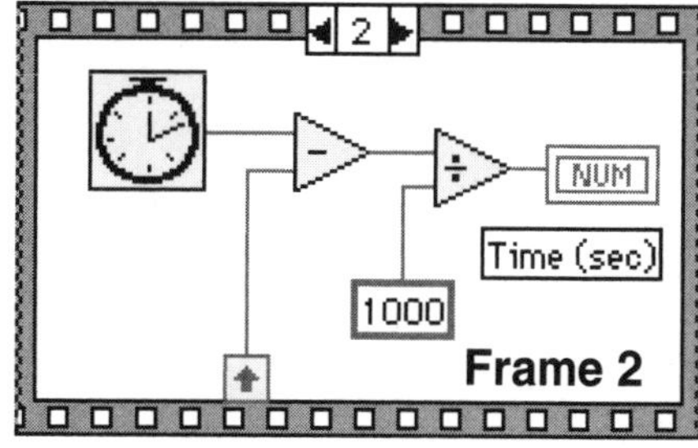

Stop the Counter
Calculate the Elapsed Time
Display the Time

How long does it take to execute one loop?

There is a difference between dividing the cycle time of the 255 PRNG numbers by 255 and executing the PRNG2.vi once. What is the cause of this difference? Can you think of a way of compensating for this difference, so that the time for n loops time is always n times the time for one loop?

■ LabVIEW Challenge: Random?

If you knew the initial value of the PRNG, and it completed n iterations, could you use the response time of the PRNG to predict what number the PRNG is currently displaying?

Overview

Ports are the computer's link with the real world. There are parallel ports, serial ports, instrument ports and network ports but the simplest port is an eight bit parallel port. This chapter uses 8-bit data formatted as a numeric or as a parallel combination of Booleans to drive the outputs on a parallel port. Examples include controlling a seven segment display, a two way stop light intersection and a stepping motor. In the later case, a driver is built to first simulate rotary motion and in later chapters execute rotary motion.

Goals

- Understanding the Parallel Port
- Data types (Boolean and numeric)
- Build a virtual seven segment display
- Design your own stop lights controller
- Build a stepping motor controller

Key Terms

- Parallel port
- Binary, hexadecimal, and BCD numbers
- 7-segment displays
- Modulo-4 counter
- Up/down counter
- Stop lights
- Stepping motor transducer

4 Calling All Ports

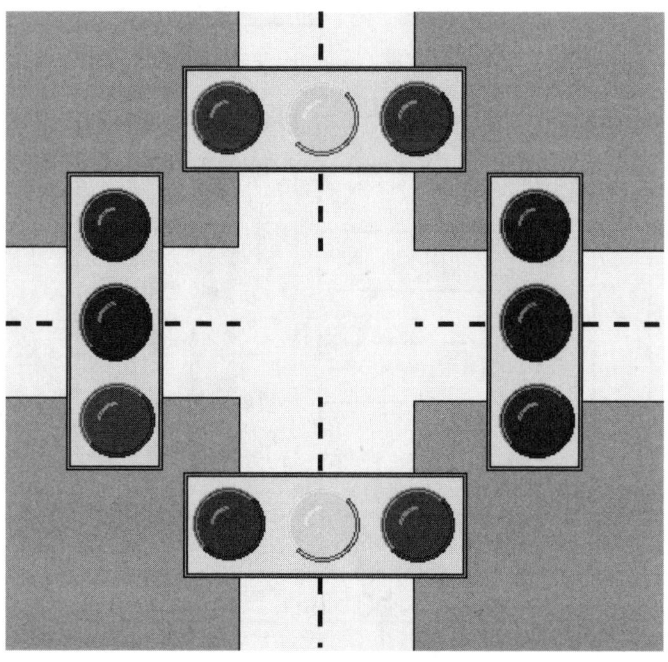

Stop Lights at a 2-Way Intersection

Ports transfer the internal calculations for measurement or control algorithms into and from the real world. The port converts the internal data types (Boolean, numeric, or ASCII) into the specific codes required by an external device. Historically, computers have provided some form of parallel and serial ports for the user. Most PCs offer two serial ports and one parallel port. One of the serial ports is commonly used for the mouse interface, leaving one free serial port. The parallel port is often configured to interface with a printer or other external devices. Macs have two serial ports and a parallel SCSI interface for external drives. Today higher speed interfaces such as the Ethernet port make computers much more useful and flexible. The port is also a bridge between asynchronous events in the real world and the synchronous world of the central processing unit or CPU.

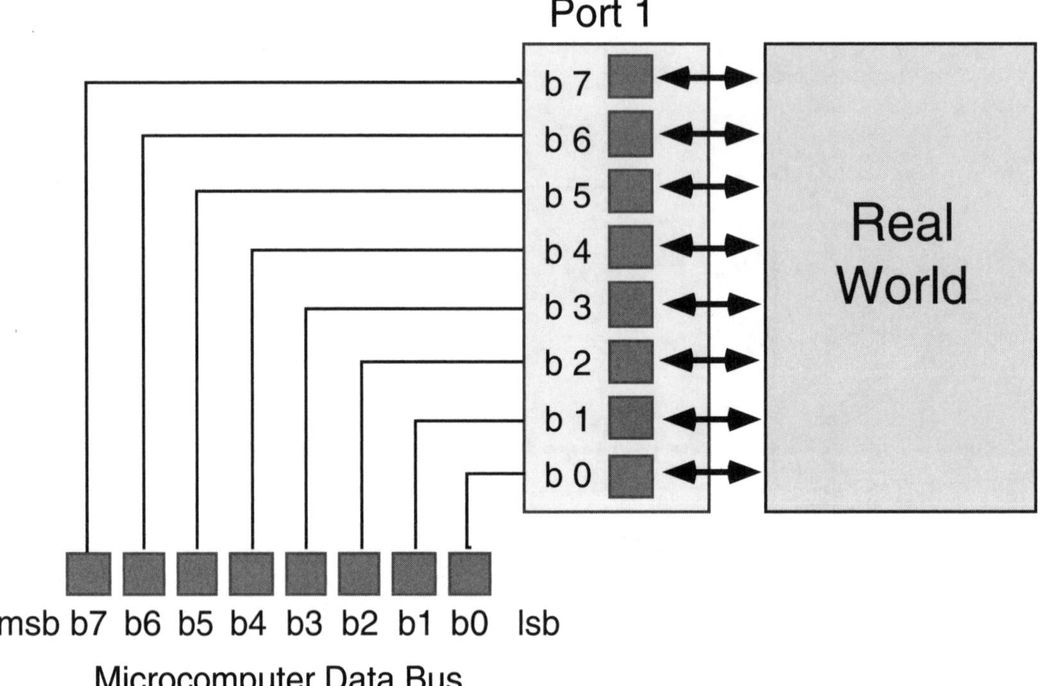

In this chapter, parallel ports are studied and simulated. A parallel port driven by Boolean controls is used to demonstrate the operation of a 7-segment LED display. In a second application, the operation of stop lights at a two-way intersection is simulated. A third example uses four bits from a parallel port to drive a stepping motor in a clockwise or counter-clockwise direction. In the following chapter, the serial port is used with an inexpensive hardware interface to bring a parallel port to LabVIEW and the simulations of this chapter come alive with lights, action, and camera.

The Parallel Port

A parallel port is just a computer memory location that is connected to the outside world. It has an unique address and can be configured to read (input) or write (output) eight bits of information. The output bits are labelled bit 0 (b0) to bit 7 (b7) and reflect the same 8-bit binary data structure used by the central processing unit.

What makes a port different from an internal memory location is that the port information can be projected into the real world to drive displays, read meters, move motors, check the status of a relay, and so on. Of course, the ability to interface a specific device often requires an electronic interface or driver circuit to satisfy timing and power requirements. The port provides the interface. With the driver electronics and CPU control, ports provide computer control of an external device.

Bits, Bytes, and a Few Words about Binary Data Types

Humans have been counting from the beginning of time and it is no coincidence that we have ten fingers and toes and also use a base 10 numbering scheme. However, a computer has, so to speak, only two "fingers and toes" and its numbering scheme is base 2. Physical properties are often binary: electronic charge is positive or negative, magnetic fields have a north and a south pole, electronic spin is up or down, gender is male or female. When humans conspired to use nature to help build computers, then computer components exploited these natural physical properties. Presence or no presence of a magnetic field codes a binary 1 or 0 in floppy disks or the hard drive. A high or low voltage codes a binary 1 or 0 in random access memory. Even the presence or no presence of current in a loop used to set or reset a mechanical switch has two binary states, open and closed. Each technology or industry has its own convention or names for these two states. For historical reasons, the communication industry refers to these two binary states as "Mark" or "Space," while sensor technology uses on or off. LabVIEW labels the two binary states as True or False.

The power of binary numbers lies in the ability of parallel binary bits to represent large numbers, either real or floating point. Since modern computers can do millions of binary operations per second, complex calculations or number manipulations can be executed in the blink of an eye.

Two bits in parallel are called a pair. There are only four possible pair combinations (00, 01, 10, or 11). Four binary bits in parallel are called a nibble and there are 2^4 or 16 possible combinations of the four parallel bits. Eight bits in parallel are called a byte and there are 2^8 or 256 possible combinations. We could continue with larger parallel combinations such as 16-bit words, or 24-bit long words or 32-bit super words. Each parallel combination give more and more possible combinations with larger and larger numbers able to be represented. For N parallel bits, there are 2^N distinct binary combinations. To convert these parallel bit structures into a numeric number, each column has a weight two times the previous one.

The decimal equivalent value, DEV can be calculated with the formula

$$\mathbf{DEV} = \Sigma\, \mathbf{g_i}\, \mathbf{b_i}$$

where $b_i = 0$ or 1, the binary weight factor $g_i = 2^i$ and i is the column number 0–N. The octal numbering scheme uses 3 binary bits to represent all the numbers from 0 to 7.

Example: Octal Number Scheme (3 bits in parallel)

2^2	2^1	2^0	<---- binary weighting factor	
4	2	1	=	# <---- Decimal Equivalent Value
0	0	0	=	0
0	0	1	=	1
0	1	0	=	2
0	1	1	=	3
1	0	0	=	4
1	0	1	=	5
1	1	0	=	6
1	1	1	=	7

The hexadecimal number scheme is more interesting as it contains sixteen states and is widely used in computer interfacing. Four binary bits in parallel—a nybble—are arranged in ascending order. The lowest order number (0000) has a decimal equivalent value of 0 while the largest number (1111) has a decimal equivalent value of 15. In hexadecimal, each state is represented by a single character from 0, 1, 2, 3, 4, 5, 6, 7, 8, 9, A, B, C, D, E, to F for the numbers 0 to 15.

Note the six extra states "fingers" are

Binary	Hexadecimal	Numeric
1010	A	10
1011	B	11
1100	C	12
1101	D	13
1110	E	14
1111	F	15

Many techniques are used to signify a hexadecimal character such as **$A** or **AH**, but LabVIEW uses the backslash in front of the hexadecimal character **\A.**

A byte can also be represented by two nybbles. The least significant nybble, bits 0–3, and the most significant nybble, bits 4–7, can each be represented by a single hexadecimal character (0 . . . F). Hence, the 8-bit binary number (11111111) having a numeric value of 255 can also be thought of as two nybbles with the binary pattern written as (1111 1111) or the hexadecimal equivalent value of $FF. Hexadecimal is a convenient method of representing parallel bit patterns that are divisible by four. The binary value on an 8-bit parallel port can be completely represented by two hexadecimal characters.

Example: A set of Boolean controls forms an 8-bit data byte (01101100) or $6C, which is passed to the real world via Port 1.

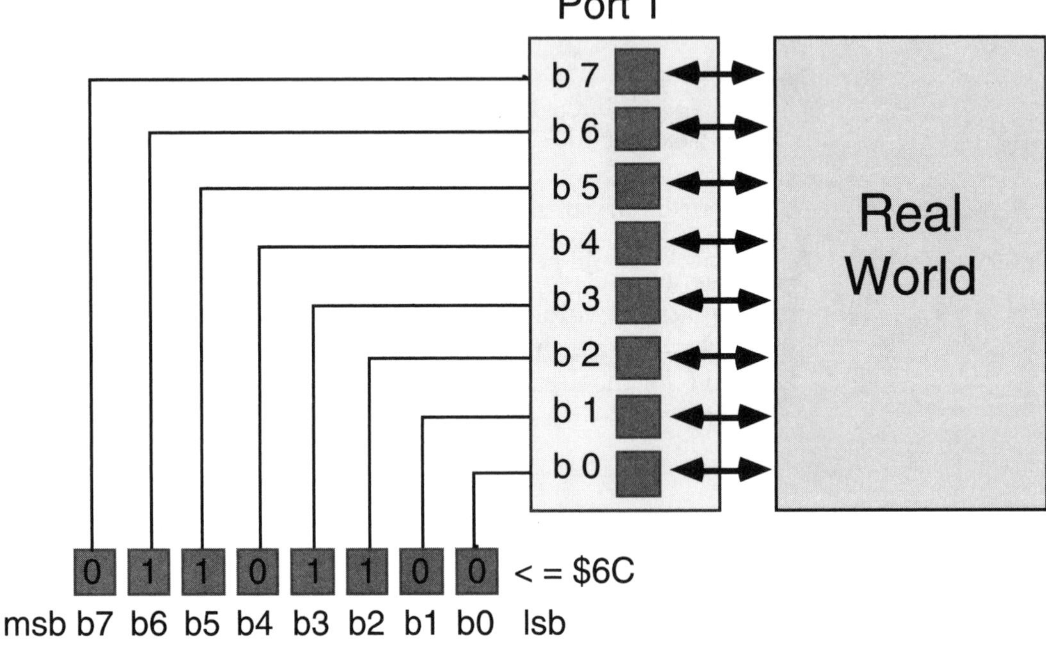

In LabVIEW, this data byte can be represented by eight Booleans, a three-digit numeric, or a two-character hexadecimal number.

Eight-bit Parallel Output Port

Consider the case where eight light emitting diodes are connected to the eight output bits of a parallel port. Seven LED indicators resized as rectangular blocks are placed in a figure-eight configuration and an eighth LED is configured as a decimal point. This arrangement, called 7-segment display, is often used in numeric displays of meters and instruments.

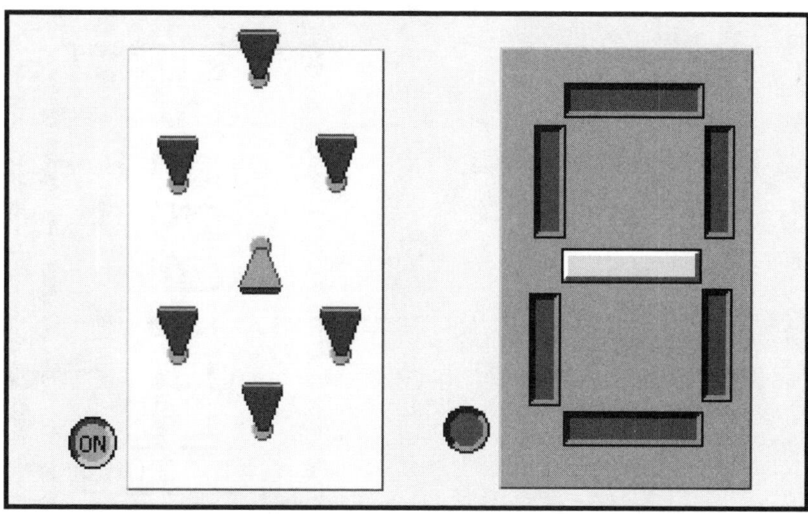

The output port bits 0 to 7 are represented by eight Boolean switches. The corresponding segments in the 7-segment display are traditionally labeled a to f. The least significant bit 0 is wired to segment a, the next bit 1 is wired to segment b, and so on. The most significant bit, bit 7, is often wired to an eighth LED and used as a decimal point. Launch the program 7-Segment.vi to observe the operation of a LabVIEW formed 7-segment display. Each time a switch is thrown, the corresponding LED indicator turns on or off. By selecting various switch combinations, some characters and all the numbers can be formed. After experimenting with the display, try outputting the message "help call 911" one character at a time.

Most 7-segment displays are driven by an encoder that converts a binary encoded nibble into a numeric number, which in turn selects the appropriate 7-segment code. In Encoder.vi, multiple case statements are used to provide the encoder function. The Case terminal [?] is wired to a numeric control whose output is formatted to select a single integer character from 0 to 9. The number zero will output the 7-segment code for 0, the number one will output the code for 1, and so on, all the way up to 9.

The Boolean constants inside each Case state are initialized to generate the correct 7-segment code.

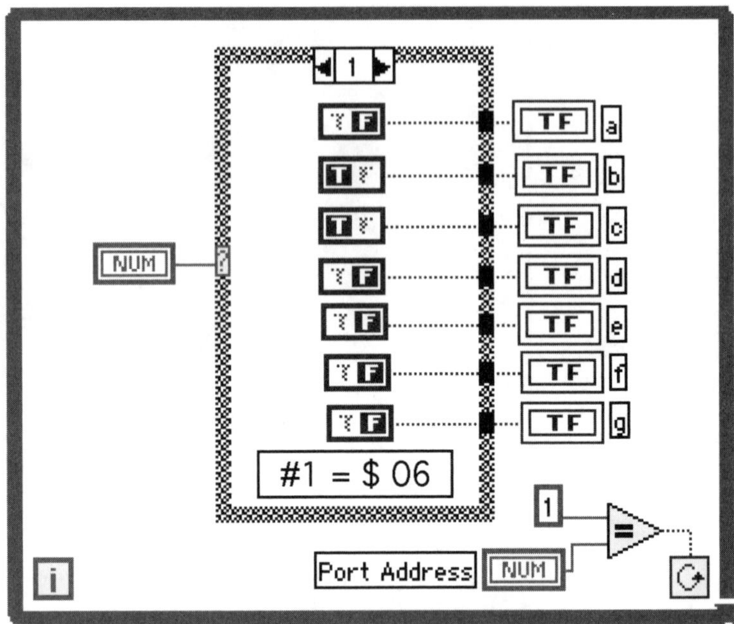

Each port has an unique address that must be selected before data can be written to or read from the real world. The correct address must be entered on the front panel to access the port and in this simulation, the Address operates the run command. Encoder.vi is also a sub-vi that can be used in other programs. Its input is a numeric and its outputs are the corresponding 7-segment codes.

Create a new program with seven LED indicators placed in the 7-segment arrangement, a numeric control for the Port address, and a slide control with integer numbers from 0 to 9 as shown on the next page.

On the diagram panel add from the chapter library the VI called Encoder.vi. The **Help** command gives the wiring diagram details. Running this program continuously allows the 7-segment codes to be displayed. Use the operating (hand) tool to drag the slider along to select the number to display.

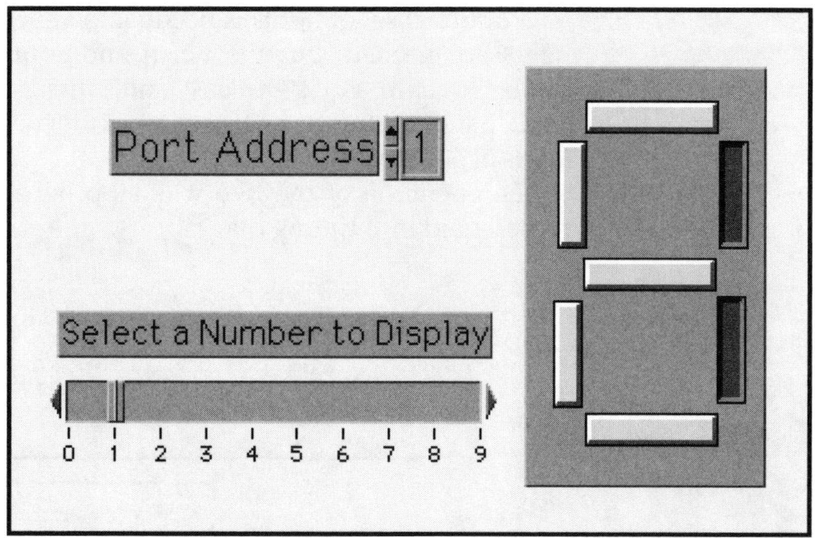

In a digital circuit, often four binary weighted lines are used to select the 7-segment pattern for the numbers 0 to 9. Such a device is called a binary to numeric encoder. Of the sixteen possible output states, only the ten number states are used. This subset is called "Binary Coded Decimal." Save your program as BCD.vi.

Exercise: Add six extra cases to Encoder.vi so that your 7-segment display can display any hexadecimal number from 0 to 15. Call this program Hex.vi.

Two-Way Stop Light Intersection

Have you ever waited at a stop light intersection and wondered how long it would take for the red light to turn green? In most intersections, the red light is on for 30 seconds, the yellow light for 5 seconds, and the green light for 25 seconds. The operation of a single set of stop lights requires two pieces of information, the on/off light code (which light is to be on) and the on-time (length of time to leave the light on). For a two-way intersection, an additional set of lights is needed. In general, four sets of stop lights are needed, but the north and south lights follow the same sequence as do the east and west lights. The LabVIEW front panel featured as the chapter opener art displays stop lights at a two-way intersection.

The operation of the two way stop light intersection can be represented by a timing diagram.

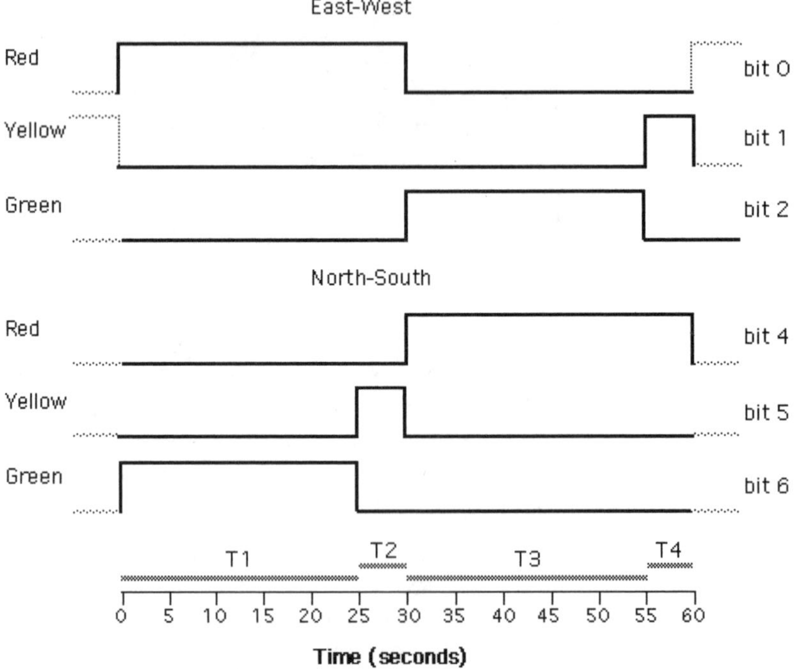

One might think that there are six states in cycling the two sets of stop lights but in fact there are only four unique states. In the 30 second timing period that the red light is on for one direction, both the green (on for 25 s) and yellow light (on for 5 s) are active in the other direction. This observation allows one to note the four operation timing periods T1,T2,T3 and T4 on the timing diagram.

Assume that six colored LED's two of red, yellow and green are wired to an 8-bit output Port. Bits 0...2 are wired to a red, yellow and green LED for the East-West direction and bits 4...6 are wired to a red, yellow and green LED for the North-South direction. To operate the LEDs, the appropriate code is sent to the Port and then held for the corresponding on timing period. After the four cases are executed, the cycle repeats. Here is the port codes for each timing period.

			T1	T2	T3	T4 ←Timing Period
	bit 6	G	0	0	1	0
East-West	bit 5	Y	0	0	0	1
	bit 4	R	1	1	0	0
	bit 2	G	1	0	0	0
North-South	bit 1	Y	0	1	0	0
	bit 0	R	0	0	1	1

Load the program Stop.vi from the Chapter 4 library and watch the lights operate.

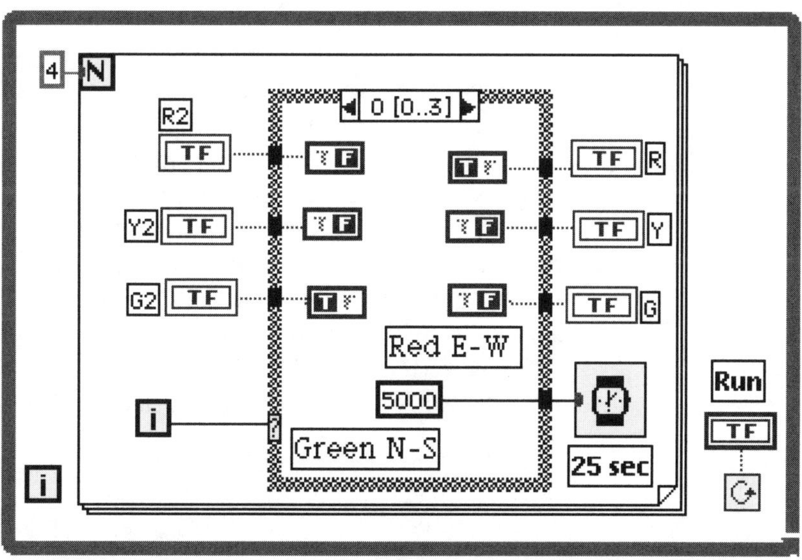

Four case statements are used to generate the Boolean control pattern required to turn the LEDs on or off. A Wait operation determines the length of display time. To speed things up, 1 second in simulation time corresponds to 5 seconds real time. A

For . . . Loop with the count terminal wired to [4] defines the stop light operation cycle and a While . . . Loop start or stops the simulation.

■ LabVIEW Challenge: Night-time Operation of Stop Lights

At night-time, the yellow light flashes in the north-south direction while the red light flashes in the east-west direction. When a vehicle arrives at the E-W approaches to the intersection, a magnetic field transducer senses the vehicle and switches the stop lights back to daytime operation for one operational cycle. *Design new program called Night/Day.vi to simulate the night-time operation of a two-way intersection. Use a Boolean switch to simulate the magnetic field car detector.*

Stepping Motors

The stepping motor converts electrical signals into rotary motion. When a pattern of appropriate voltage levels is applied to four voltage-to-current generators, the selected lines energize electromagnets inside the stepping motor to form a magnetic field pattern. The rotor being attached to a permanent magnet near the end of the electromagnets rotates in response to the magnetic field pattern. By controlling the input phases and the sequence of the phases, the stepping motor can be rotated in a clockwise or a counterclockwise direction. A four-phase stepping motor in full step mode uses the following phase patterns.

	Pattern				
Phase→	a	b	c	d	
Step 1	0	1	1	0	
Step 2	0	0	1	1	
Step 3	1	0	0	1	Clockwise
Step 4	1	1	0	0	
Step 1	0	1	1	0	

Each time the pattern is changed to an adjacent pattern, the stepping motor will rotate one step. The number of phases and the number of coils determine the angular resolution. A typical motor in the full step mode will have 200 steps per revolution. Note that the pattern repeats after four steps. A shift in pattern one way causes a clockwise rotation while a shift in pattern the other way causes the counterclockwise rotation. To cause a rotation in a specific direction, two commands are required: **Step**, which causes the pattern to shift to an adjacent pattern, and **Direction**, which defines which adjacent pattern is selected.

To simulate the stepping motor driver, four bits on an output Port are assigned to phases a, b, c, and d. A four-element Case statement is used. The selection of the phase is dictated by a numeric input (0 1 2 3) connected to the case terminal **[?]**. The four-case elements contain the phases (0110), (0011), (1001), and (1100).

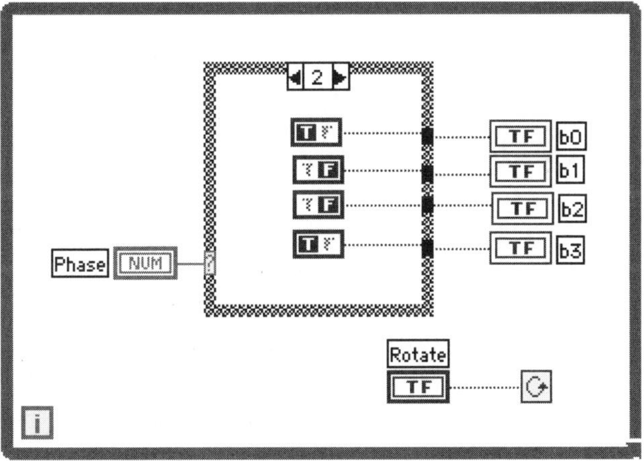

For the front panel, choose four LED indicators to represent the phases. These LEDs symbolize the current patterns sent to the stepping motor. A four-position slide control is used to select which step is active. When run, the operating tool is used to select one of the phases and the phase pattern is displayed on the LED indicators. Save your program as Phase.vi.

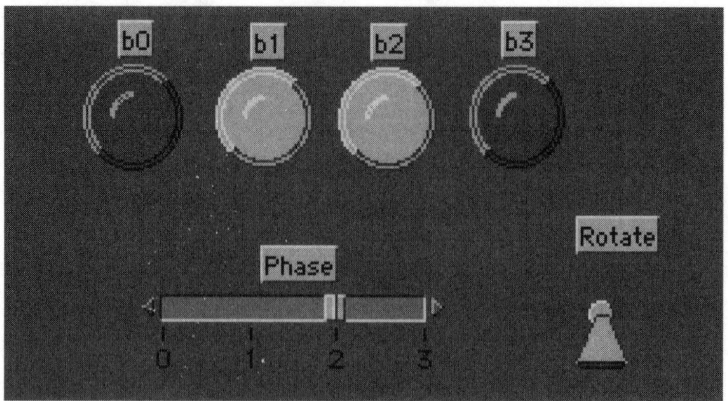

To make the motor rotate continuously, the input indicator must be incremented through the phases 0 1 2 3 0 1 2 3 0 1 2 3, and so on. Decrementing the sequence 3 2 1 0 3 2 1 0, and so on, causes a continuous rotation in the other direction. A device to generate this pattern is a modulo-4 counter.

Modulo-4 Counter

A modulo-n counter is one whose output repeats after n counts. For example, a counter with an output sequence such as (3 1 0 2 3 1 0 2 3 1 0 . . .) is a valid modulo-4 counter. An ordered modulo-4 counter would have a sequence like (0 1 2 3 0 1 2 . . .). It is this type that is most useful for the stepping motor driver.

To cause rotation in either direction, an ordered modulo-4 up/down counter is required. On command from a direction switch, the counter can run forward or backward. In the forward direction, the sequence required for the Phase.vi is just 0 1 2 3 0 1 2 3 . . . . To generate this sequence, it is necessary to detect the number 3 and then reset the counter to 0 for the next cycle. For this operation LabVIEW provides the **Select** function, found in the **Functions>Comparison** palette. The select function is a data fork where the Boolean input *s* decides which data input (true or false) is to be passed to the output.

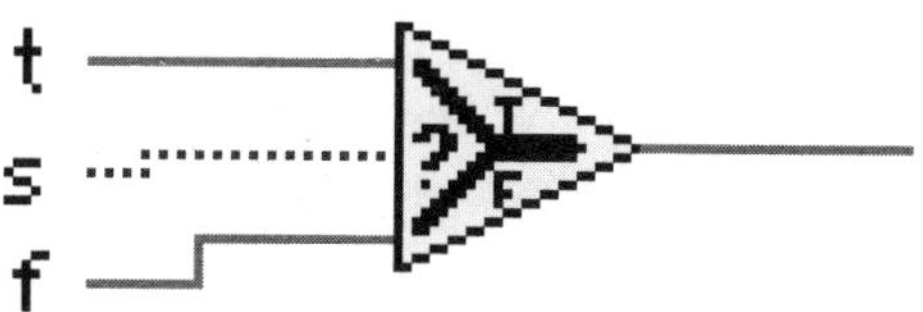

If the Boolean input *s* is true then input *t* passes to the output; if *s* is false, input *f* is passed to the output. In the following diagram, note how the shift register of the While . . . Loop is used to build a cyclic ring counter.

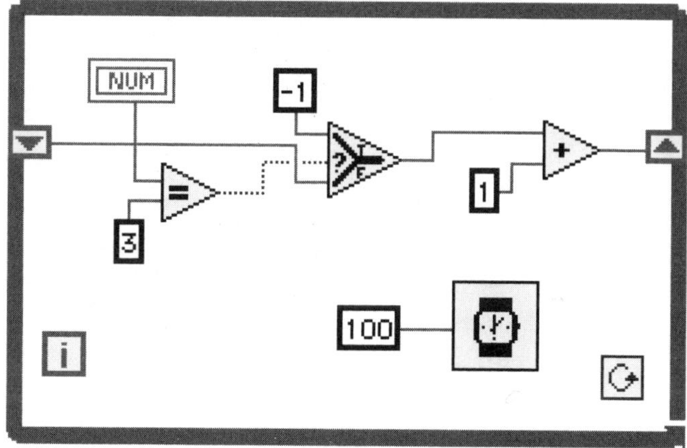

In the normal sequence when the current count does not equal 3, one is added to the count. However, when the count is 3, the data sequence is replaced with –1, which, when added with 1, becomes the reset value 0. Note the While . . . Loop control terminal is not connected. As a result, each time the sub-VI is called, it will execute the While . . . Loop only once. In this case, the next count in the modulo-4 count up sequence is returned.

The addition of two more select functions modifies the above program to produce an up/down counter with the ability to run forward or backwards through the four values of the ordered numeric sequence.

Load and run Up/Down.vi found the Chapter 4 library. Observe how the sequence changes direction whenever the switch is thrown from down to up.

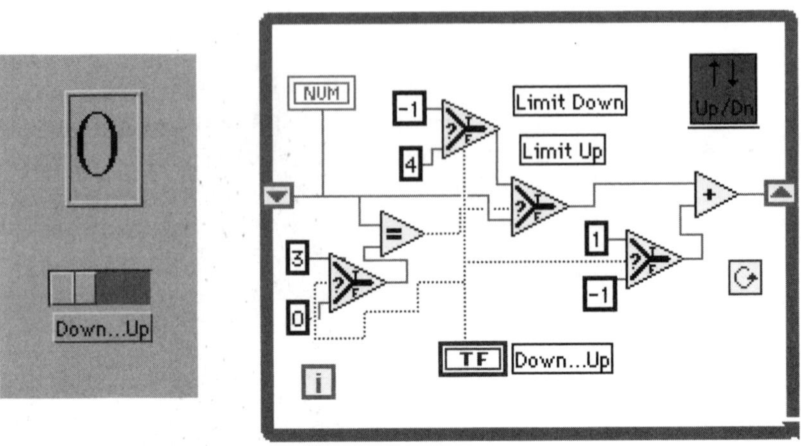

We are now in the position to put these two VIs together to build a stepping motor driver. The stepping motor simulator uses the output LEDs (bits 0 to 3) to display the phase changes. In a real application, these outputs would go to the current generators of the stepping motor interface. A single slide switch allows the direction to be changed from clockwise to counterclockwise.

To provide a virtual simulation of the motor turning, a radar plot is used to show the current position of the rotor. A numeric indicator displays the current angle of the stepping motor with respect to the horizontal axis. The front panel switch [Scan] enables the program. Load and run SteppingMotor.vi.

On the block diagram, note how the Up/Down.vi output drives the case structure to generate the stepping motor phases. The angle of rotation is calculated from the number of times the While . . . Loop executes. Incrementing or decrementing is controlled by the [CCW—CW] front panel switch. The polar plot sub-VI displays the current angle in a graphical format. Details of this VI are outside the scope of this chapter, but it will be discussed in a later chapter on arrays. By all means, open up the VI and take a look at its operation.

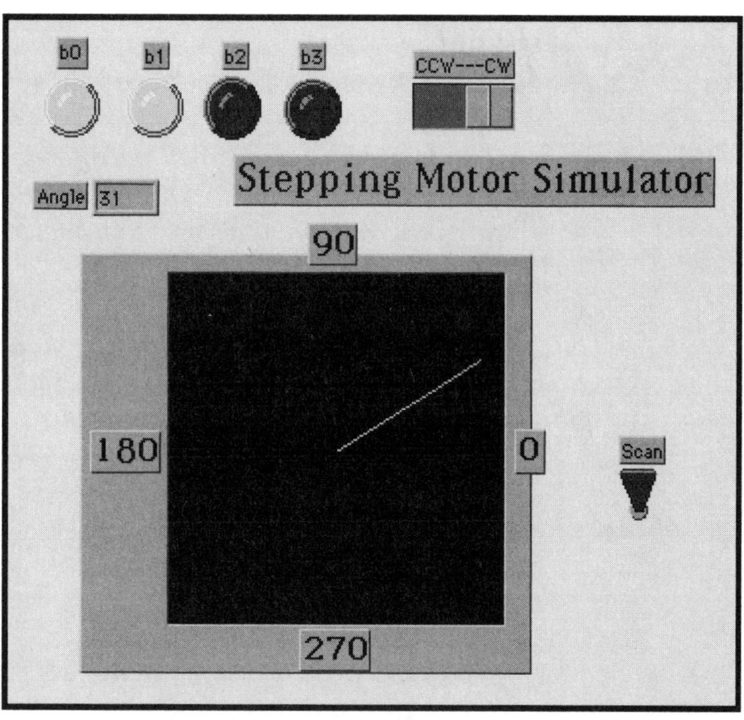
b0
b1
b2
b3
CCW---CW
Stepping Motor Simulator
Angle 31
90
180
0
270
Scan

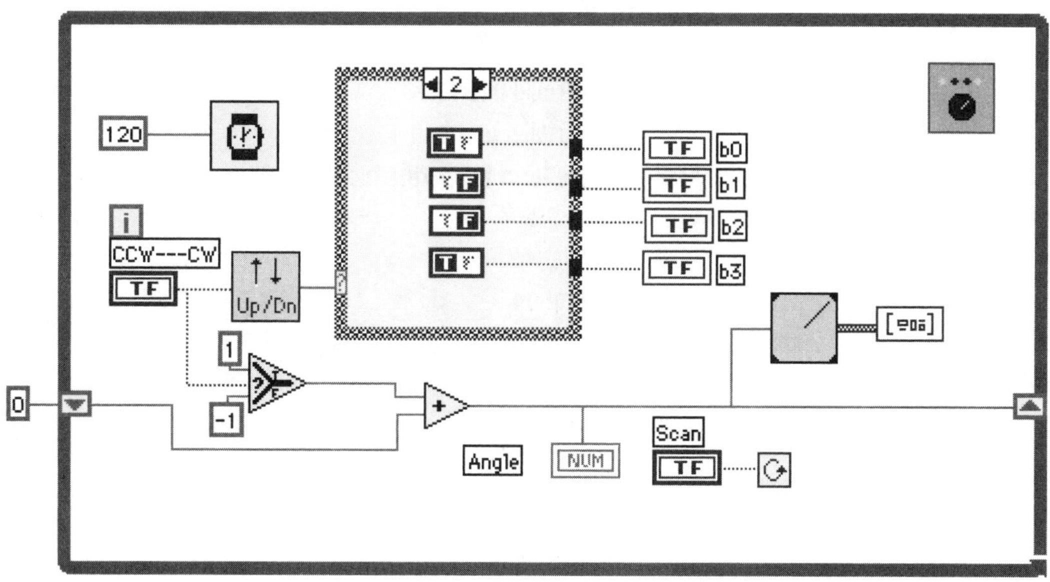
2
120
i
CCW---CW
TF
Up/Dn
b0
b1
b2
b3
1
-1
0
Angle
NUM
Scan

Overview

Serial ports are available on all microcomputers and form one of the simplest interfaces to the real world. A detailed understanding of serial waveforms and protocols opens up a wide range of instruments and projects to the LabVIEW enthusiast. A low-cost microcontroller is presented that converts serial messages into parallel data bits. Stepping motors, LED displays, relays, and so on can all be controlled through this interface. Finally, a serial driver for a temperature controller is presented to hone the reader's skills.

Goals

- Understand the serial port
- Understand RS232 waveforms
- Build a general purpose serial driver
- Understand serial to parallel interface
- Learn how to use the parallel port in the real world
- Design a serial interface for a temperature controller

Key Terms

- RS232 serial communication
- Baud rate, number of data and stop bits
- ASCII characters
- LabVIEW string data type
- Read and Write serial icons
- Microcontroller
- Temperature controller

Serial Communication

5

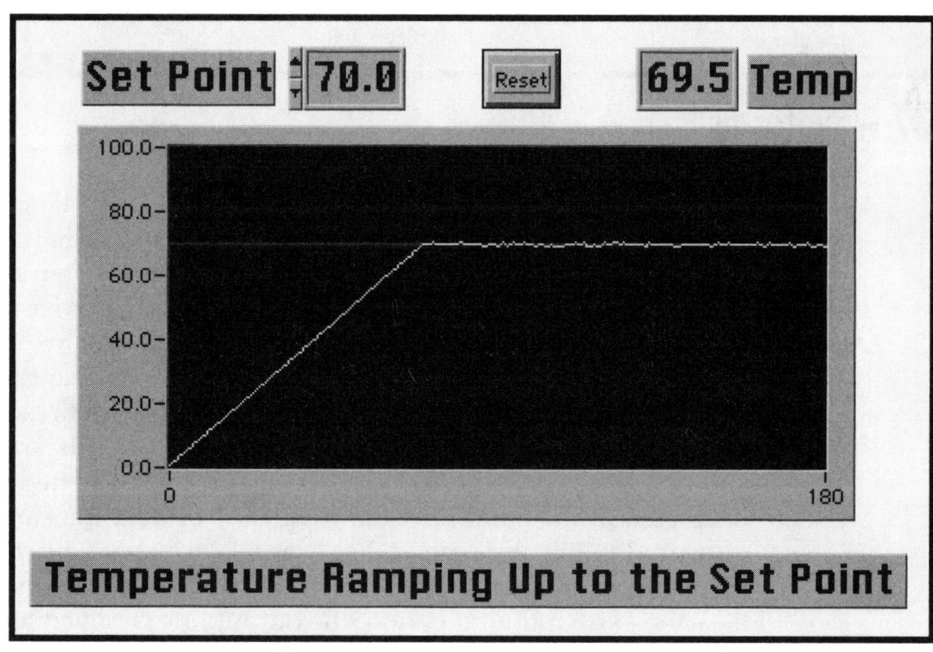

Temperature Profile of a Furnace as It Is Heated up to a Set Point

Many instruments, controllers, and computers are equipped with a serial interface. Our ability to generate device specific VIs opens up a whole new world of measurement and control. In LabVIEW, it is a simple matter of writing a character string to a serial driver and the data comes out the serial connector in a standard bit serial format called RS232C. This protocol defines the order of the bits and the waveform shape in both time and amplitude. Some optional hardware handshaking lines and the connector pin assignments are also specified. Understanding serial protocol is essential for successful interfacing serial devices. On the surface, it may look complex. However, underneath all the pages of specifications is a very simple interface. At a minimum only three communication lines are needed for data transfers between a computer and an external device: transmit, receive, and a reference ground.

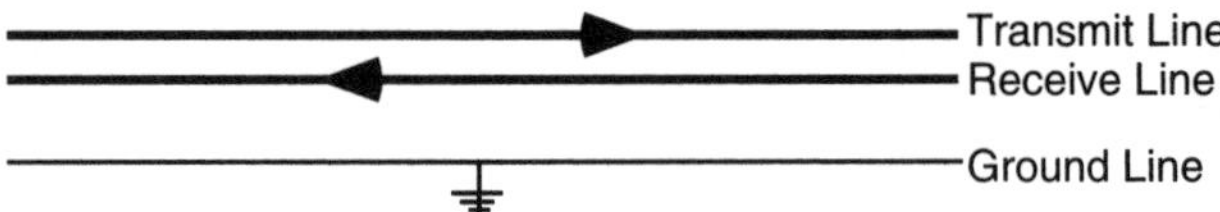

RS232 Waveforms

In serial communications, a high level is called a Mark state, while the low level is called the Space state. In normal operation, the output line is in a high state, which will often be denoted as a 1 or in LabVIEW as a Boolean True. The transmitter signals the receiver that it is about to send a data byte by pulling the transmit line low to the space state (0). This falling edge or negative transition is the signal for the receiver to get ready for incoming data. In RS232 communication all data bits are sent and held for a constant period of time. This timing period is the reciprocal of the baud rate, the frequency of data transmission measured in bits per second. For example, a 300 baud data rate has a timing period of 1/300 of a second or 3.33 milliseconds. At the start of each timing period, the output line is pulled high or low and then held in that state for the timing period. Together these transitions and levels form a serial waveform.

Consider an 8-bit data byte ($3A) or in binary representation (0011 1010). For serial communication, the protocol demands that the least significant bit (b0) be transmitted first with the most significant bit (b7) last. By convention, time is represented as moving from left to right; hence, the above data byte would be transmitted as (01011100), in reverse order.

The protocol also requires that the data byte be framed by two special bits, the start bit (a Space state) and the stop bit (a Mark state).

The start bit follows the falling edge signal to the receiver amd thus is a low (0). The stop bit signaling that the data byte transmission is completed is always high (1). The addition of these framing bits requires ten timing periods to send one byte of data.

If each byte represents an ASCII character, then ten serial bits are sent for each character. For example, a 9600 baud modem would be capable of sending 960 characters per second. The addition of the start and stop bits to the serial data byte generates a 10-bit data stream. In terms of a timing diagram, the RS232 serial waveform for ($3A) looks like the following.

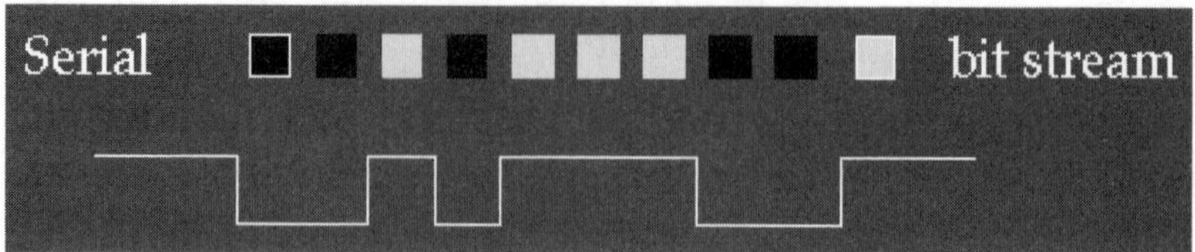

In summary, for a serial waveform, the start of transmission is signaled by a falling edge. The serial line is held low for one timing period. The least significant bit is sent first followed by bits b1 to b7. Each bit [a (1) or (0)] holds the line high or low respectively for one timing period. The stop bit held high for one timing period signals the end of transmission. The serial communication line is now ready for the next data byte.

ASCII Characters

RS232C does not specify what the data byte will represent. The data could be an 8-bit binary number, a 7-segment code, stepping motor phases, or anything the designer wishes the data to represent. For standard serial communications between computers and serial devices, the 8-bit data codes consists of a 7-bit ASCII code (bits 0 to 6) with the eighth-bit (bit 7) reserved for error detection in the form of a parity bit. The 128 possible codes of 7-bit ASCII data have been assigned to the upper and lower alphabet characters, the numbers 0 through 9, punctuation marks (? ,! . . .), special codes (@#$%^ . . .), and some non-printable control codes such as carriage return, linefeed, bell, and so on. A summary of the ASCII codes and their hexadecimal representation is found in the Appendix A at the end of this chapter. The parity bit, if used, has two states, called even (0) or odd (1). If it is agreed that a data communications link shall use even parity, then the received data bits should add up (modulo-2) to a (0). If any received data byte has a bit sum not equal to 0, then a transmission error has occurred. For odd parity, the binary sum modulo-2 should be (1). Many serial interfaces choose to use no parity and the eighth bit, even if present, is ignored.

LabVIEW Serial Drivers

In LabVIEW, a message is transmitted on the serial port by sending an ASCII character or string to the **Serial Port Write.vi.** The write VI looks after all the formatting and waveform generation to produce an RS232 waveform at the transmitter pin of the serial connector. Like any port, the serial port must have a valid port address, or in LabVIEW, a valid port number.

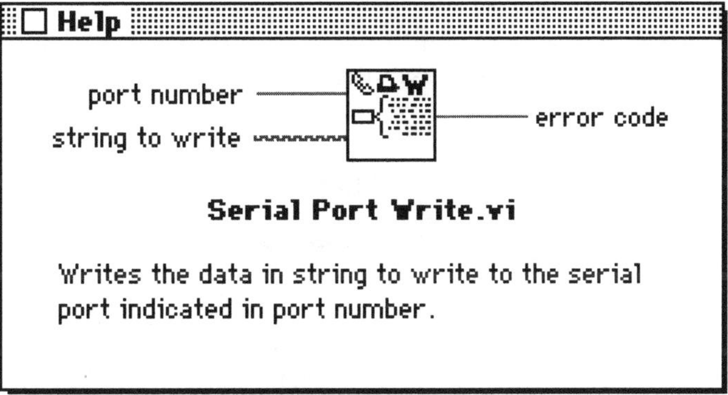

The **Serial Port Write.vi** also returns a numeric error or diagnostic code. A zero code (0) indicates no error. Other error codes can be found in the LabVIEW manuals.

The simplest serial driver consists of an input string and the port address.

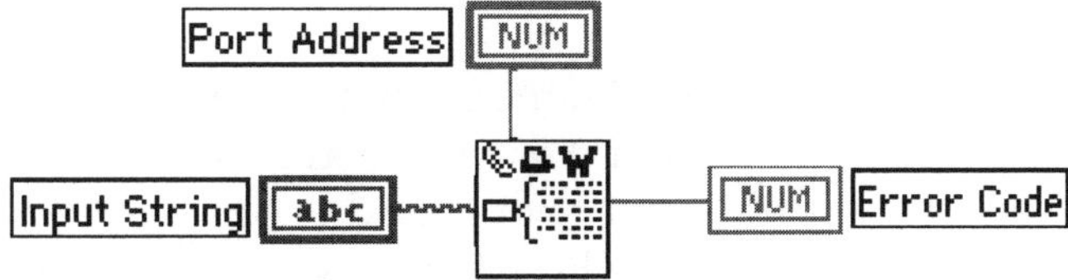

The error code can be carried to the front panel as a numeric code for error reporting. Each time the VI is executed, the input string will be serialized and sent to an external device as a RS232 waveform on the serial transmitter pin. Since the serial bit transmission rate may be slower than the speed with which characters are written to the serial port, a serial buffer holds the characters temporarily until the port is ready to send them.

For successful serial communication, the serial protocol on the computer must match that expected by the external device. Several parameters must be initialized:

Port address [0, 1 . . . 13] depends on computer type

Baud rate [110, 300 . . . 36000]

Number of data bits [6, 7, or 8]

Number of stop bits [1 or 2]

Parity [0 for even, 1 for odd, or 2 for no parity]

In LabVIEW, an initialization function, **Serial Port Init.vi**, is used to set these parameters.

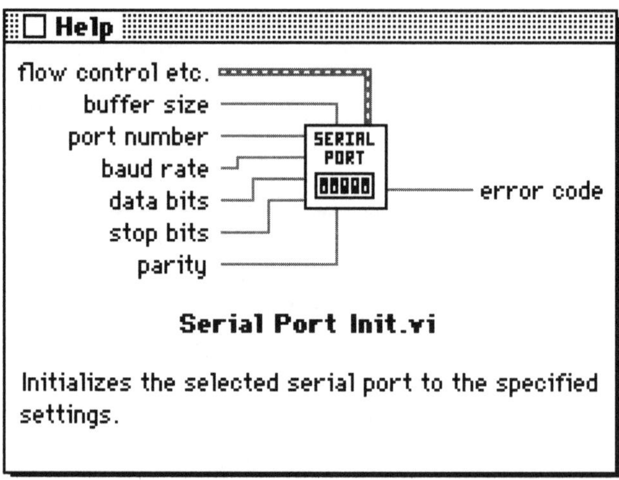

The parameters need only to be set once, at the beginning of a program. For most serial devices, the buffer size and flow control need not be used because the default values are often satisfactory. These parameters will be discussed later.

Reading data from the serial port is almost as simple as sending a data byte. Two icons—serial write and serial byte count—are used to collect characters on the serial port. **Bytes at Serial Port.vi** reads the serial port buffer addressed and returns a numeric: the number of bytes waiting in the input buffer.

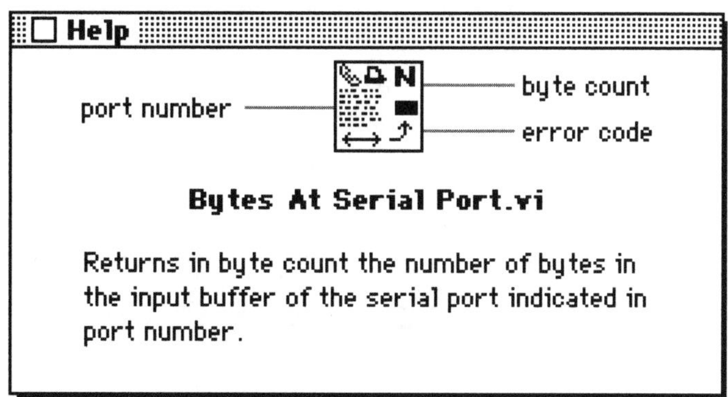

It is essential that this VI be executed first since the **Serial Port Read.vi** needs not only the port address but also the number of bytes to read.

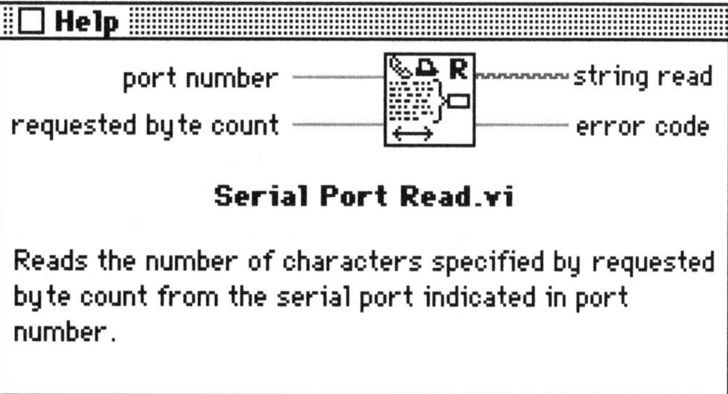

The **Serial Port Read.vi** fetches from the input buffer the number of ASCII characters requested by the byte count. You can now see why you may have to specify the size of the serial buffer. If the incoming string length exceeds the default value (1024 characters), information may be lost. For large incoming strings, the serial buffer size can be made as large as the system RAM permits. However, buffering does have a dark side. Sometimes characters may be sent to the serial port when the program is not expecting serial data. When the VI looks at the serial port, it would find these characters followed by the expected characters. LabVIEW dutifully collects all the characters in the serial buffer and sends them to the front panel. To eliminate these unwanted characters, an additional step called flushing the buffer may be necessary.

General Purpose RS232C LabVIEW Driver

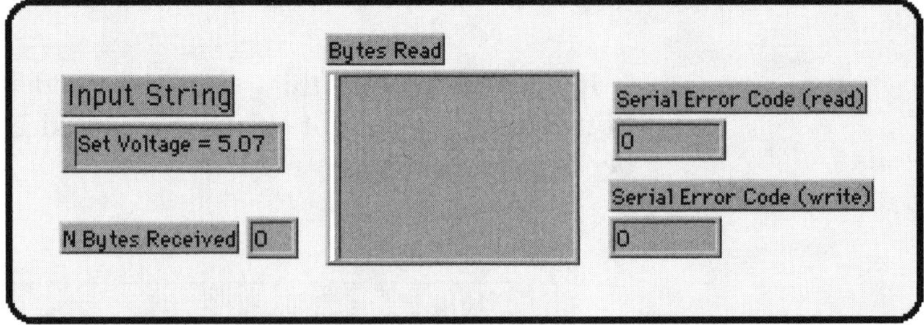

A general purpose RS232C serial driver forms the basis of an interactive serial diagnostic tool. You can send and receive messages at the driver level to test a serial link between LabVIEW and an external serial device. Once the device protocol is determined, the driver can be modified to become a device driver unique to that interface. The driver is based on the premise that for every action (output string command) there is a reaction (input string response). Inputs to the serial driver are the serial port address and a control string. Response from the serial interface is the ASCII characters and the number of characters. For diagnostics, the read and write error codes are also displayed on the front panel. Serial communication is a sequential operation, hence a four-step sequence structure is appropriate.

Frame 0 carries all the initialization information.

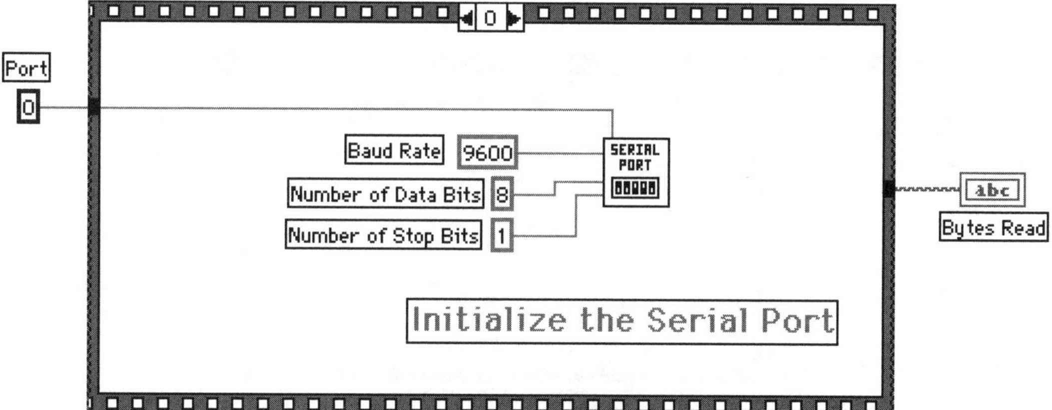

In general, the initialization VI is only executed once, usually at the start of the program. In the diagnostic program presented here, it is included as the first frame to ensure the port is initialized. Once the device specific sequence is determined, the frame can be removed and the initialization VI placed at the beginning.

Frame 1 sends the input command string.

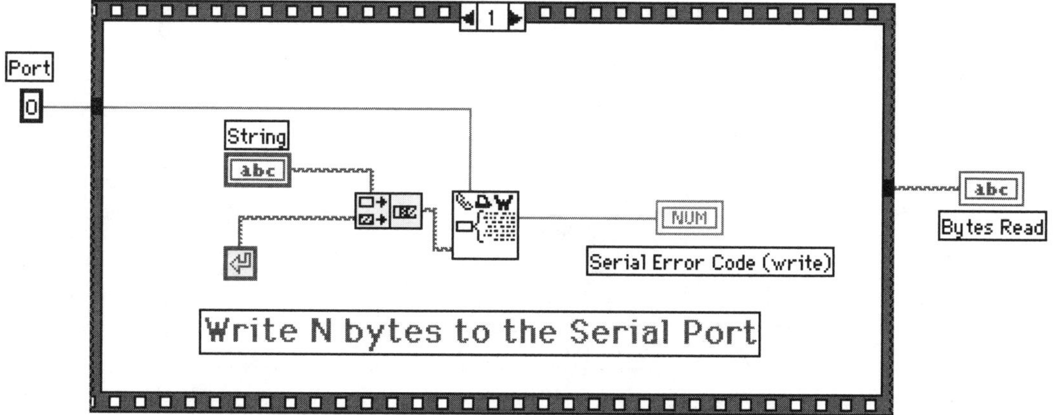

Most serial data messages are terminated by a carriage return so one has been added to the input string using the **String Concatenate.vi.**

Frame 2 flushes the buffer.

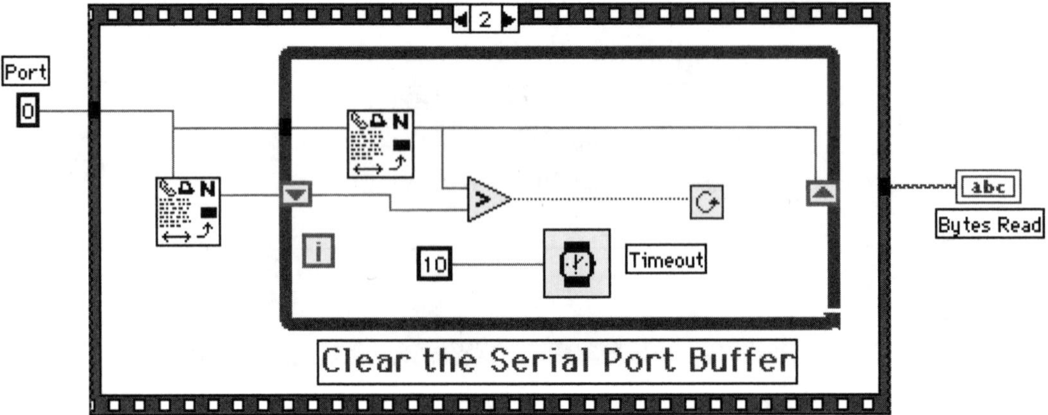

This frame removes unwanted characters from the serial buffer if present. This is not always necessary but a good precaution in the beginning. If not needed, it can be removed.

Finally, frame 3 reads the number of bytes received in the input buffer and retrieves that number of bytes.

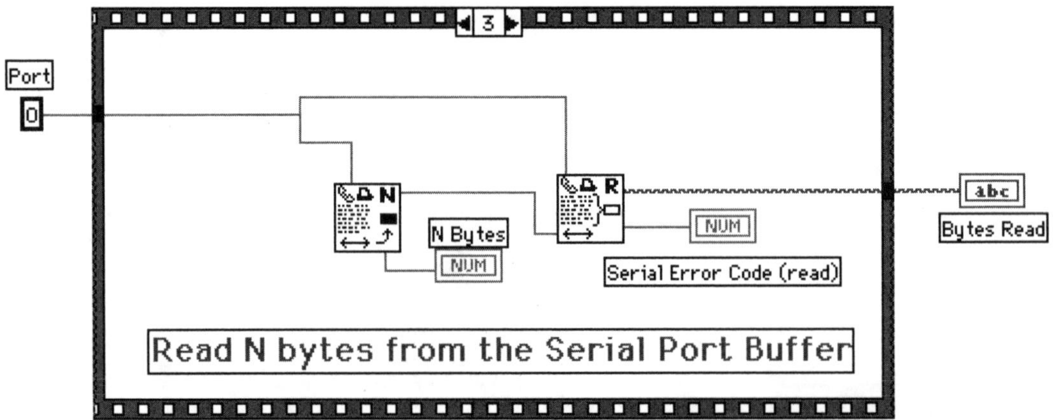

To test this driver program, the transmit and receive pins on the computer's serial port can be shorted together (see Appendix B for details). This simulates the simplest of all serial interfaces. The transmitted character string is echoed to the computer monitor. What is particularly interesting is to view the data flow of the VI during execution. Run the driver program with the execution highlighting button (light bulb) enabled. You will be able to watch the data flow action as the data stream is sent to the serial port, reflected by the shorting wire, and returned to the message pad on the front panel.

Giving the Serial Port a Parallel Life

LabVIEW does not support the PC parallel port nor does the Mac have a parallel port. But all computers do have serial ports. Even though the PC may use one of its ports for the mouse interface, there is often at least one free serial port. The MAC has two serial ports, the modem and printer port. The following circuit uses a inexpensive microcontroller chip to convert an RS232 encoded ASCII message into an 8-bit parallel output. Inside a computer, bit voltages corresponding to a high or low level are often not standard. A voltage line driver at the serial port converts these levels into standard RS232 levels. That is a high state is signaled by a voltage greater than 3 volts and less than 28 volts. A low is signaled with a negative voltage greater than –3 volts and less than –28 volts. In the following circuit a MC1489 serial line driver does the opposite. It converts the RS232 levels into the voltage levels that the integrated circuit microcontroller SM101B expects.

The parallel outputs are TTL compatible and can be used to drive a LED display or relays as shown. This simple interface opens up a whole new world of low-cost computer interfacing to LabVIEW.

To control any bit on the parallel port, a four-character ASCII message must be sent through the LabVIEW serial port to the microcontroller. The message consists of a microcontroller address ($0 to $7), the output bit level (N for on or F for off), the bit address (0 . . . 7) and a carriage return character (<CR>). The following program provides control of bit 7 on the parallel port of

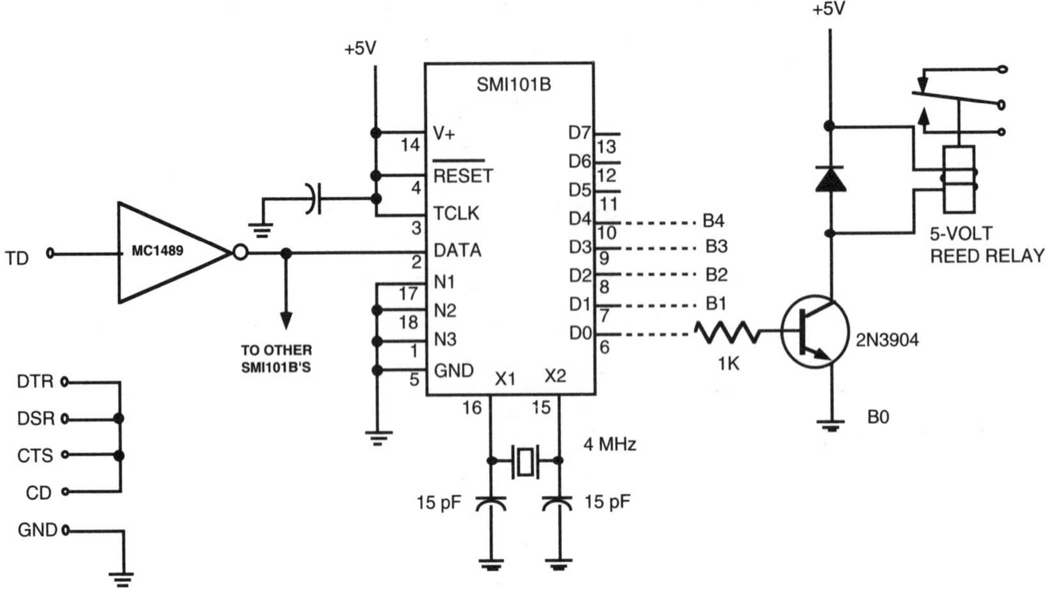

the microcontroller having device address $0. A single Boolean switch on the front panel provides the input command. This state is echoed on a LED display and the ASCII message sent to the microcontroller is shown in a string indicator.

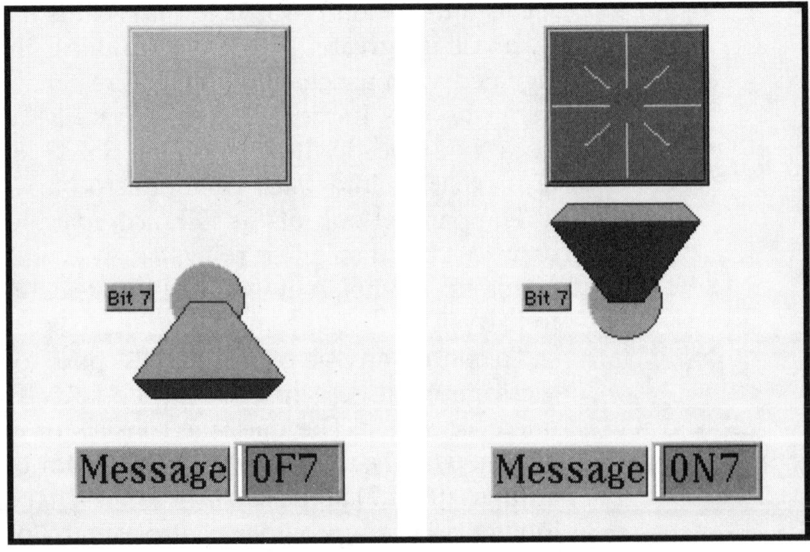

The single bit controller program uses a case construction and a string concatenate operation to generate the ASCII command message. The Boolean state of the switch control selects the on case (<|True|>) or the off case (<|False|>). For convenience, a LED display echoes the switch (on/off) state on the front panel. The end of message character <CR> is also the execute command for the microcontroller. On receipt of the string command, the bit command embedded in the ASCII string is decoded and passed on to the addressed output, Pport1.vi.

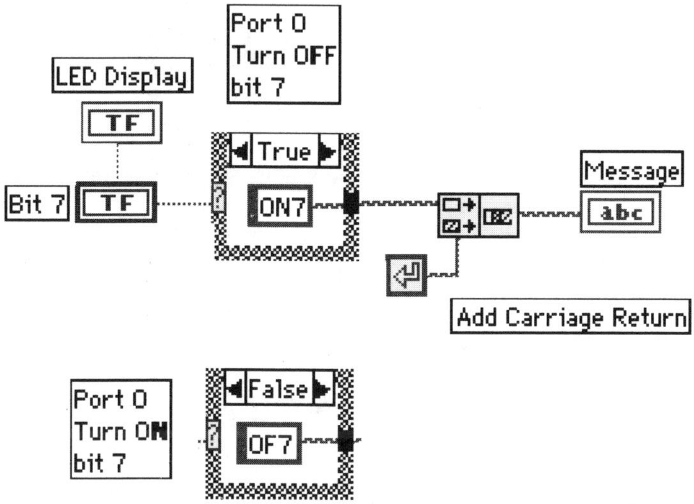

As a sub-VI, this program is called Pport.vi with an icon [Set Reset bit]. Its input and its output are a Boolean and an ASCII string. It is now a simple matter to replicate this VI seven times to build a complete 8-bit parallel port controller. In a second program, called Pport8.vi, all bits on the parallel port can be programmed interactively from the front panel.

Eight command strings are required to form the complete message. These strings are concatenated together and then displayed on a front panel message board as well as being transmitted out the serial port by the RS232.vi driver.

The diagram panel uses two concatenate operations to generate the complete message. This message string is then passed to the [RS232] driver icon as well as echoed to the front panel. No

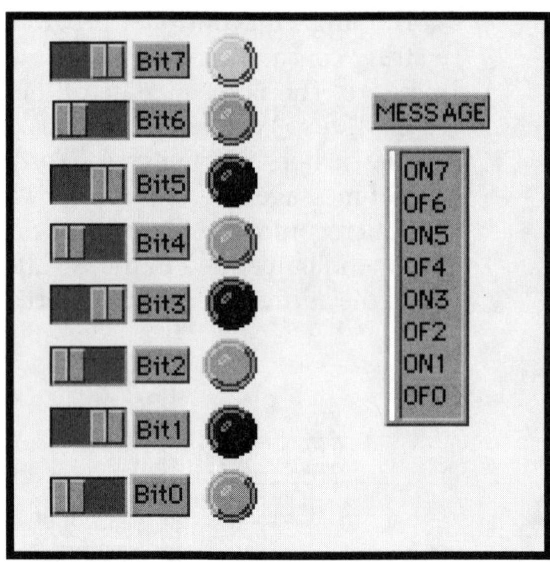
Bit7
Bit6
Bit5
Bit4
Bit3
Bit2
Bit1
Bit0
MESSAGE
ON7
OF6
ON5
OF4
ON3
OF2
ON1
OF0

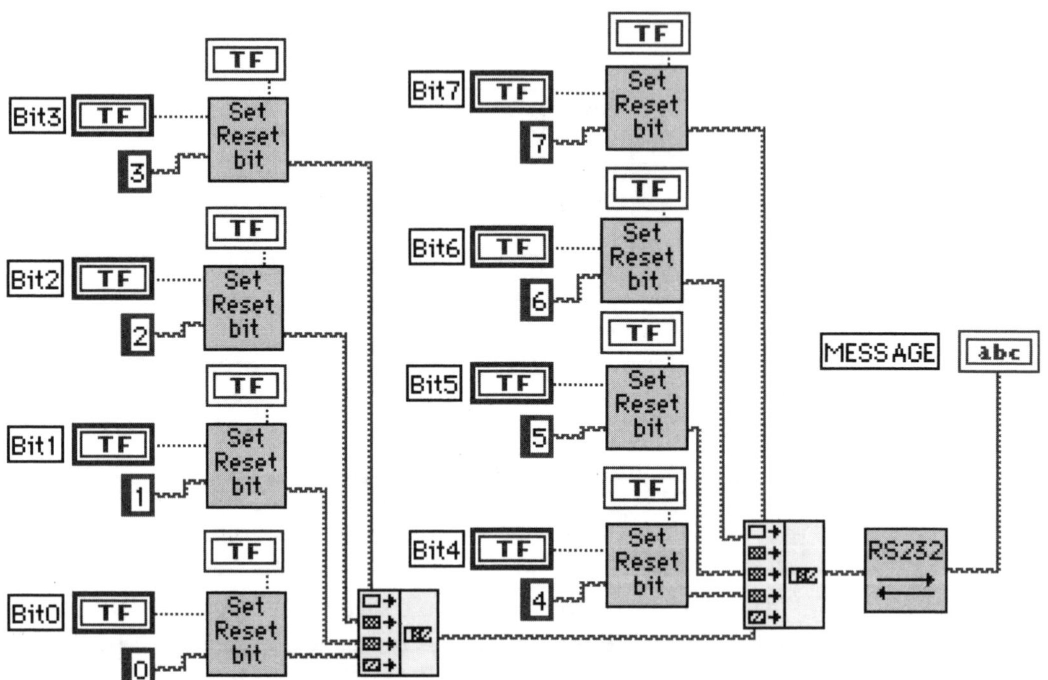
Bit3
Bit2
Bit1
Bit0
Bit7
Bit6
Bit5
Bit4
TF
Set Reset bit
3
2
1
0
7
6
5
4
MESSAGE
abc
RS232

response message is returned by the microcontroller when a command string is issued. Hence, the serial driver is very simple, consisting only of the Write.vi. The initialization needs only to be executed once, and it can be removed from the serial driver (frame 0) and placed outside the main programming loop.

Load the program entitled Pport8.vi and observe its operation. It can be run in simulation or with a serial to the parallel port interface circuit to control real devices.

The parallel port can then be connected to real LEDS, relays or TTL devices to make your world come alive. Here are a few activities.

Exercise 1: VU-meter

Many stereo systems have VU-meters to display the audio power level. These indicators measure the output power level on a log scale (decibels). Design a VI that converts an 8-bit number (0–255) into eight power levels. As the power level is raised, the eight LEDs are lit in sequence.

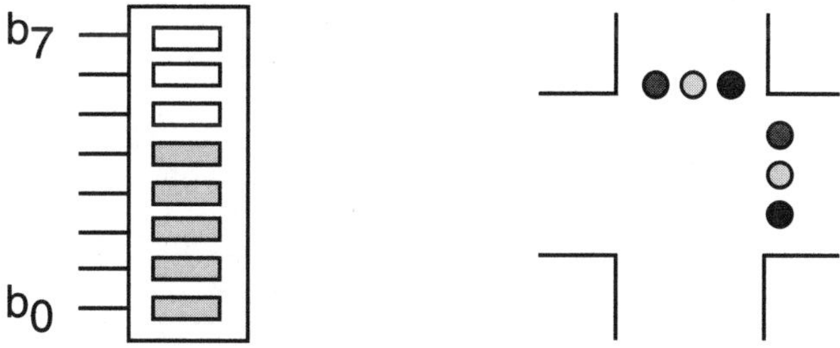

Exercise 2: Stop Light Intersection

Place six LEDs colored red, yellow, and green in a stop light intersection arrangement and connect the LEDs to six parallel output lines. Modify the virtual stop light program of the last chapter to echo the state of the virtual LEDs on LabVIEW's front panel to the real world.

Exercise 3: Seven-Segment Display

Connect the eight output lines of the parallel port to a 7-segment display. Send a sequence of characters one after the other with about a 1/10 of a second delay between updates. Try sending your name or a simple message like "call 911."

Note that some LED displays require more current than the microcontroller can supply. In this case, a buffer amplifier is needed for each line.

Exercise 4: Stepping Motor

Four bits on the parallel port can be used to generate the phases of a stepping motor and you can make your world spin. In this case, current amplifiers are needed to provide an output current of 30 to 100 ma for each phase of the stepping motor.

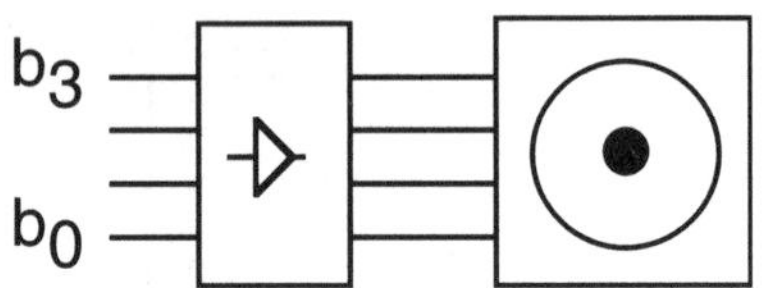

Exercise 5: Controlling Large Power Devices

Many real-world devices require large currents to operate electrical or mechanical devices. The simplest interface is with an electromagnetic or electronic relay. Four relays on the output port could be connected to a variety of electromechanical devices to put a little action into your projects. Or try controlling four colored floodlights to set the mode.

Eight-Bit Input Port

Using a different microcontroller (the SM102) and a slightly modified program provides digital inputs for your computer again via the serial port. This time, however, the full serial driver is used. Addressing the microcontroller causes it to read the input parallel port. The microcontroller responds with an ASCII number whose binary value is representative of the bit pattern on the parallel input port. For example, the command sequence to read the input port of the microcontroller with address "0" is simply 0<CR>. Assuming all the inputs were high, then the response would be 255<CR>. Recall the number 255 is the decimal equivalent value for (11111111).

The simplest input interface would be a series of eight switches connected to the parallel inputs pulled high.

The Temperature Controller Version 1.0

Temperature controllers are transducers commonly found in commercial applications where the setting of temperature, the programming of temperature or the maintanance of temperature is critical. Commercial controllers often use RS232 serial links as the communication channel. In many cases cryptic ASCII messages form the communication protocol. In the following example, one such temperature controller is simulated. Command messages consist of a single ASCII character. Version 1.0 allows for the reading of the current temperature, the temperature set point, the status message, and allows the controller to be reset. The temperature set point is adjusted on the front panel of the controller. Four functions shown in square brackets [ ] are currently implemented:

[V] Requests the software version number

[S] Asks for the current set point temperature

[R] Resets the controller

[T] Requests the current temperature.

On receipt of the version command, the controller responds with the software version number. On receipt of the set point command, the controller responds with temperature set on the front panel of the controller. On receipt of the temperature command, the controller responds with the current temperature in degrees °C. On receipt of the reset command, the temperature controller reads from the front panel the set point and then ramps the temperature from room temperature up to the set point at a rate of 1 degree C per second. When the temperature reaches the set point, the temperature is held until the controller is reset. A simulation program, Temp Controller.vi, is found in the program library. Run the VI and observe the ASCII response for each single character command. After Reset, run the VI continuously to observe the ramping up of the temperature.

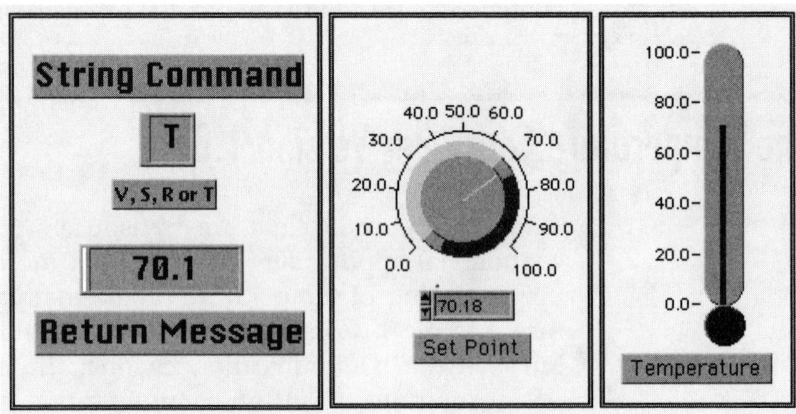

LabVIEW Challenge: Temperature Controller Interface

Design a program that exercises the temperature controller VI. On the front panel, display the response to the cryptic commands [V], [S], [R], and [R] in user friendly format (c.f., the graphic at beginning of the chapter). To show the full operation of the temperature controller, use a numeric indicator for the temperature and the set point, an ASCII message indicator for command responses, and a temperature-time chart to observe the controller in action.

Appendix A ASCII Character Set

■ Hexadecimal Codes/ASCII Codes

00	NUL	10	DLE	20	SP	30	0
01	SOH	11	DC1	21	!	31	1
02	STX	12	DC2	22	"	32	2
03	ETX	13	DC3	23	#	33	3
04	EOT	14	DC4	24	$	34	4
05	ENQ	15	NAK	25	%	35	5
06	ACK	16	SYN	26	&	36	6
07	BEL	17	ETB	27	'	37	7
08	BS	18	CAN	28	(	38	8
09	HT	19	EM	29	)	39	9
0A	LF	1A	SUB	2A	*	3A	:
0B	VT	1B	ESC	2B	+	3B	;
0C	FF	1C	FS	2C	,	3C	<
0D	CR	1D	GS	2D	–	3D	=
0E	SO	1E	RS	2E	.	3E	>
0F	SI	1F	US	2F	/	3F	?

40	@	50	P	60	`	70	p
41	A	51	Q	61	a	71	q
42	B	52	R	62	b	72	r
43	C	53	S	63	c	73	s
44	D	54	T	64	d	74	t
45	E	55	U	65	e	75	u
46	F	56	V	66	f	76	v
47	G	57	W	67	g	77	w
48	H	58	X	68	h	78	x
49	I	59	Y	69	i	79	y
4A	J	5A	Z	6A	j	7A	z
4B	K	5B	[	6B	k	7B	{
4C	L	5C	\	6C	l	7C	|
4D	M	5D	]	6D	m	7D	}
4E	N	5E	∧	6E	n	7E	~
4F	O	5F	–	6F	o	7F	DEL

Non-printable ASCII Characters

00	NUL	Null	10	DLE	Data Link Escape
01	SOH	Start of Heading	11	DC1	Device Control 1
02	STX	Start of Text	12	DC2	Device Control 2
03	ETX	End of Text	13	DC3	Device Control 3
04	EOT	End of Transmission	14	DC4	Device Control 4
05	ENQ	Enquiry	15	NAK	Negative Acknowledge
06	ACK	Acknowledge	16	SYN	Synchronous Idle
07	BEL	Bell (Audio sound)	17	ETB	End of Transmission Block
08	BS	Backspace	18	CAN	Cancel
09	HT	Horizontal Tab	19	EM	End of medium
0A	LF	Line Feed	1A	SUB	Substitute
0B	VT	Vertical Tab	1B	ESC	Escape
0C	FF	Form Feed	1C	FS	File separator
0D	CR	Carriage return	1D	GS	Group Separator
0E	SO	Shift Out	1E	RS	Unit Separator
0F	SI	Shift In	1F	US	Delete

Appendix B: Serial Connections

PC Computers

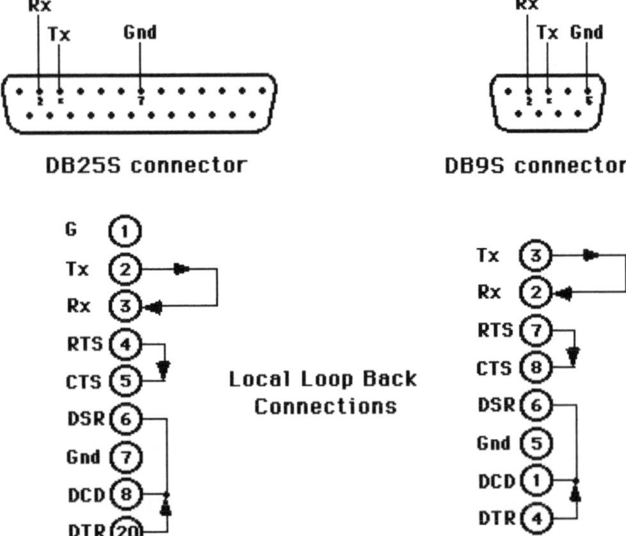

■ MAC Computers

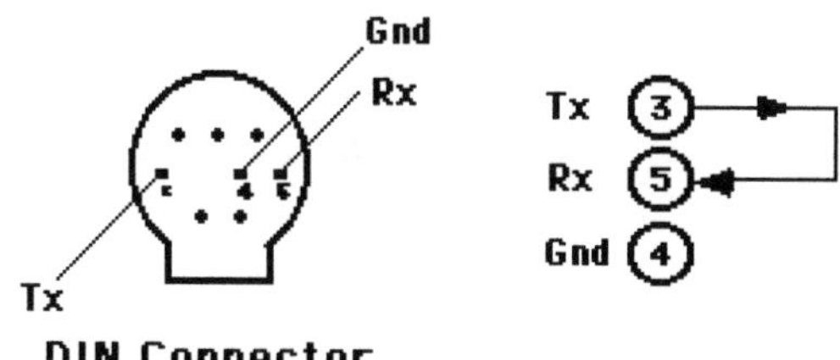

Overview

All interface messages between computer control programs and external devices over the popular interface ports are in the form of ASCII strings. LabVIEW supports the string data type with a wide range of operations and functions. Strings can be "sliced and diced" and grouped together with other strings to form string messages. ASCII strings are a common form of data type used in file operations, and thus string messages as commands, programs, or data can be easily stored to memory with LabVIEW's file operations. Data logging is one of the most common tasks in data acquisition and two examples are provided to build the readers skills. A study of digital plotters reveals the many ways string operations are used in an interfacing project.

Goals

- Understand strings constants, variables, controls, and indicators
- Understand string functions
- Review string message protocols
- Build a general purpose data logger
- Design a data logger for chemical luminescence
- Simulate a digital plotter
- Exercise a LabVIEW Etch-a-Sketch

Key Terms

- String messages
- Parsing a string
- Building a string
- Write characters to a file
- Digital plotter
- Charts, graphs, and XY plots
- Data logger

String Along With Us

6

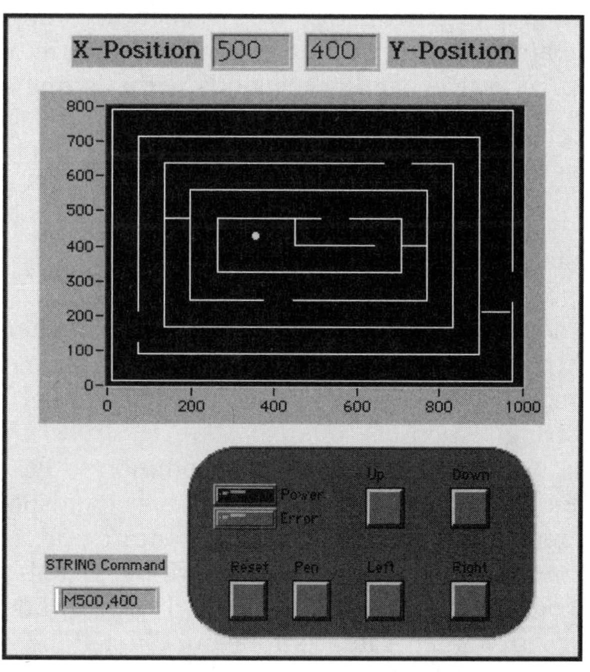

LabVIEW Simulation of a Hewlett-Packard Digital Plotter

A string is a collection of ASCII characters. Together these characters can form a command, status, or data message. In this chapter, message transfer protocols for specific I/O devices are studied as a means to understand strings, string operations, and message transfer techniques for CPU-to-device communication. In each case, the input to a device is a string and the device responds with a string. For example, a command message to an external digital voltmeter might be "Get Data." The input/output device driver receives this message and executes the acquire data function. It then sends back to the I/O driver the message "OK, DVM reads 5 volts."

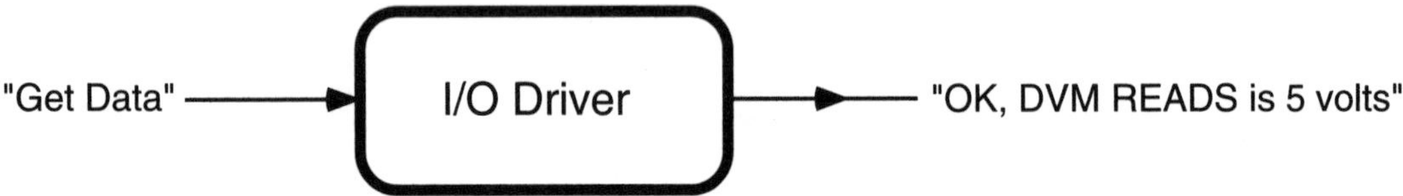

In the general case, device handlers do not act until commanded by an input string. After execution of the device process, the handler responds with a string acknowledge. Take a look at a few examples.

An 8-bit parallel port, when commanded by a string message, takes a character embedded in the input string and passes it on to the output port. The 8-bit character encodes the on/off state of each of the eight bits that are to be set or reset on the port. The output operation in general requires no acknowledgment. However, an input parallel port, when commanded by a string message to read an 8-bit port, acknowledges the command with the bit pattern on the parallel port encoded as a string response.

A Centronics printer port common to PC computers uses an 8-bit parallel output port and two handshaking lines. Data is transferred over the parallel output port to the printer. One handshaking line from the printer signals the CPU that the printer is ready to receive data. The other line from the Centronics driver generates a strobe pulse to tell the printer that the data on the output port is valid and can be transferred into the printer's character buffer on the falling edge of the pulse. When the character buffer is full, the printer signals the CPU with a flag (printer not ready) not to send any more characters until the printer catches up. There are two types of messages signaled by these handshaking lines: status messages from the printer (Busy/Not Busy) and data transfer commands.

The CPU to printer dialog might read

```
Printer are you busy?
   If Yes, then I will call back later
   Else I am transfering a data character
end.
```

The serial port of the previous chapter has already been used to send simple and some not-so-simple messages. ASCII strings are the standard message data type for serial communication. The serial port is bidirectional, and as a result, messages can be coming in and going out at the same time. Since the CPU has a much faster response than the serial ports, it can keep the outgoing and input buffers full and empty respectively. Command, status, or data string messages are often followed by an End of Message character, EOM. For command messages, EOM is often the carriage return character <CR>. The external device interprets this character as an execute command. A simple outgoing message might be "Hello digital voltmeter. What range and function are currently set? <CR>." The voltmeter reads the messages, formulates a reply and on receipt of the <CR> might respond "Volts DC, 10 V full-scale <CR>." The computer also uses the <CR> from the reply to signal the end of the voltmeter message.

IEEE488 data transfers use two 8-bit ports. One port is a bidirectional parallel bus used for the transfer of data and command bytes. The other port is a mixture of input and output handshaking lines. All IEEE488 messages, whether input or output, are in the form of device specific ASCII messages. For example, a digital voltmeter might respond to a read command with the message "$NDCV + 1.82E+01." Loosely translated, the message reads: The current voltage reading on the front panel is 18.2 volts DC. The character "$N" is a manufacturer's code. All IEEE488 task messages and device responses are in the form of ASCII strings.

The internet communicates with an ASCII encoded Internet Protocol, IP datagrams that contain sender address, receiver address, and a block of data. A higher level Transfer Control Protocol (TCP) that supervises the IP datagrams also uses ASCII messages to link the user program with the network. Together these two protocols form the TCP/IP internet protocol.

A detailed understanding of how string messages are built and dissected is essential for successful communication with external devices.

Strings: Constants, Variables, Controls, and Indicators

A string is a collection of ASCII characters formulated to represent a message. Strings are just a different data type with their own set of functions and operations. Strings can be sliced and diced, glued to other strings, converted to other data types, and much more. LabVIEW has a wealth of string operations and functions.

A string is said to be a constant if it does not change during a process. A simple string constant message in this text is written as a sequence of characters contained inside double quotes, for example, "Hello." In LabVIEW, the string is typed into a string control box colored gray. Output strings are displayed within a gray indicator box with a vertical white bar. The corresponding terminals on the diagram panel have boxes with the letters "abc" contained inside the terminal. String terminals and pathways are colored purple. The following is an example of a string control and indicator.

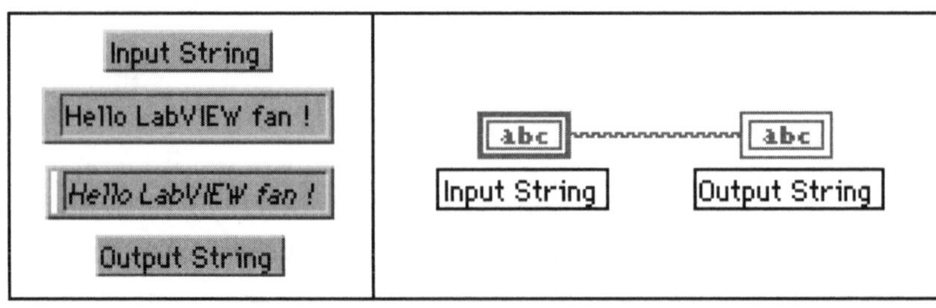

LabVIEW uses the backslash (\) character to indicate non-printable characters, followed by a lower case character. For example, "\r" is LabVIEW's way to indicate the carriage return <CR> character. Upper case letters are reserved for the hexadecimal codes "ABCDE and F." The carriage return character in hexadecimal is "$0D" or in LabVIEW can also be written as "\0D."

The following are some common "\" LabVIEW codes

\00–\FF	Hexadecimal value of an 8-bit character	
\b	Backspace	(ASCII BS $08 or \08)
\f	Formfeed	(ASCII FF $0C or \0C)
\n	Newline	(ASCII LF $0A or \0A)
\r	Return	(ASCII CR $0D or \0D)
\t	Tab	(ASCII HT $09 or \09)
\s	Space	(ASCII SP $20 or \20)
\\	Backslash	(ASCII "\" $5C or \5C)

Common string characters such as carriage return, line-feed, tab, and null can also be found as [icons] in the **Functions>String** palette.

String Functions

Many string functions require the length of the input string before the function can be processed. LabVIEW provides a String Length function that outputs the number of characters in a string as a numeric value. This operation is found in the **Functions>Strings** palette.

String messages are often composites of string constants and variables concatenated together to produce the message. Consider the case of a command message "Set DAC to 12.5 volts" sent over the IEEE488 interface to an digital-to-analog converter. The header "Set DAC to" is a string constant. The value 12.5 is a numeric variable that must be converted into a variable character string, "12.5." The trailer is another string constant,

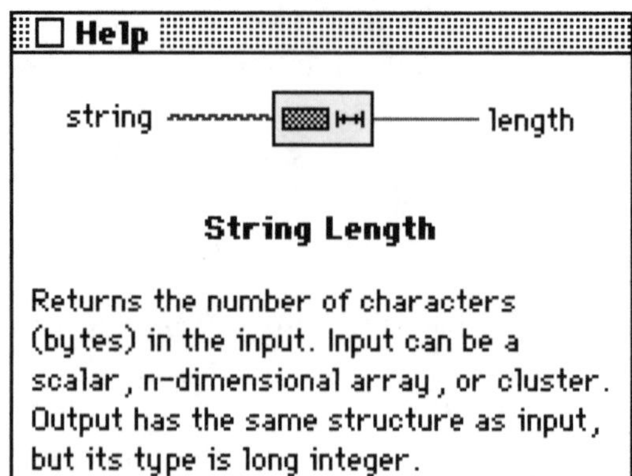

"Volts." In many cases, an execute character (<CR>) is appended to the end of the message to signal the device to execute the command message.

The LabVIEW function **Format & Append** found in the **Functions>Strings>Additional String to Number Functions** subpalette converts the numeric value into a character string according to a format string. The string constant "%.2" provides a formatted character string message with two characters after the decimal point. The **Concatenate Strings** function glues the input strings together into the string message. Note that the resize tool can be used to expand the icon downwards to reveal additional string inputs. The following example, Build String.vi, shows how the string message is built on the block diagram.

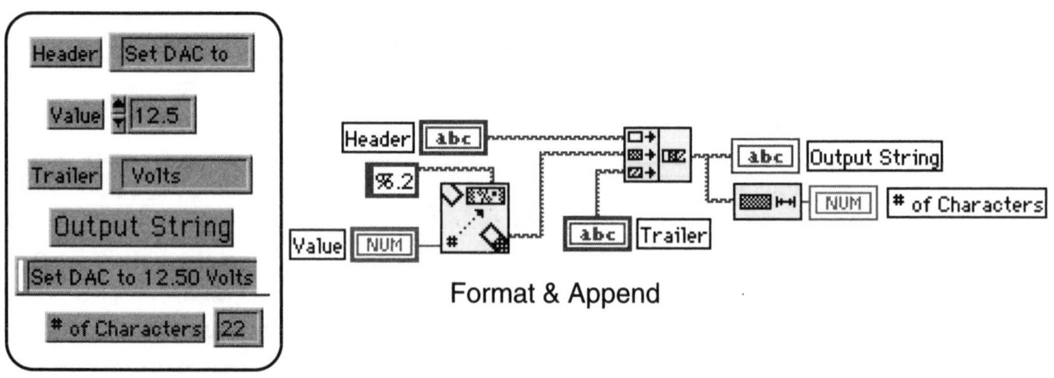

Can you explain why the number of characters from the **String Length** function reads 22?

Parsing a String

An external device often responds with a string message consisting of string constants and variables. Future calculations or displays may require that the string characters corresponding to numeric values be isolated from the message and converted into a numeric. Such an operation to slice and dice a string is called parsing. Consider the case where a digital voltmeter responds to a read command with the message "VOLTS DC +1.34E+02." The character substring "+1.34E+02" is to be converted into a numeric. LabVIEW provides a **String Subset** function that returns a subset beginning at an offset (counting from 0) and containing length, number of characters. Both offset and length are numerics. For the type of voltage (DC or AC) from the return message, the offset would be 6 and the length would be 2. To retrieve the DVM voltage, an offset of 9 is required and the length is given by the number of characters in the message minus this offset. Study the following parse diagram, Parse String.vi.

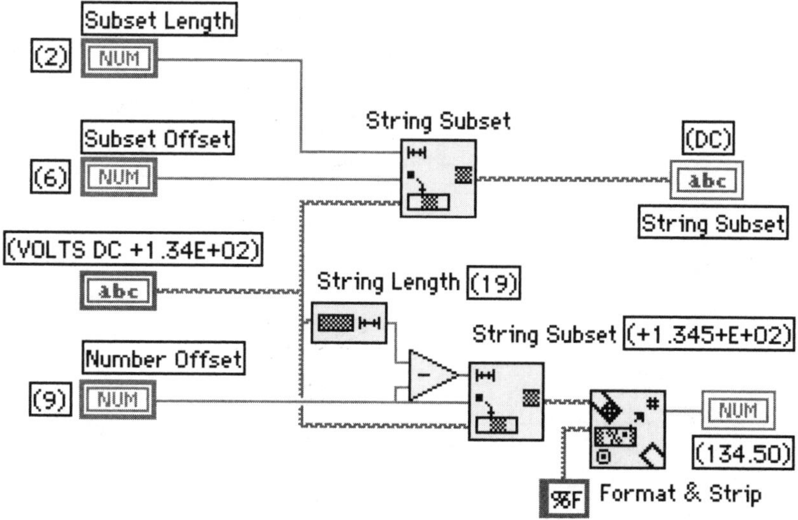

Note that a **Format & Strip** function is used to convert the string number into an numeric number. It is found in **Functions>Strings>Additional String to Number Functions** subpalette. This is a very useful function that can also be used to find a matching string, convert a substring into a number, and return all characters after the number.

Exercise

A programmable power supply is remotely controlled by a ASCII string message with the format

SET X.XXX {VAC, VDC, AAC, or ADC}

The word "SET" signals the power supply to set its output to X.XXX volts or amps (AC or DC). The cryptic three-character code choses the function.

Create a power supply command message program. A front panel slide switch selects the function, Volts or Amps (DC or AC). Building the string message follows the same method shown above. You need to concatenate the header, number, and trailer strings. Format and Ap-

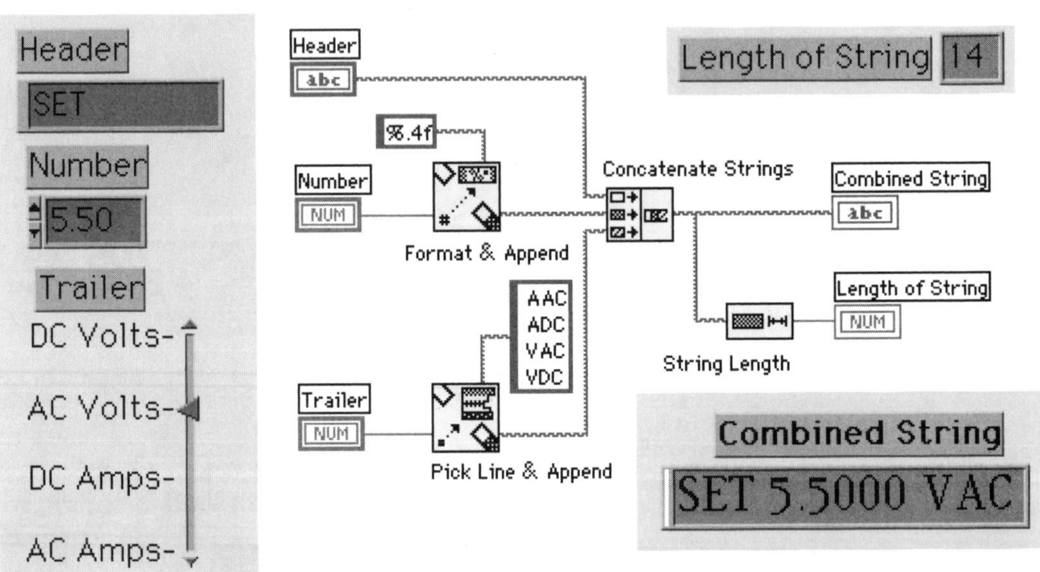

pend converts the number to a string, Concatenate glues the strings together, and String Length reports the length of the string.

The choice of function is selected by another useful string function called **Pick Line & Append.** This function has a list of string constants (AAC ADC VAC & VDC). Each element in the list has a unique numeric integer index (0, 1, 2, or 3). The line index picks from the multiline list of strings the selected string to output. Note that elements in the list are separated by a carriage return. In operation, the header string is typed into the header box. The number is entered in the numeric control box and the function is selected by clicking and dragging the operating tool on the slider. Save your program as Build String2.vi.

One of the most common data acquisition tasks is the logging of a physical parameter or parameters in real time and saving the data into a file. Tab delimited text files can be read by many programs including plotting, word processing, and of course spreadsheet programs. This format style requires columns to be separated by the <Tab> character and rows by a carriage return <CR> character. The next example builds a data logger using a data set from a chemical luminescence reaction. When executed, a tab delimited data file will be created.

Chemical Luminescence Data Logger

Chemical luminescence is produced when a chemical reaction yields an electronically excited species, which emits light as it returns to the ground state. The simplest type of reaction can be formulated by

$$A + B \longrightarrow C^* + D$$
$$C^* \longrightarrow C + h\nu$$

where A and B are the reactants, D is one of the products, and C^* represents the excited state product that after some period of time decays to a stable state C with the release of a photon of energy $h\nu$. Certain chemicals, when mixed together—for example, the popular glow sticks—display luminescence. In nature, this reaction is called bioluminescence and examples include fireflies, certain jellyfish, and bacteria.

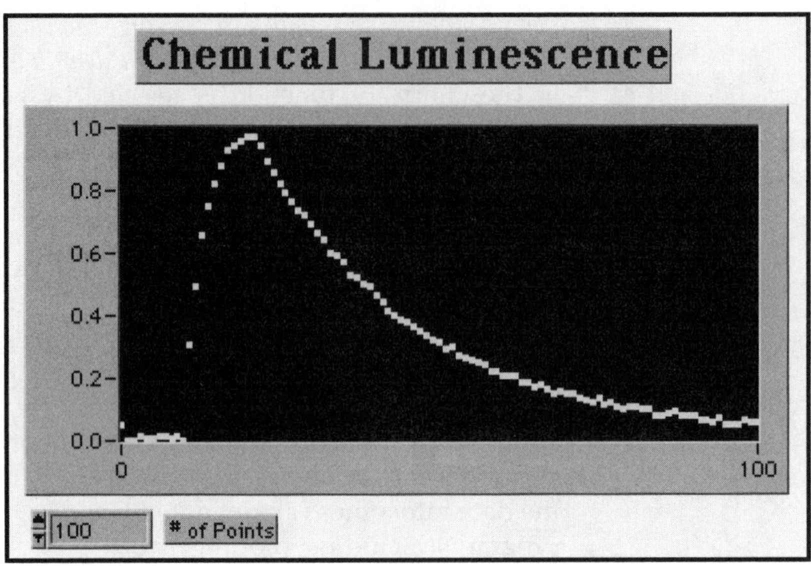

Instrumentation is remarkably simple, consisting of a suitable reaction chamber, a photomultiplier tube to measure the light emitted in the reaction, and a LabVIEW data logger. In all cases, the intensity of the light is directly proportional to the amount of reactants. A typical chemical luminescence signal rises rapidly as soon as the reactants mix. As more and more molecules react, the signal grows in strength. A maximum is reached when the reagent and analyte are fully mixed. The light then fades slowly in a more or less exponential fashion until all reactions are complete. Lumin.vi simulates a chemical luminescence reaction. Each time this VI is called, a new value of the luminescence curve is returned. Successive calls generate a data set whose graph is shown above.

In the following program, the loop index [i] is used to address the VI. The first value when [i] = 0 is used to reset the simulation. Thereafter, each call generates the next measurement.

The data logger program shown above calls Lumin.vi N times and reports the luminescence signal to the front panel as well as producing an ASCII character string using the **Format & Append** string function. A time stamp for each measurement is generated by the string function **Get Date/Time String** from the **Functions>Time & Dialog** palette. This function will output an

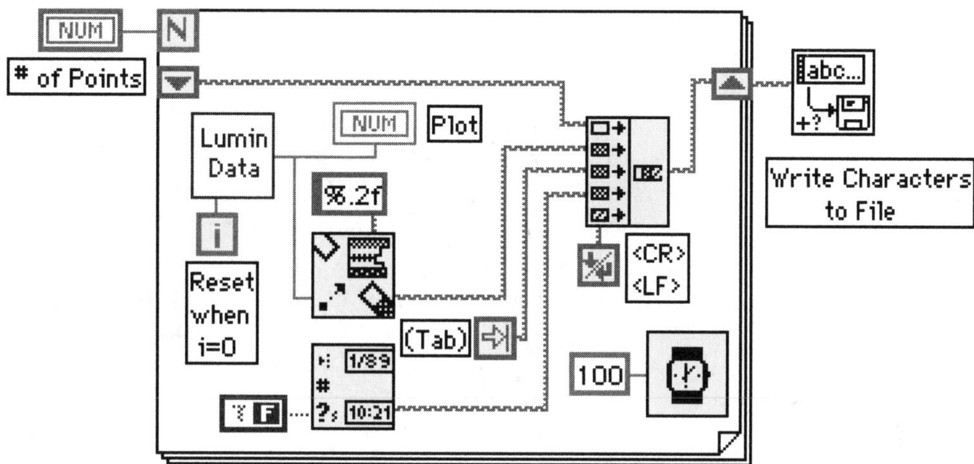

ASCII string containing the date, time, or both in a variety of formats. In our case, only the reaction time is required and accuracy to the nearest second is satisfactory. The concatenate string function builds a single line of data consisting of the measured luminescence, a tab character, the time stamp and carriage return/line feed characters. The latter characters appended to the end of the message signal the end of that row or line. Note that the entire line goes to a shift register input. On the next iteration, the new line of data is added to previous lines. The file keeps getting larger and larger until the loop counter [N] equal the [# of Points]. All lines are then passed to the **Write Characters to File** function to generate the file. A dialog box pops up on the front panel with a request for a file name and location. Since the file was prepared as a tab delimited text file, it can be read by many programs including spreadsheets, word processors, and plotters. Try it! Plothum.vi is found in Chapter 6 library.

Digital Plotters

Computers send plot commands to digital plotters over a Centronics parallel port, a RS232 serial port, or an IEEE488 interface. In all cases, the commands and data messages are in the form of

ASCII strings. A digital XY plotter is simulated to demonstrate how these devices are controlled by string messages. The plotter has a platen whose plot area is covered with a XY grid. Each position (X,Y) has an absolute set of coordinates. In this example, the X coordinate ranges from 0 to 1000, while the Y coordinates range from 0 to 800. The HOME position is designated as (0,0). The pen can be moved anywhere on the platen in the allowable range. A point with a coordinate outside the range will cause the pen to stick at the last allowable limit in that direction. For example, the point (1500, 400) will stick on the right edge at (1000,400). An error message is given and the pen will stay stuck to that edge until the X coordinate comes back into range.

Version 1.0 of the plotter program XYPlotter1.vi has only two commands implemented, HOME and MOVE. The ASCII string message format is

H	**moves pen to (0,0)**
Mx,y	**moves pen to (X,Y)**

Load XYPlotter1.vi and take the pen for a test drive. When running the program, make sure you have turned the power switch on. After typing a command into the ASCII message pad, execute the command by clicking on the [Send] button. Note that a {return} character need not be typed when entering a command string. The [Send] button provides the execute command.

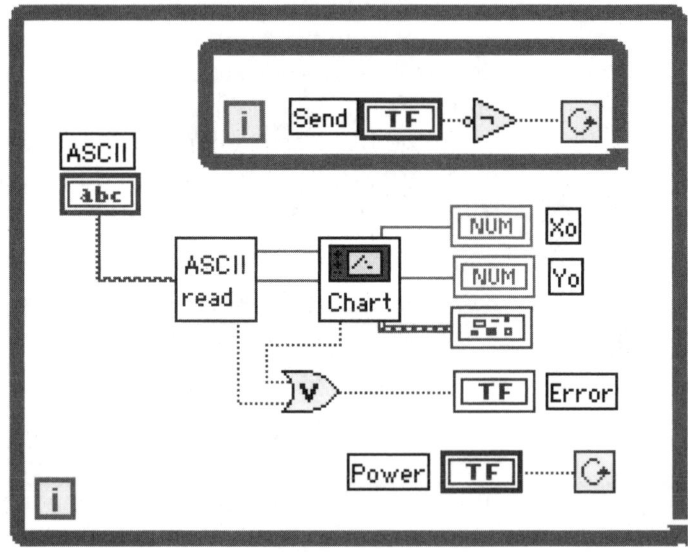

The string command is converted into a pair of numbers (X,Y) in the sub-VI called ASCII read.vi. This point is then plotted on the front panel with the sub-VI [Chart]. If the requested point is outside either of the ranges, then an error LED is indicated on the front panel. The sub-VI called XYPlot.vi uses an interesting set of nested case statements to simulate a plotter with sticky limits. In addition, **Build Array** and **Bundle** functions (discussed in detail in the next chapter) are used to form the cluster input for the XY graph. Note also how the error flag is created with multiple OR inputs.

The ASCII Read.vi uses complex nested case statements and various string functions to extract the plotter commands (H or Mx.y) and calculate the absolute plotter coordinates. Study its block diagram in detail.

Digital plotter Version 2.0 adds the draw function to the plotter routines. Draw lowers the pen and draws a line from the current position to the new position.

Dx,y draws line from (X_0,Y_0) to (X,Y)

where (X_0,Y_0) is the current position of the pen and (X,Y) is the new position. Draw is essentually an incremental plot (Move) command with the pen down.

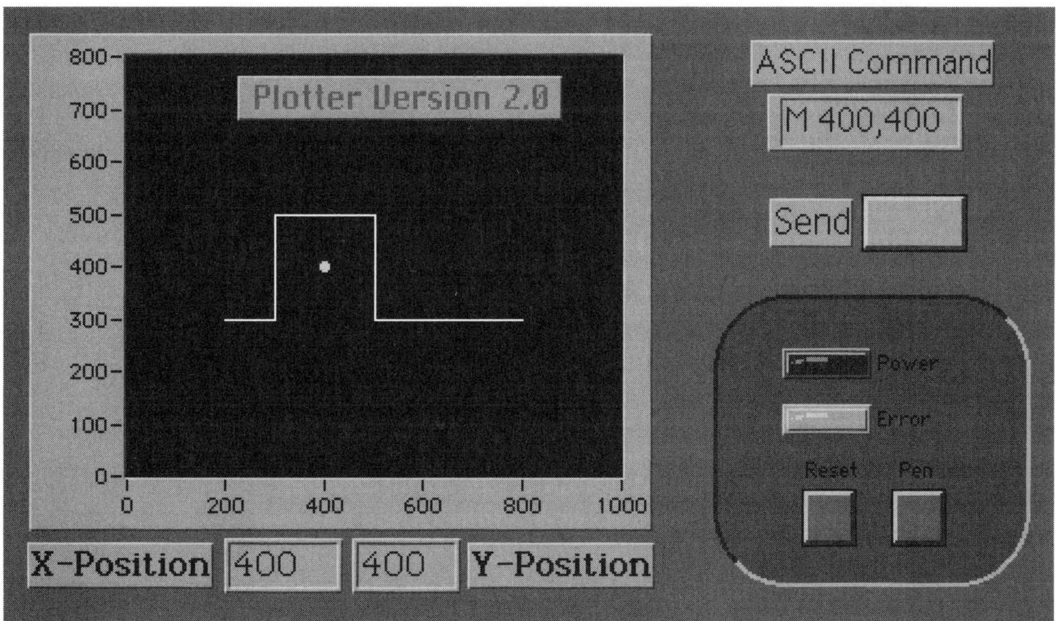

Load and launch the program XYPlotter 2.vi. Try drawing a box in the middle of the plot field. Note that the draw command also sets a button indicator to show whether the pen is up or down. When the move command is issued the button is up. When the draw command is issued the button is down.

LabVIEW Challenge: How Did They Do That?

Study the block diagram to follow the data flow that generates the up/down pen action.

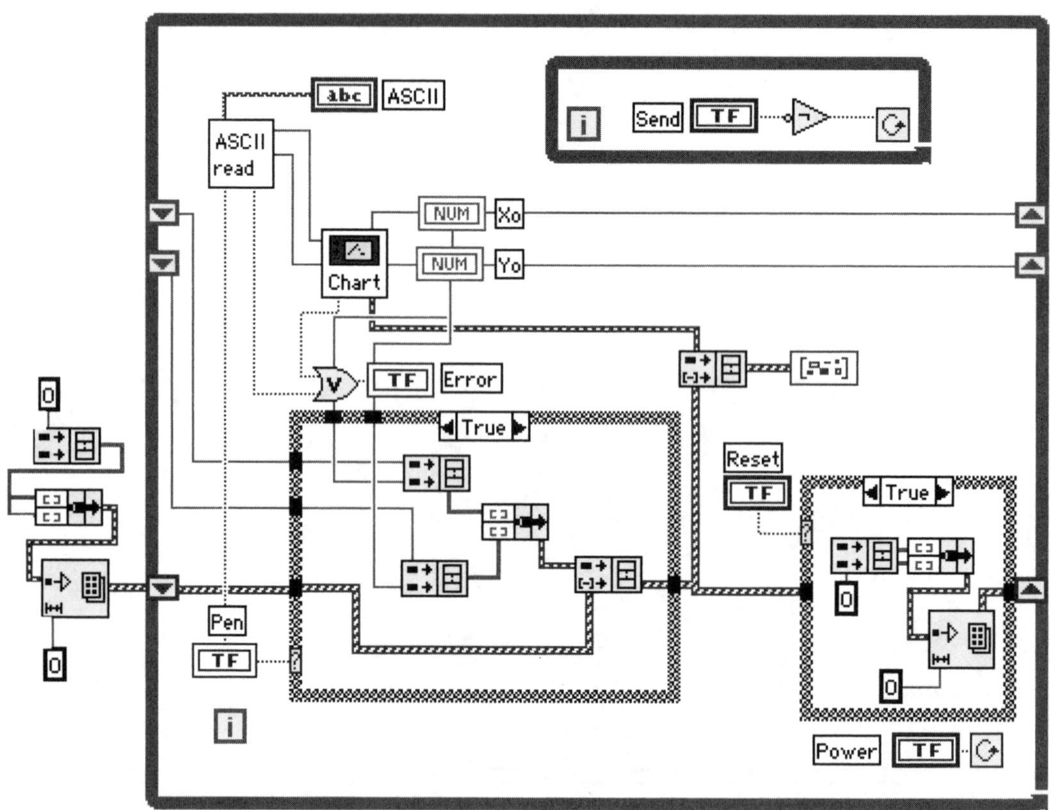

A reset function is also provided to clear the screen and set internal counters back to zero. The diagram panel for Version 2.0 is considerably more complex, since the draw command requires both the current position and the new position before it can calculate the plot command. Several shift registers are added to implement this function.

Etch-a-Sketch Plus

Many digital plotters have a joystick on the front panel, so the operator can set the initial position of a plot field. Version 3.0 shown on the chapter opener art, provides this feature by adding four button controls—[Up], [Down], [Left], and [Right]. Each time the button is pressed the plotter pen moves forward five plotter units in the direction selected. The pen button is now a control. If the button is up, the move command is executed. If the button is down, the draw command is executed. Load XYPlotter 3.vi and take that "pixel" for a test ride. Simple and complex rectangular structures can be drawn on the plot field.

If the artwork is not up to your standards, there is always the [Reset] button to clear the field and start over. Note that the plotter commands in string format are shown in a front panel display so the artist can follow the ASCII trail in creating a masterpiece.

Overview

One, two or n-dimensional set of numbers can be represented by a single compact data type called array. LabVIEW contains a wealth of array functions which allow arrays to be manipulated in the most complex manner while still maintaining a simple and elegant programming style. A virtual experiment to study the heating curve of a liquid is used to illuminate the programming style with arrays. Along the way a least squares fit to a straight line is used to include the calibration table of a platinum resistance thermometer into the heating curve experiment. A visualization of random numbers and the design of polar plots are used to display some of the unique array operations.

GOALS

- Understand arrays, array operations and array functions
- Build one- and two-dimensional arrays
- Study measuring temperature with a platinum resistance thermometer
- Design a polynomial fit to a data set
- Study the heating curve of a liquid
- Visualize random number distributions
- Design a VI for polar plots

KEY TERMS

- Temperature sensors
- Platinum resistance thermometer
- Arrays and indexes
- Building arrays
- Polymorphism
- Bundle operation
- Spreadsheet files
- Radar plots

Arrays of Light

7

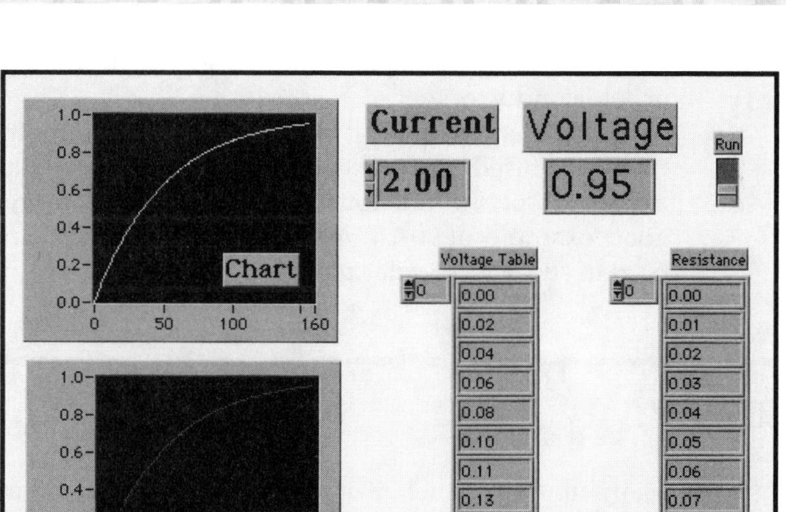

Heating Curve for Bringing a Pot of Water to the Boiling Point. The chart output displays the temperature as a function of time as the water is heated. At the end of the experiment, the data is converted into an array and displayed on the front panel in array and graph format.

Temperature is perhaps the most commonly measured parameter in science and engineering. The primary temperature standards are fixed temperatures produced by physical phenomena. The triple point is an unique point in temperature and pressure where all three phases (solid, liquid, and gas) can coexist in equilibrium. These are usually expressed in an absolute thermodynamic temperature scale called Kelvin. Zero degrees Kelvin corresponds to the absolute minimum heat energy. The triple point of water is 273.16 K. at 4.58 mm Hg. To correct to standard pressure, at 0 mm Hg the triple point is reduced by 0.01 K. In science and engineering, the Celsius temperature scale is the most common scale, having the freezing and boiling point of water at 0°C and 100°C respectively at standard pressure. Hence, the conversion factor to convert Celsius to Kelvin is

$$°K = °C + 273.15$$

There are many types of thermometers but the most commonly used sensors are platinum resistance, thermocouples, thermistors, and solid-state devices. All these thermometers depend on the thermal variation of a physical property of metals or semiconductors.

In this chapter, the platinum resistance thermometer is featured and used as a means of studying arrays, charts, graphs, and other special functions. Random numbers are revisited as a good example of array magic. Finally, the stepping motor is used to study how polar plots are created from XY graphs.

What Is An Array?

Many times a sequence of measurements is taken and information is contained not only in the magnitude of each measurement but also the change in magnitude over time. An array is simply an ordered set of numbers where each number has both a magnitude and an index. In mathematics, a one-dimensional array might be denoted by $X(i)$, where X is the representation for the whole array, i denotes the ith element in the array. In LabVIEW, a one-dimensional array is displayed as an icon with two connected boxes on the front panel. The smaller box with the scroll bars contains the index *i* and the larger box represents the array. Arrays can be numerics, strings, or even Booleans. To define the data type, select a control or indicator and place it over the larger box. When the mouse button is released, the larger box is filled with the selected data type and it turns gray. Now the larger box contains the value of the element selected by the index box. The following diagram shows numeric and string arrays for both controls and indicators on the front panel and on the diagram panel. Study the differences carefully.

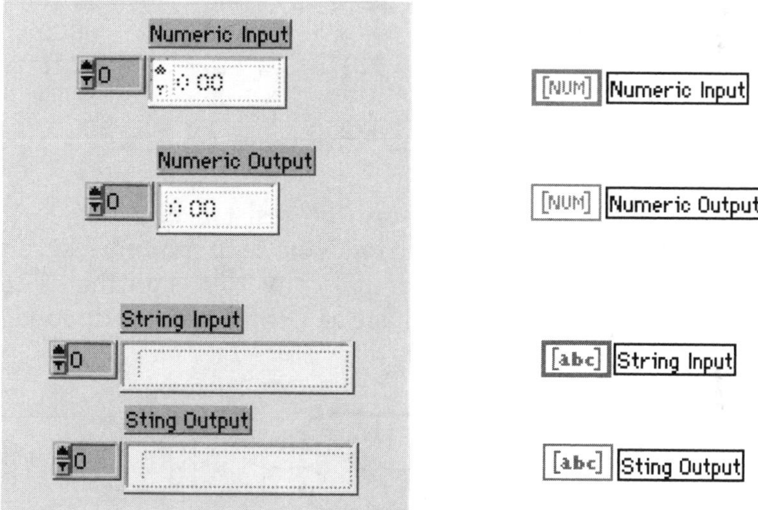

Platinum Resistance Thermometer

Platinum is a noble metal that can withstand high temperatures and harsh environments with good stability and repeatability. The temperature sensor is a coil of thin platinum wire mounted on a strain-free quartz support and sealed inside a ceramic case. Other models use a thick film of platinum on a substrate of alumina (Al_2O_3). The standard platinum thermometer is carefully trimmed to have a resistance of 100.0 Ω at 0°C. In North America, the temperature coefficient is about 0.392 ohms per degree

Centigrade. The material is so reproducible that standard tables of resistance versus temperature can be published.

T (°C)	R (Ω)	T (°C)	R (Ω)	T (°C)	R (Ω)
−200	18.49	−100	60.25	0	100.00
−190	22.80	−90	64.30	10	103.90
−180	27.08	−80	68.33	20	107.79
−170	31.32	−70	72.33	30	111.67
−160	35.53	−60	76.33	40	115.54
−150	39.71	−50	80.31	50	119.40
−140	43.87	−40	84.27	60	123.24
−130	48.00	−30	88.22	70	127.07
−120	52.11	−20	92.16	80	130.89
−110	56.19	−10	96.09	90	134.70
−100	60.25	0	100.00	100	138.50

To measure the resistance, a simple circuit consisting of a constant current source, platinum resistance thermometer, and analog-to-digital converter is all that is required. Resistance is just the voltage across the resistor divided by the current.

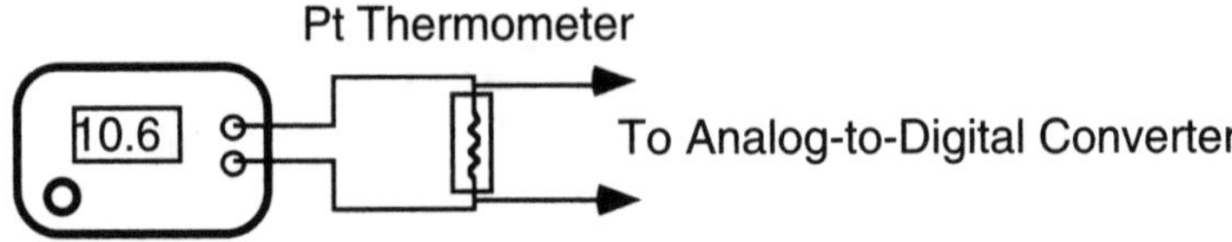

The analog-to-digital converter can be in the form of a digital multimeter connected to a computer with an RS232 or IEEE 488 interface or a DAQ card plugged into the computer bus. LabVIEW has drivers for each interface. The final result is that when the driver VI is called, the analog voltage measured across the thermometer is returned. One note of caution is that the constant current must be small enough so that intrinsic Joule heating within the sensor does not a cause an error. In practice, this means that the driving currents are in the microamp range and the voltage across the platinum thermometer is in the millivolt range.

Forming a Data Array

Consider an experiment to measure the rate of heating for a liquid. A beaker of water is placed on a retort stand just above a Bunsen burner. A platinum resistance thermometer is placed into the liquid. The voltage across the thermometer is monitored and sampled by LabVIEW every 30 seconds. A plot of the data stream displays the heating curve of the liquid.

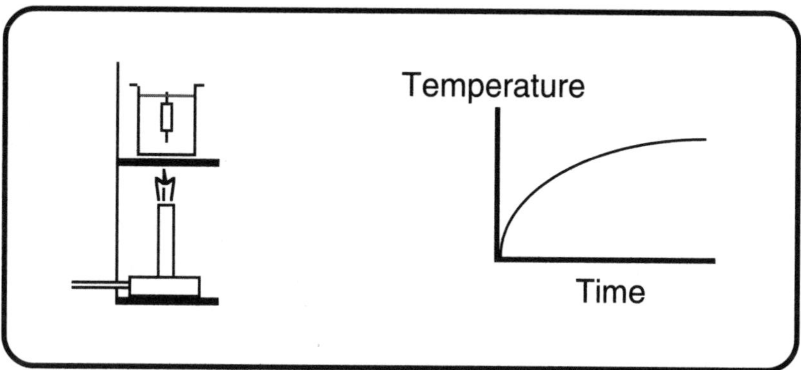

To simulate the experiment, an analog driver called Pt Therm.vi provides the data set. Each time this VI is called, the temperature for the next 30-second time interval is returned. To measure the temperature profile, Pt Therm.vi is placed inside a While . . . Loop. The loop index is used to trigger the VI to produce the next measurement each time the loop executes. A digital indicator on the front panel is used to observe the voltage signal as it is measured. To see the profile in real time place a chart on the front panel. From the **Controls>Graph** menu select **Waveform Chart.** Make sure to put the chart terminal inside the loop. Run the program to observe the heating curve.

How does one collect the complete data set so that it can be used as a single entity to be saved or used in another part of the program? Arrays catch the data and store it in sequence as a contiguous block of memory. Each array has a unique name or label and as such data can be passed on to other functions as a single block.

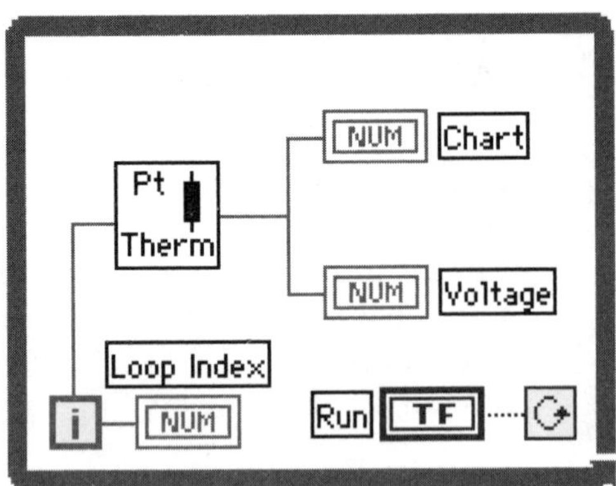

From the **Controls>Array & Cluster** menu select **Array.** Now from the **Controls>Numeric** menu select **Digital Indicator.** Place the indicator on top of the right hand box of the Array icon. The box will display a dashed insert line when the indicator is on top of the array. Release the mouse (unclick) and the indicator drops inside the array, now shaded light gray. On the diagram panel place the array terminal outside the loop. Wire the output of Pt Therm.vi to the array terminal. A tunnel will be created through the loop as the wiring proceeds. The link between the tunnel and the array may be dashed. If so, pop up on the tunnel and select **Enable Indexing.** The dashed line will change to a thick colored line. The thicker link is LabVIEW's way for displaying array data paths. Both the While . . . Loop and the For . . . Loop can index and accumulate arrays at their boundaries automatically. Each iteration creates a new element, in this case the next voltage. When the loop is terminated, the array values are passed on to the indicator. Run the program for a few cycles and using the scroll bars investigate how the data is stored in the front panel array Icon. Select from the **Controls>Graph** menu **Waveform Graph** and place it on the front panel. Its terminal is connected to the array link outside the loop. Run the program and observe what happens.

The chart function is just like a strip chart recorder that displays data as it is received, hence in real time. On the other hand, graphs wait until all the data points are received before

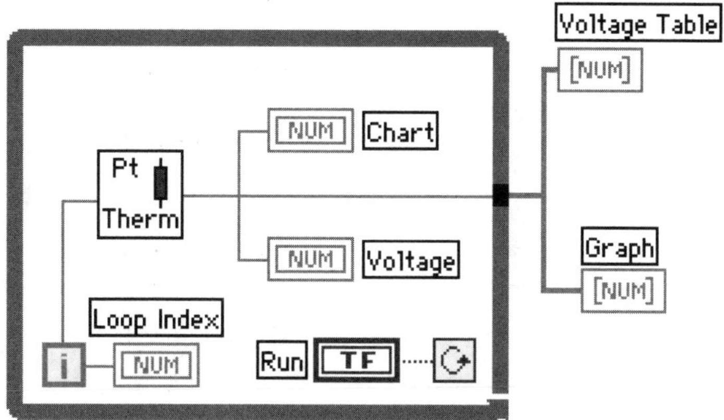

displaying the graph. This is an important distinction to remember. Charts use a numeric data stream, while graphs use an array data stream.

Array Functions

Arrays are just another data type and there are many things one can do to or with an array. Elements can be counted, new elements can be added, and elements can be extracted, replaced, or modified. Arrays can be split, sorted, or transposed. The next section uses the heating curve data to demonstrate some of the basic array functions and techniques.

In general, the number of elements or measurements in an array is not known. The **Array Size** function returns the number of elements inside an array. This is important because many array functions need to know the number of elements in an array before the function can execute. For one-dimensional arrays, the array function returns a single numeric, the number of elements in the array. For two-dimensional arrays, the **Array Size** returns an array with two elements, one for the numbers of rows and the other for the number of columns. Add the **Array Size** function and observe it in action.

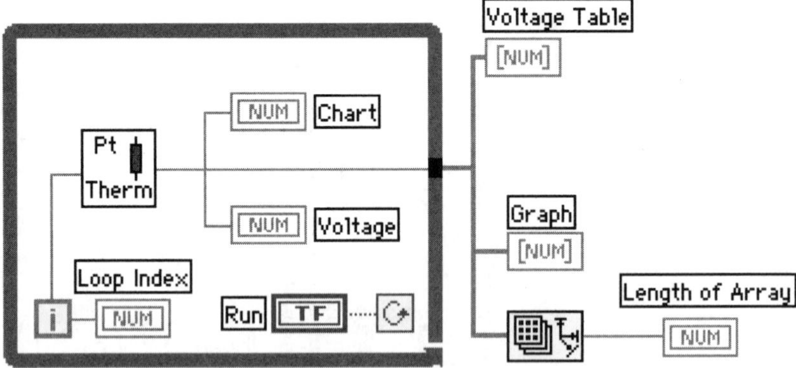

Note the value of the loop index and the length of array value. The loop index counts from 0 while the length of array counts from 1.

Another interesting question is How do you clear an array? The simplest method is to select **Reinitialize All to Default** from the **Operate** menu. This selection sets all elements in the arrays to the default value zero. On the other hand, charts and graphs are cleared by popping up on the chart or graph and selecting **Data Operations>Clear Chart.**

Returning to our experiment at hand, the array now contains a record of the voltage measurements. However, the physical measurement is resistance. To calculate the resistance, each voltage measurement is divided by the current flowing through the thermometer. Watch the polymorphism-magic!

Create on the front panel a new array of numeric indicators called Resistance. Add a numeric control for the constant current source. On the diagram panel add a divide function. Wire the top input to the array link and the bottom input to the current. The divide output is wired to the resistance array terminal. What is this? Two different data types wired up to the same function and LabVIEW understands what to do. In this case, each element of the array is divided by the numeric input producing a new array. It is an example of operation polymorphism. Run the program to observe the results.

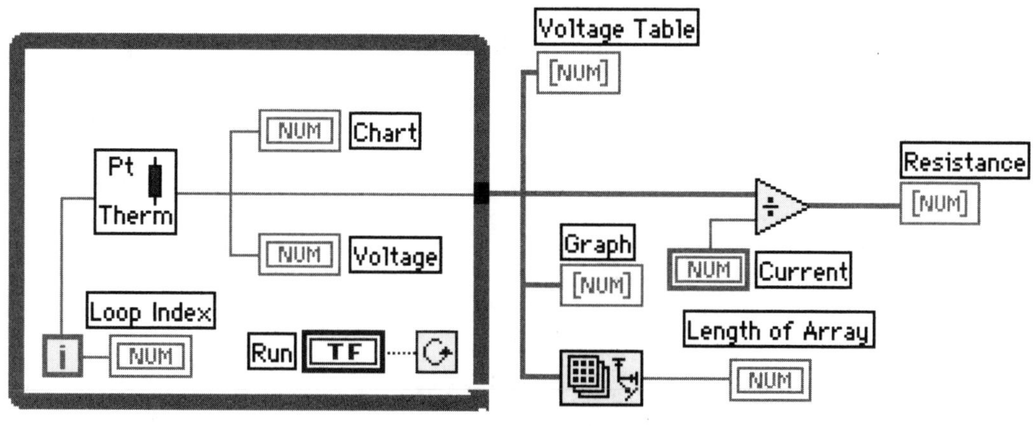

Saving Your Data

Another useful array function is the **Write To Spreadsheet File.** This little jewel converts one- or two-dimensional data sets to a text string and saves them on a floppy disk or hard drive in a spreadsheet file format. A tab character is used to signal a new column and a carriage return character to signal a new row. You can even transpose the data file from horizontal to vertical format if required. Most word processors, spreadsheets, and plot programs can read these files directly. From the **Functions>File I/O** menu select **Write To Spreadsheet File.vi.** Wire the resistance array link to the 1D data input of the file icon, see next page. Run your program or the library VI, Exp2.vi. When the pop up dialog panel appears, save the data file at a convenient location because it will be used later.

■ Calibration Thoughts

The platinum resistance versus temperature table shown earlier provides a calibration table through which the measured resistance can be converted into a temperature value. The table is an example of a two-dimensional array having 2 columns and 31

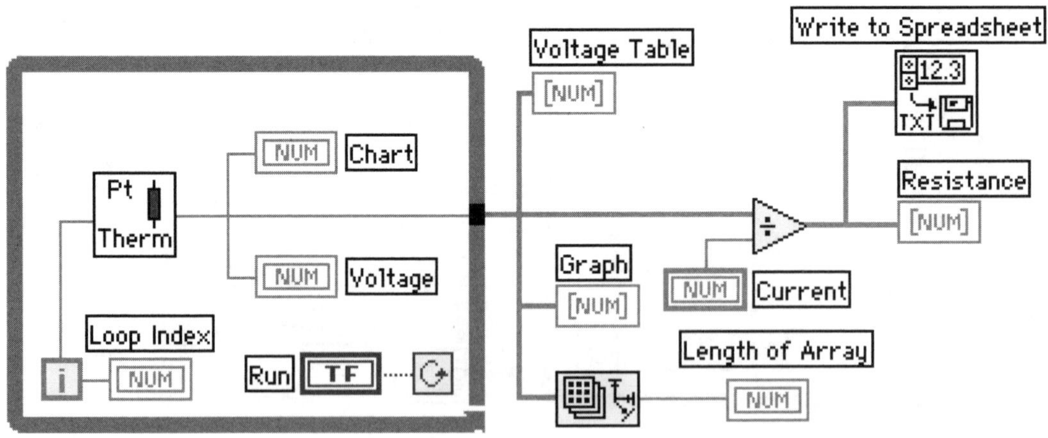

rows, a total of 62 elements. LabVIEW works just as well with 2D data sets as 1D data sets. On the diagram panel, 2D data links are shown as a colored double line. In fact, a 2D array can be built from two 1D arrays using the **Build Array** function, found in the **Functions>Array** menu. The following program creates a 2D array of sine and cosine values and saves it as a spreadsheet formatted file.

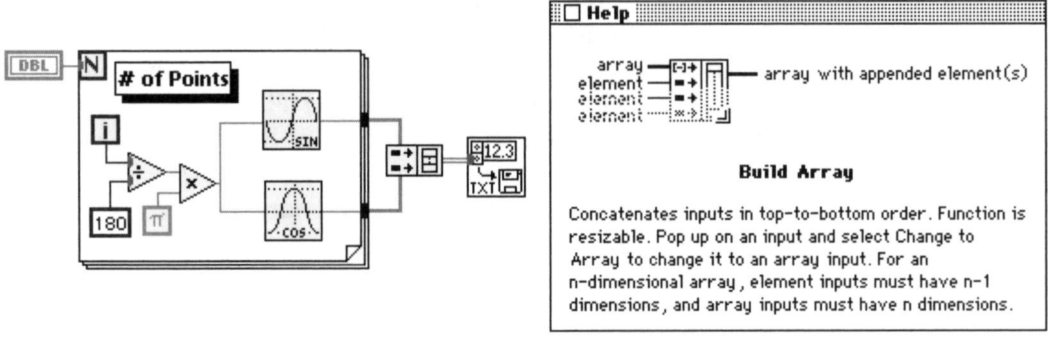

An algebraic relationship between the temperature and resistance is required to convert the measured resistance values into temperature. To help us, another array function, **Curve Fit**, is used. This function executes a polynomial regression to a 1D array of X values and a 1D array of Y values. The fit type can be linear, exponential, or polynomial. If polynomial, the order must be less than the number of data elements. The output generates an array of coefficients, the mean standard error between the raw and fitted data, and a fit array that can be used to visualize the quality of fit.

The platinum resistance thermometer calibration table is stored as a spreadsheet file called Platinum Calibration.data. You can look at it with any text processor or spreadsheet program. The file is in the form of a two-dimensional array. Curve fit requires two 1D arrays: one for X values and one for Y values. The first step is to unbuild the 2D array into two 1D arrays. This is accomplished using the using the **Index Array** function.

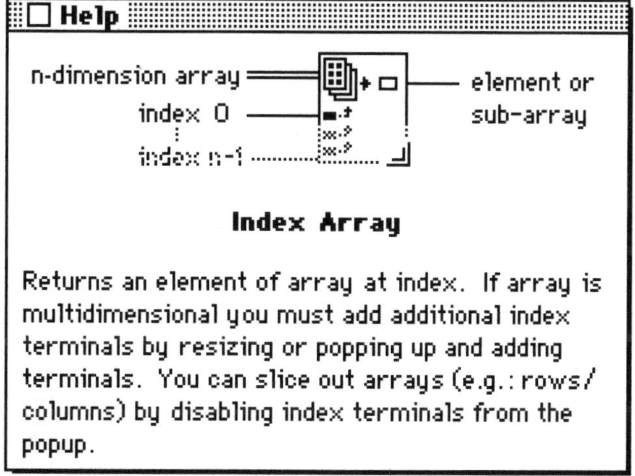

Select two **Index Array** functions from the **Functions>Array** menu. Using the position tool, resize the icons downward to add a second index. Pop up on the upper index box and select *Disable Indexing.* The icon will change to the form shown below. The lower index terminals are wired to the constants 0 and 1. This is the array address for the first and second row. The output will be a 1D data array containing either the resistance (0) or temperature (1) values. You can now add the **Curve Fit** function and wire up the array data paths to front panel indicators. This example displays the original 2D array, the coefficient array, and the fit array. The latter can be used to show the values calculated from the fit at the x (resistance) values. If the fitted values and the original data points are plotted on the same graph, a goodness of fit can be displayed.

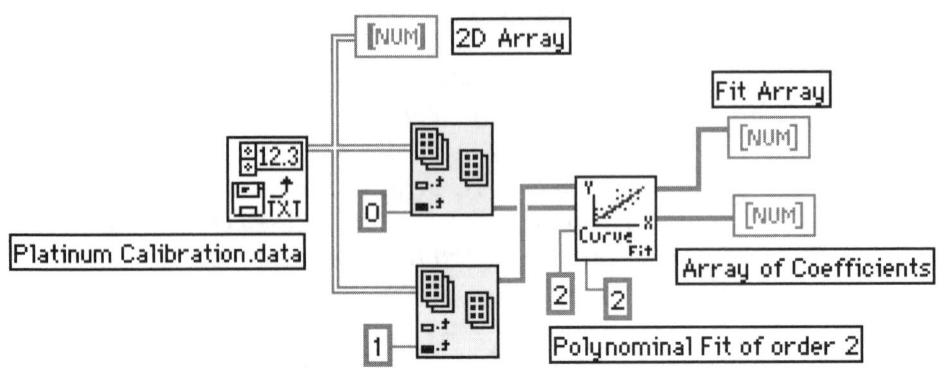

After running this program, you can write down an expression for the calibration curve.

$$\text{Temperature} = C_0 + C_1 * \text{Resistance} + C_2 * (\text{Resistance})^2$$

Here C_i are the coefficients found in the Array of Coefficients.

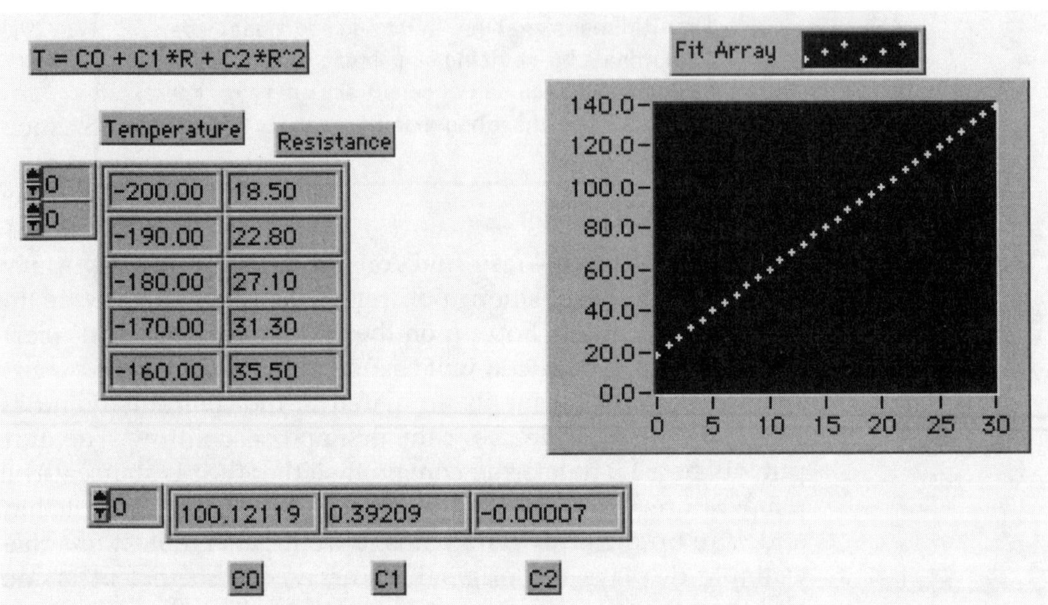

The final step is to add the calibration expression into the measured data set and plot the actual temperature versus time graph for the heating curve of a liquid. This is left as an exercise for the reader.

Random Numbers Revisited

A simple check on the randomness of random numbers is to plot the number of times a given number occurs for each number. For a large number of trials, the distribution should be uniform across the range of choices. In this investigation, a die is rolled and LabVIEW counts the number of times each number appears in N throws. Random 6.vi from Chapter 3 is again used to simulate a six-sided die, and it becomes our source of random numbers. Recall that each time this VI is called, it generates a random integer between 0 and 5. To keep a record of the number of each side thrown, a 1D array consisting of six elements is used. The 0th element contains the number of ones, 1st element contains the number of twos, and so on. In the following program an **Initialize Array** icon is placed outside the For . . . Loop. It initializes the array so each of the six elements contains the number zero. The accumulation of counts is accomplished by the shift register in the For . . . Loop. Note again polymorphism—the shift register contains array data types.

When the die is thrown, Random 6 is called and the number rolled is used as an index into the count array. **Index Array** allows you to extract a given element from the count array. For example, if the die was three, then the number of threes in the count array is at the array index two. **Index Array** makes available the current count as its output. The current count is incremented by one, then put back into the array at the same index location. This second feat is accomplished with the **Replace Array Element** function. Its output is the updated count and it in turn is passed on to the shift register in the count array. When the For . . . Loop has completed N loops, the count array contains the number of ones, twos, threes, and so on and is passed outside the loop for display. The data path goes in parallel to six

Index Array functions. Each function returns the contents of the element whose index is wired to the index terminal. The number of counts is then passed onto a digital indicator for display.

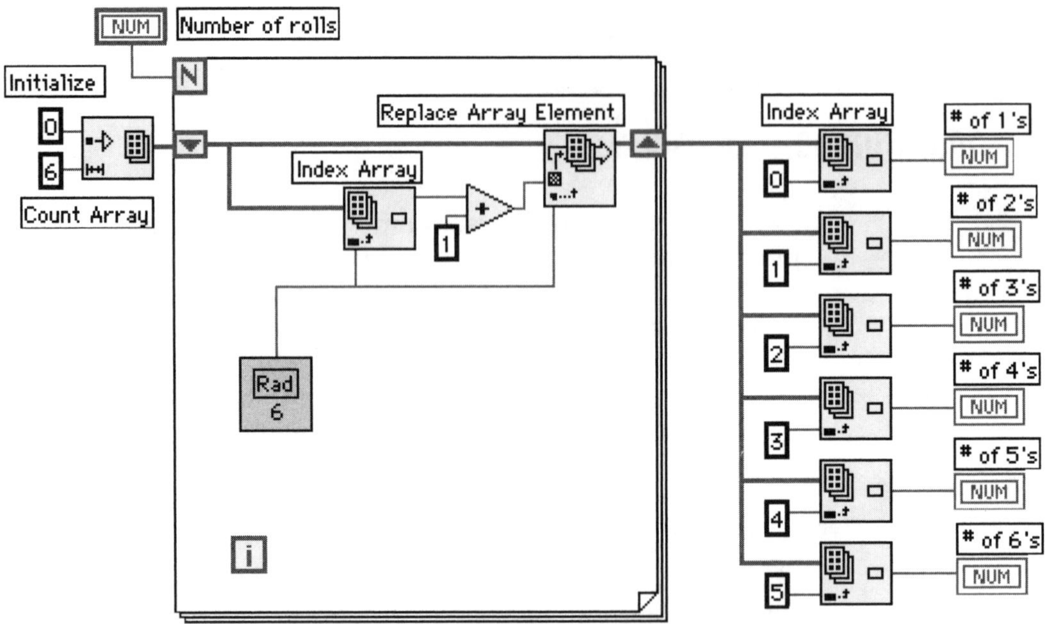

Roll the die, Count6.vi about 200 times to get a good appreciation for the variation found in the distribution of occurrences of each number. What is your favorite number? Does LabVIEW have a favorite?

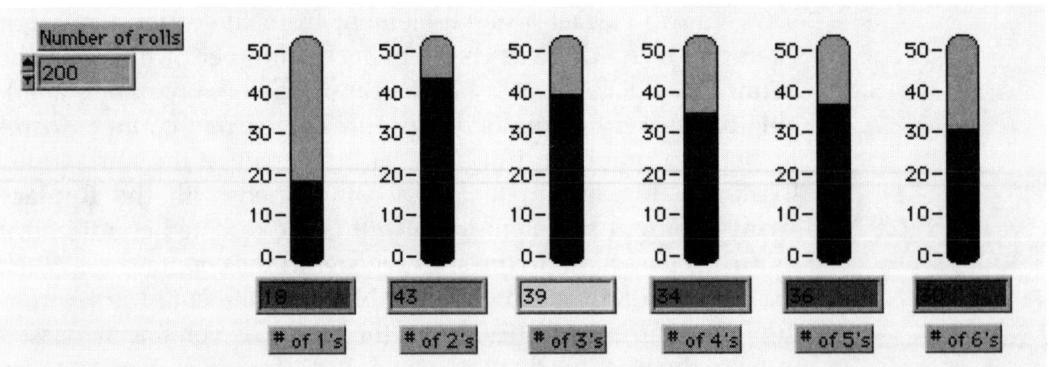

Polar Plots

Radar or swept angle polar plots are useful in displaying angular data or in simulating rotating machines. LabVIEW provides another graph type that can be used to create radar plots. Waveform graphs take data from an array and assumes that the data is sampled at equal intervals of time. XY graphs plots pairs of points (x,y) and as such the relationship between points need not be uniform. The **XY Graph** function requires that the X and Y arrays be bundled into a cluster input. A cluster is just a group of data paths combined to form a new data type. A cluster can contain Booleans, numerics, or arrays. The **Bundle** function is found in the **Functions>Cluster** menu.

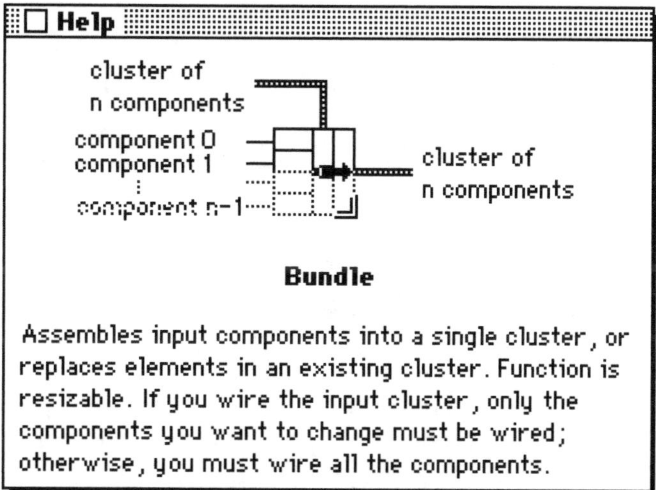

The radar plot will use a cluster input consisting of the two 1D arrays bundled together. The first step in understanding the radar plot is to plot a circle graph. The following program features the **Sine** and **Cosine** functions to create *X* and *Y* components of a unit vector .

$$X = \text{Cos}\,(\theta) \quad Y = \text{Sin}\,(\theta)$$

The For . . . Loop index is used for the angle and each loop corresponds to one degree. Multiplying the index by a constant defines the angular interval. This value is in turn converted into radians, a requirement for **Sine** and **Cosine** functions. The output of the sine and cosine functions tunnel through the For . . . Loop to be autoindexed and form the 1D arrays of X and Y. These are in turn bundled together into a cluster for the XY Graph terminal.

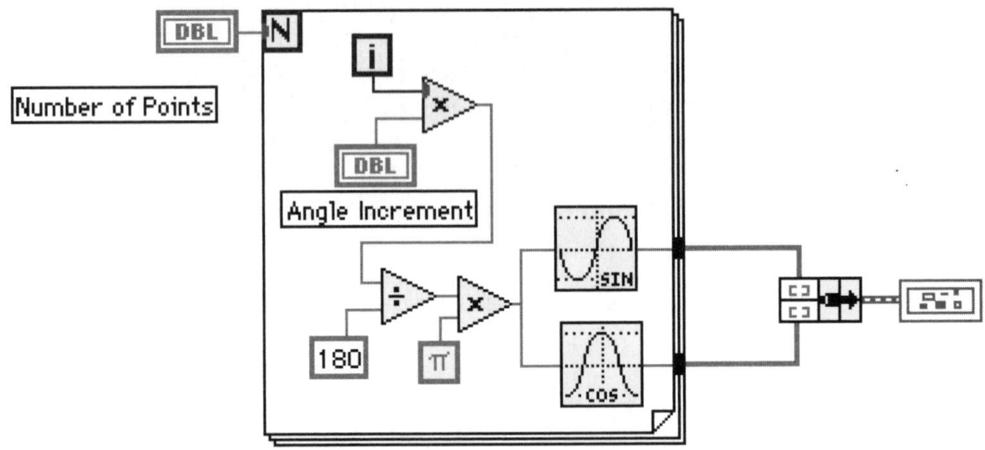

To see the circular plot on the front panel (next page), it is necessary to remove the autoscale feature found in the pop-up menu of the **XY Graph** under **X Scale** and **Y Scale.** Set the X and Y scales from–1.1 to 1.1. Try different line and point types on the pop-up menu of the legend. Especially interesting are the interpolation types from the same pop-up menu.

The graph output waits for all the values to be processed before plotting the points, Circle1.vi. How would you plot one point at a time?

In this case, each time a point is calculated a new graph is plotted. The autoindexing feature of the For . . . Loop is not used. The X and Y components from the sine and cosine functions are simple numerics that are combined using the **Bundle** function into a cluster point. The **Build Array** does the trick by converting the cluster point into a 2D array, albeit a very small array. Place the program inside a For . . . Loop and watch the unit vector tip trace out a circular path, Arc.vi.

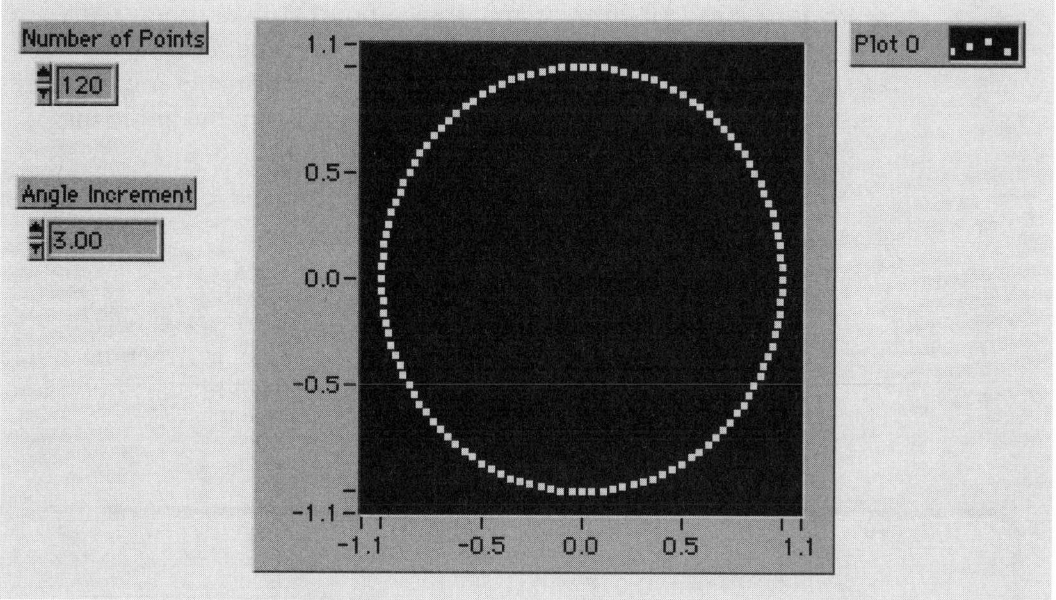

For the radar plot, it is necessary to plot a rotating unit vector in real time. The vector is created by plotting a straight line from the center (0,0) to the vector tip (1,θ). XY Graph uses Cartesian coordinates, hence a conversion from polar to rectilinear coordinates is necessary.

$$x = r \cos(\theta) \qquad y = r \sin(\theta)$$

LabVIEW provides this function in the **Functions>Analysis> Array Operations** menu using **Polar to Rectangular.vi.** The cen-

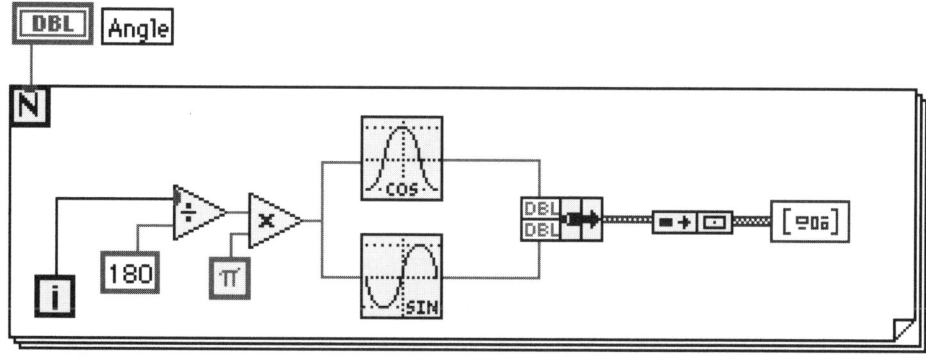

ter point (0,0) and circumference point (1,θ) are then passed to the **Build Array** function and bundled for the XY Graph.

To observe the unit vector on the front panel graph, it is necessary to select from the pop-up legend menu the following

Point Style → (none)
Line Style → (solid line)
Interpolation → (line)

Each time the program, Polarplot.vi is run or called, a unit vector at the input angle is plotted. To make a sweeping radar plot, place this VI inside a For . . . Loop or While . . . Loop and watch the action.

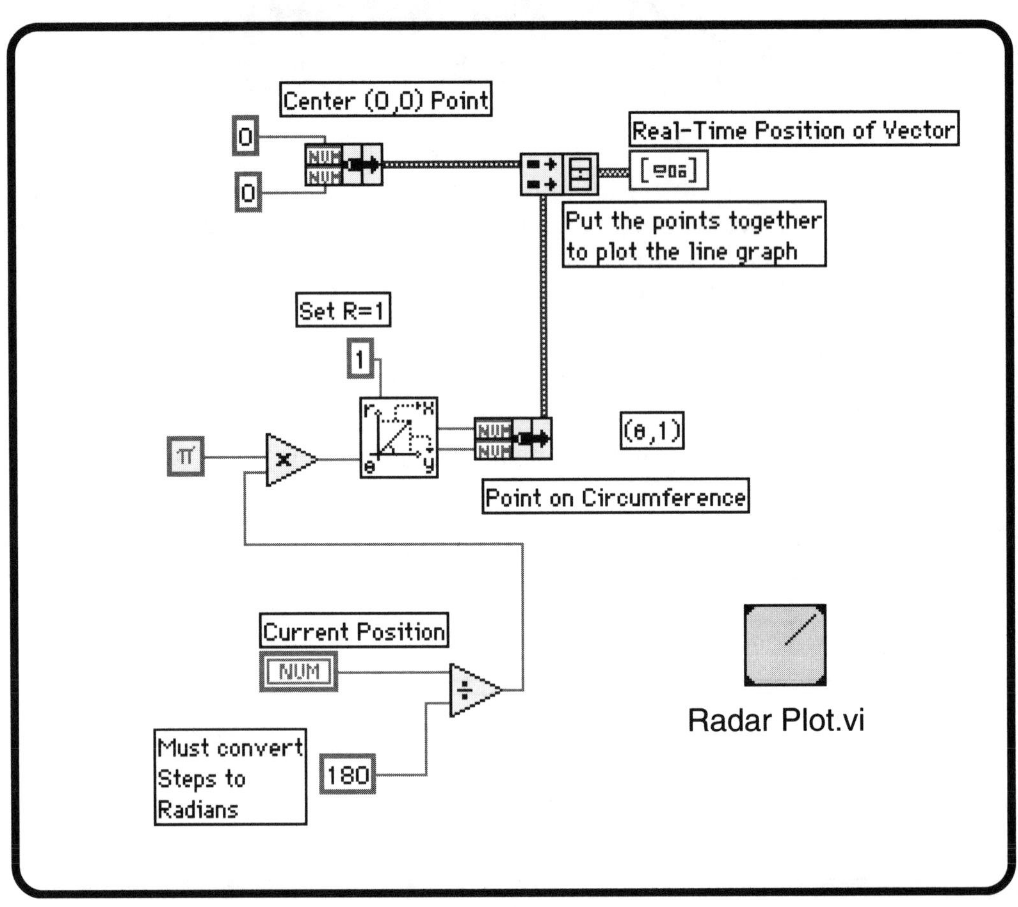

Overview

Semiconductor thermometers offer a low cost solution to accurate and reliable temperature measurements. Several real world experiments demonstrate how temperature measurements can be used to reveal a deeper understanding of a physical process. Several approaches to the calibration of thermocouples demonstrate the flexibility of LabVIEW programming and how built-in functions (icons) make using thermocouples in an industrial environment a simple task.

Goals

- Understand semiconductor thermometers and thermocouples
- Build a data logger to measure cooling curves
- Determine the cooling constant from a cooling curve
- Study calibration algorithms for thermocouples
- Understand hardware, software, and LabVIEW solutions
- Observe the changes of state in the melting of an ice ball

Key Terms

- Semiconductor thermometers
- Current-to-voltage converter
- Heat transfer by radiation losses
- Cooling constant
- Thermocouples
- Thermocouple standards
- Cold junction compensation

Some Like It Hot: Semiconductor Thermometers

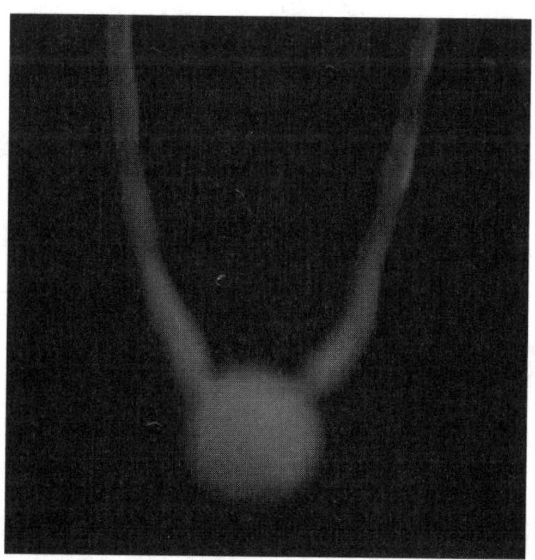

Changes of State: The tip of a chomel/alumel thermocouple is successively dipped into water then liquid nitrogen until an ice ball is formed. As the ice ball warms up, it displays a unique heating curve.

The current flowing in a semiconductor pn junction can be represented by the diode equation

$$I = I_0[\exp(eV_d/kT) - 1]$$

where I_0 is the reverse bias current that depends weakly on temperature and the material properties of the p and n type semiconductors, V_d is the voltage across the pn junction diode, T is the absolute temperature measured in degrees Kelvin, and the atomic constants k/e = 86.1709 mV/°K. When the diode is forward biased, the exponential term dominates and the above equation can be rearranged and written as

$$V_d = \{86.1709 \ln(I/I_0)\}\ T\ (\mathrm{mV/°K})$$

If a diode is placed in a constant current circuit, then the term in the curly brackets is a constant and the voltage across the diode is directly proportional to absolute temperature.

This expression is also valid for transistor junctions and in the special case where a matched pair of transistors pass different currents I_1 and I_2, the difference in the base-emitter voltages is related to the absolute temperature in a similar fashion

$$\Delta V_{BE} = 86.1709 \ln(I_1/I_2)\ T$$

The Analog Devices temperature sensor AD590 uses eight identical transistors in parallel with a single transistor to set the ratio I_1/I_2 equal to 8. The voltage difference is denoted in the following figure as V_r and is directly proportional absolute temperature.

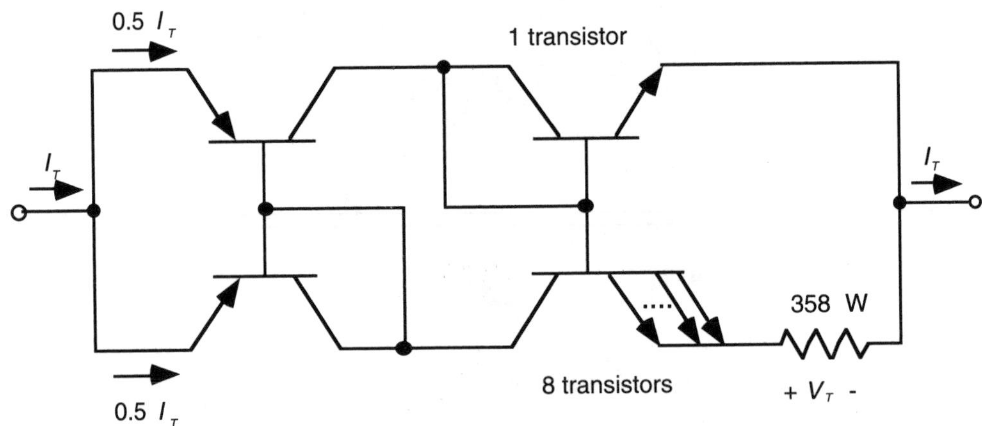

The transducer equation becomes

$$V_r = [86.1709 \ln(8)]\ T = 179\ (\mu\text{V}/°\text{K})\ T$$

One-half of the input current I_T passes through the single transistor while the other half of I_T passes through the eight transistors connected in parallel. The 358 Ω resistor is laser trimmed so that the current in the lower loop $V_r/358$ is exactly 0.5 I_T. The total output current is then

$$I_T = 2^*\ (179/358)\ T = 1.0\ (\mu\text{a}/°\text{K})\ T.$$

The AD590 has a direct relationship between the current flowing through the device and absolute temperature with a proportionality constant of 1.0 μa/°K. For example, an ambient temperature of 0°C yields a current of 273 μa.

The AD590 can be driven with a power supply in the range +4 to +30 volts and operates in over wide temperature range of –55°C to +150°C.

The simplest and most flexible temperature transducer is to pass the sensor output current into an OpAmp circuit configured as a current-to-voltage converter. The transform equation is

$$V_{\text{out}} = -\,i\,R_f$$

where i is the input current and R_f is the feedback resistor. A resistance in the megaohms allows the microamp current levels from the sensor to be converted into voltage levels suitable to be input into analog-to-digital circuits using only one integrated circuit.

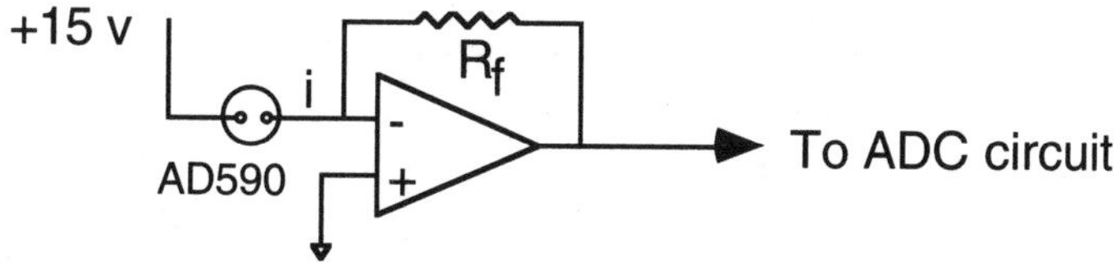

$$V_{\text{out}} = -\,1.0\ (\mu\text{a}/\text{K})\ T\ .\ R_f\,(\Omega)$$

For example, at 20°C the circuit would output –2.93 volts using a feedback resistor R_f equal to 10 KΩ. By placing the sensor in an ice bath, any offset voltage can be trimmed to zero so that the voltage reads precisely 2.7317 volts (T = 273.17°K). By subtracting 273.17 μa from the summing point of the OpAmp, the circuit would output a voltage directly proportional to degrees Centigrade. Further, by placing the sensor in a bath of boiling water and adjusting the feedback resistor to produce an output of 10.00 volts corresponding to 100.0°C, a simple and calibrated temperature transducer is built. It can be read by any digital voltmeter or analog-to-digital circuit with an output calibrated in degrees Centigrade and the absolute error will be less than 0.4°C over the entire operating range. However, if the sensor voltage is to be logged under computer control, then the calibration offset and gain constants can be included in software. With such a simple temperature sensor many experiments are possible.

Some Like It Hot! The Cooling Coffee Cup

We all know that a hot cup of coffee stays hotter longer in a covered cup than a cup with no lid; but how much longer, two, three, or four times?

Two cups of hot coffee in Styrofoam cups are placed on a table at room temperature. AD590 temperature sensors are placed into each cup near the center of the liquid. One cup is covered with the usual plastic lid provided by the retailer and the other is left open to the air.

LabVIEW Challenge: Cooling Curve of Hot Cup of Coffee

Design a LabVIEW program to sample the temperature of each cup at 10-second time intervals and plot the cooling curve for each cup on the same chart. How much faster does an uncovered cup of coffee cool than a covered cup of coffee?

Before conducting the experiment, let's do a little reasoning. First, Styrofoam is a good insulator, so little heat will be lost by conduction through the sides or bottom. Convection effects will certainly aid in the cooling once currents are established inside the liquid. If you stir your coffee, it cools faster. But unstirred, natural convection is not the major cooling factor. Most of the heat lost will come from radiation through the lid on the covered cup and from the top surface on the open cup. All objects radiate energy continuously in the form of electromagnetic waves. In the coffee cup problem, heat transfer is primarily in the form of infrared radiation. The rate at which an object emits radiant energy is proportional to the fourth power of absolute temperature. Stefan's law is expressed by

$$P = \sigma A e T^4$$

where P is the power radiated by the body in Watts, σ is Stefan's constant equal to 5.67×10^{-8} ($\mathrm{W/m^2 \cdot {}^\circ K^4}$), A is the cross-sectional area (m^2) of the emitting surface, and e is the emissivity, a factor that ranges from 0 to 1 depending on the radiating material. In addition, there is also a similar factor for the heat absorbed by the coffee and its cup from the room, $P = \sigma A e T_0^{\,4}$, where T_0 is the room temperature. Again, we assume that no heat is conducted from the room temperature environment through the Styrofoam cup, only from the upper surface. The net heat lost is the difference.

$$\Delta P = \sigma A e (T^4 - T_0^{\,4}) \text{ Watts}$$

With a knowledge of the initial and final temperatures, the quantity of coffee, the specific heat and the insulation factor of the cover, one can deduce the cooling curve for each case. It is not so easy to calculate an answer, but it is very easy to use LabVIEW to measure the temperature curves and find the answer empirically.

Temperature Logger

Two temperature measurement channels are simulated in data sets called Coffee+.vi and Coffee.vi. The plus sign indicates the covered cup. Each time these VIs are called, the next temperature measurement is returned. The data sets were recorded at 10-second intervals using digital voltmeters interfaced over the IEEE 488 data bus (see Chapter 15). Only the uncovered data set is used in the following example. It is left as an exercise for the reader to add the other channel to answer the original question.

A priori, one has no idea how long it will take the coffee to cool to room temperature because most people drink it long before the temperature reaches that point. Each data point is separated from the previous one by 10 seconds. A While . . . Loop is used to sample the data set. Each data point is then plotted on a front panel chart to observe the cooling in real time. The data collection can be stopped at any time using the Run button.

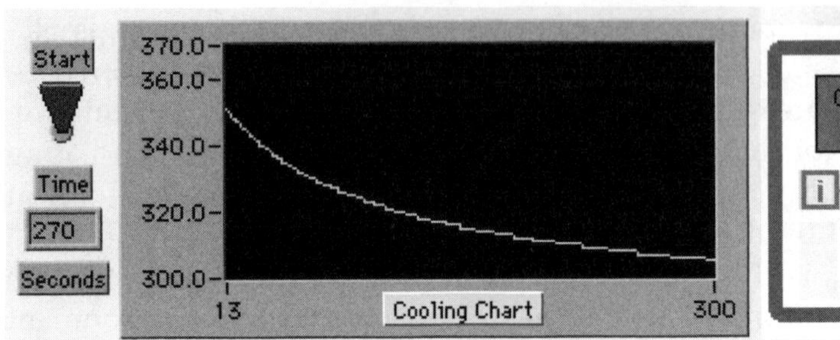

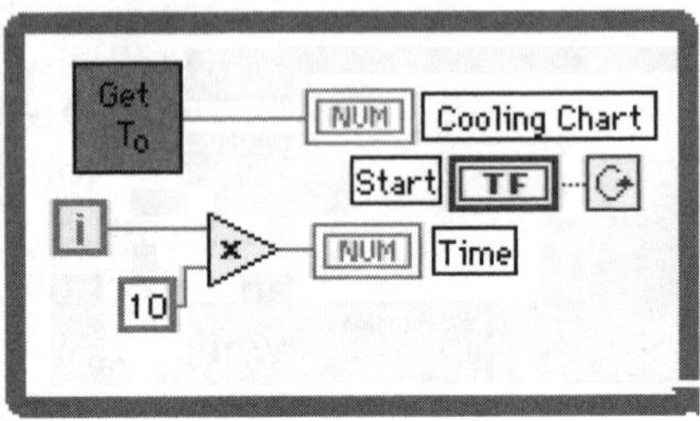

Assuming the cooling curve follows an exponential curve, the temperature can be the expressed as

$$T = (T_0 - T_f)\exp(-t/\tau) + T_f$$

where T_0 is the initial temperature and T_f is the ambient room temperature. The cooling constant τ is given when the coefficient of the exponential term is equal to (–1), that is, $t = \tau$ and the temperature at this time is given by

$$(T - T_f) = (T_0 - T_f)\exp(-1) = 0.63\,(T_0 - T_f)$$

From the slope of a plot of $\log[(T - T_f)/(T_0 - T_f)]$ versus t, the cooling constant, τ can be determined from $\tau = -1/\text{slope}$.

Exercise 1 Design a LabVIEW program to determine the cooling constant τ.

It is a little more interesting to see the cooling curves for each data set plotted together on the same graph where the differences are readily evident.

Exercise 2 Add the second data set to the program in Exercise 1. Plot both cooling curves on the same chart and answer the question, How much faster does an uncovered cup of coffee cool than a covered cup of coffee?

In the same spirit, there is another interesting question to ponder about drinking hot coffee. After I purchase a cup of coffee in the morning, it is a 5-minute walk back to my office. If I want the coffee to be as hot as possible when I start to drink it in my office, should I put the cream in it immediately after purchasing the coffee or should I wait until just before I am ready to drink it?

You can use the same techniques described above to tackle this problem. Better yet, with the above equations and a few assumptions about the specific heat of coffee and cream, size of the container, and so on, one can predict what will be the temperature just before the coffee is drunk and use the LabVIEW Temperature Logger to confirm your calculations.

Lost in the Desert

There has been an age-old debate about which color of clothing to wear in hot climates. Common sense seem to say wear light-colored and lightweight clothing. However, in some desert climates, many natives wear heavy dark clothing. Dark colors absorb the heat better, hence in the bright sun, dark-colored cloth will heat up to a higher temperature than light-colored cloth. At night, dark absorbers become good radiators and provide rapid cooling. The answer to this question is not just the color, but also in the style of clothing. Flowing robes allow natural evaporation from the skin to drive miniature air conditioning currents inside an outfit. All air conditioners need a source of power and in the desert dark clothing is a generator.

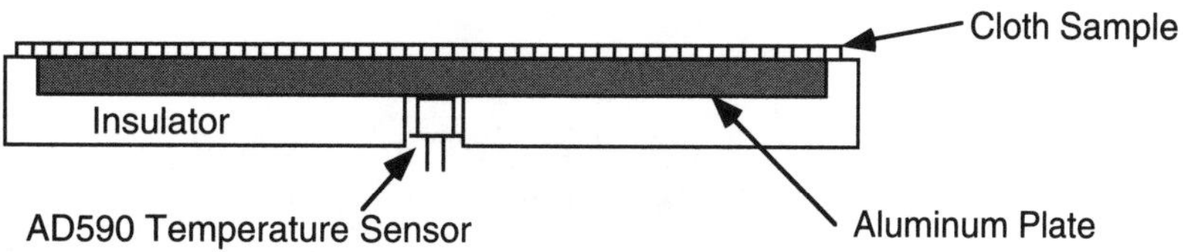

To answer at least part of the question, a small temperature sensor attached to a conducting metal becomes a test bed for absorption studies. A sheet of aluminum 4 inches by 4 inches is set into an insulating frame. An AD590 temperature sensor in a TO-5 metal can is pressed against the back side of the aluminum plate. A small amount of thermal grease between the sensor and plate aids in the heat transfer from the plate to the sensor. A piece of dark clothing is pressed to the front surface, using some transparent thermal paste. For reference a similar plate is covered with a light-colored piece of clothing. The sensors are attached to current-to-voltage amplifiers and their outputs connected to two channels of a LabVIEW data logger. The test plates are placed equidistant from a 200 Watt sunning lamp and the heating curves recorded.

LabVIEW Challenge: Absorption of Sunlight by Light and Dark Clothing

Design a two-channel data logger to observe the absorption curves of light and dark clothing. A typical data set is contained in two spreadsheet formatted files entitled dark.data and light.data. After plotting the data sets, answer the following questions. What is

the maximum temperature the light and dark clothing samples reach?

the temperature difference between the light and dark samples?

the initial rate of heating for the two samples?

Thermocouples

Thermocouples (TCs) are economical, rugged; they possess good long-term stability and can be used from cryogenic to jet-engine exhaust temperature ranges. In addition, their small size provides fast response to sudden temperature changes. A thermocouple consists of two dissimilar metals joined together at one end. A voltage called the Seebeck EMF is generated across the thermocouple if the two ends are at different temperatures. In the following figure two different metals are bonded together at the measurement point T_1 and the other ends are attached to two copper leads that are located in a reference temperature bath, T_0.

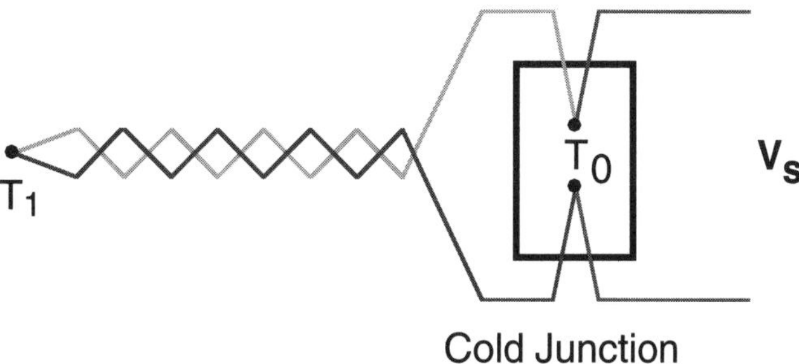

The Seebeck voltage V_S is a product of the Seebeck coefficient α and the temperature difference $(T_1 - T_0)$

$$V_S = \alpha\,(T_1 - T_0)$$

The Seebeck coefficient is a property of the metal or alloy used in the thermocouple construction, and for small temperature differences α is a constant. T_0 is commonly called the cold junction because of a common practice to immerse the reference junctions into a slurry of ice and water at 0.0°C. This ensures T_0 is independent of room temperature variations. Common thermocouple materials such as chromel/constantan (Type E), iron/constantan (Type J), copper/constantan (Type T), chromel/alumel (Type K), and platinum/platinum-10% rhodium (Type S) cover a wide variety of temperature ranges.

Over a wider temperature range, V_S is not quite linear with the temperature difference. In this case, calibration tables for the common thermocouple materials are readily available. The following table gives the thermoelectric output (mV) for chromel/constantan thermocouples for the temperature range –260 to 800°C.

Suppose a chromel/constantan thermocouple displayed a voltage V_S = 30.456 mV. What is the temperature at the measurement junction? A simple linear interpolation between the two adjacent calibration points 400°C (28.943) and 450°C (32.960) would yield a value of 418.83°C. If the thermocouple calibration is not linear in this region, then the calculation would be in error.

Thermoelectric Output of Chromel-Constantan Thermocouples

T (°C)	Output (mV)	T (°C)	Output (mV)
–260	–9.797	60	3.683
–240	–9.604	80	4.983
–220	–9.274	100	6.317
–200	–8.824	120	7.683
–180	–8.273	140	9.078
–160	–7.631	160	10.501
–140	–6.907	180	11.949
–120	–6.107	200	13.419
–100	–5.237	250	17.178
–80	–4.301	300	21.033
–60	–3.306	350	24.961
–40	–2.254	400	28.943
–20	–1.151	450	32.960
0	0	500	36.999
20	1.192	600	45.085
40	2.419	700	53.11
60	3.683	800	61.022

A better method would be to use all the calibration points to generate a polynomial expression for the temperature T_1 versus the measured voltage V_S. The National Bureau of Standards uses a ninth-order polynomial to represent the chromel/constantan calibration curve. A file entitled Chrom/Const.data contains the chromel/constantan calibration table as a 2D array of numbers.

LabVIEW Challenge: Chromel/Constantan Calibration Table

Design a LabVIEW program that fits a polynomial to the chromel/constantan calibration table and calculates the thermocouple temperature using the polynomial and the measured Seebeck voltage. How does the polynomial value compare with the linear interpolation when $V_S = 30.456$ mV?

The problem with using thermocouples to calculate absolute temperature is that TCs are basically a differential temperature sensor. A known reference junction is required if the temperature of the other junction is to be inferred from the voltage measurement. Most temperature tables of common thermocouples are referenced to 0°C, the water-ice point. The maintenance required to keep the reference point at 0°C is both tedious and continuous. It would be much more useful if the ambient temperature could be used as a reference. Two problems present themselves: (1) The calibration tables are published for 0°C reference, and (2) the ambient temperature drifts over time. A solution is to measure the ambient temperature with an absolute thermometer such as the AD590 temperature sensor. The sensor output is then added to the thermocouple output to compensate for the ambient temperature. Electronically the composite signal is referenced to 0°C and the standard calibration tables and polynomials can be used to calculate the temperature.

Cold Junction Compensation

The AD590 is an ideal ambient cold junction compensation (CJC) sensor with its high precision, good reproducibility, and simple circuitry. LabVIEW provides a library of thermocouple VIs found in the **Functions>Data Acquisition>Signal Conditioning** subpalette to convert a TC voltage into an actual temperature using a reference temperature from an ambient sensor and the standard thermocouple transfer function for the chosen thermocouple Type. **Convert Thermocouple Reading.vi** uses two numeric inputs, the TC voltage and the ambient (CJC) temperature signal.

The thermocouple type (B, E, J, K,. . . . T) is chosen as an integer from 0 to 6 and the CJC Sensor can be either an electronic RTD sensor or a thermistor.

The cold junction sensor voltage is measured and converted into a temperature inside the case statement. This temperature

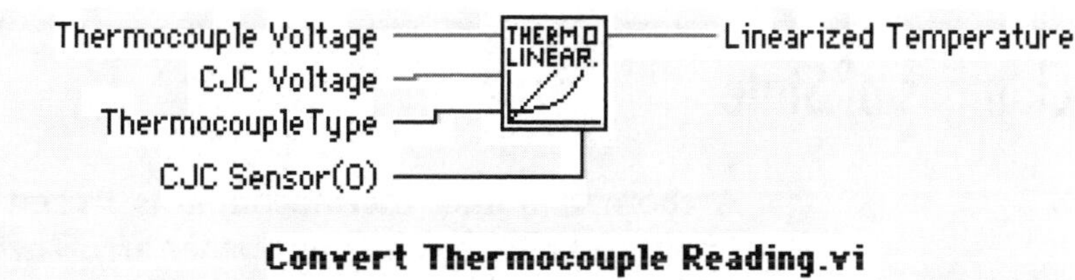

Convert Thermocouple Reading.vi

Takes a voltage value and returns the corresponding temperature in ∞C.

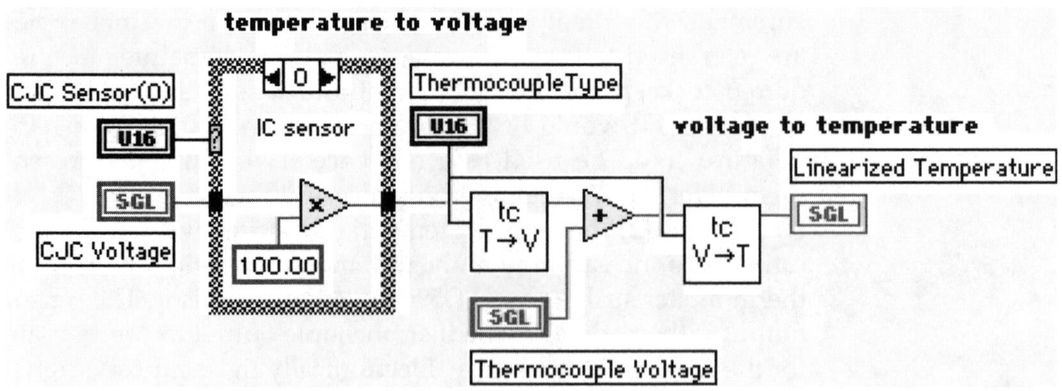

is then converted into the equivalent thermocouple voltage that would be read on a TC (Type ?) if it was used to measure the ambient room temperature. This value is then added to the real TC voltage signal and used to calculate the temperature at the tip of the thermocouple.

A hardware solution (opposite page) for cold junction compensation is in the form of an integrated circuit, the AD594/596. This monolithic IC is both a thermometer and a thermocouple amplifier. The AD594 is precalibrated by laser trimming to match the characteristics of Type J (iron-constantan) thermocouples. AD595 is matched for Type K (chromel-alumel) thermocouples. The TC leads are soldered directly to input pins 1 and 14 on the integrated circuit. Inside the IC, an AD590 measures the ambient temperature and the appropriate compensation voltage is subtracted from the TC voltage to electronically reference 0°C. The output is then scaled to produce a signal level of 10 mV/°C using the standard calibration tables. This 14-pin IC package requires only a supply voltage in the range of +5 to +/–15 V and a thermocouple to measure temperature to an accuracy of +/–1°C.

Changes of State

A chomel/alumel thermocouple is frozen into the center of a small ice ball (see chapter opener art). The ball is suspended in free air at an ambient room temperature of 24°C. Cool vapors

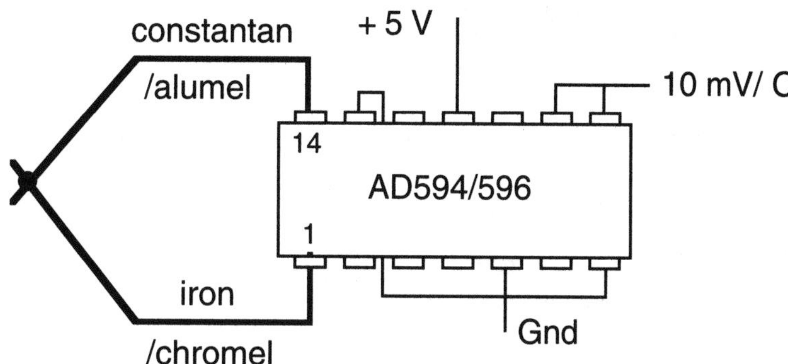

fall from the ice ball as the room temperature radiators heat the outer layers. Inside, the temperature slowly rises. As the outer layers melt into water, gravity pulls the liquid to the base, where drops form and fall from the ice ball. Eventually the liquid melts to the thermocouple position at the center of the "ball" where water and ice exist in equilibrium, at least for a few moments. It is at this point where the latent heat of fusion holds the temperature constant until all the ice has vanished. The ice ball is now replaced with a liquid drop and the temperature rises more quickly. The drop size diminishes as evaporation occurs. A change of state from liquid to gas can sometimes be seen in the heating curve near the dew point. Once the liquid drop is gone, the temperature rises to the ambient temperature of the room.

A file called Ice Ball.data contains a data record of the thermocouple measurements and the sample time taken during the lifetime of the ice ball. These measurements were taken using a LabVIEW DAQ card and Analog Input VI.

■ LabVIEW Challenge: Ice Ball Melting

Design a LabVIEW program to observe the heating curve of the ice ball as it melts and evaporates. How many changes of state (ice to water, water to air) can you find in the data set?

Overview

We are all familiar with the point-and-click remote infrared controllers for consumer electronic products. But how do they work? Can a LabVIEW program provide computer control of our television, stereo receiver, CD player, or power amp? This chapter introduces line of sight communications in the infrared region of the optical spectrum. A IR transceiver module allows the design of algorithms for an IR proximity sensor, an optical chat line between computers, and an optical transponder. In each case IR encoded data is sent from a transmitter and the reflected, modulated, or encoded signal is returned as a series or infrared light pulses. The chat line and transponder application can be run without IR transceivers.

Goals

- Understand infrared communication transmitters and receivers
- Build an IR transceiver for the serial port
- Use an IR transceiver as a proximity sensor
- Build an IR optical chat line between two computers
- Study an IR optical transponder application
- Study data encryption techniques using a pseudo-random number generator

Key Terms

- Infrared light emitting diodes
- IR receivers
- Proximity detector
- Chat line
- Master/slave communication protocol
- Data encryption
- Analog-to-digital converter

IR Communications

9

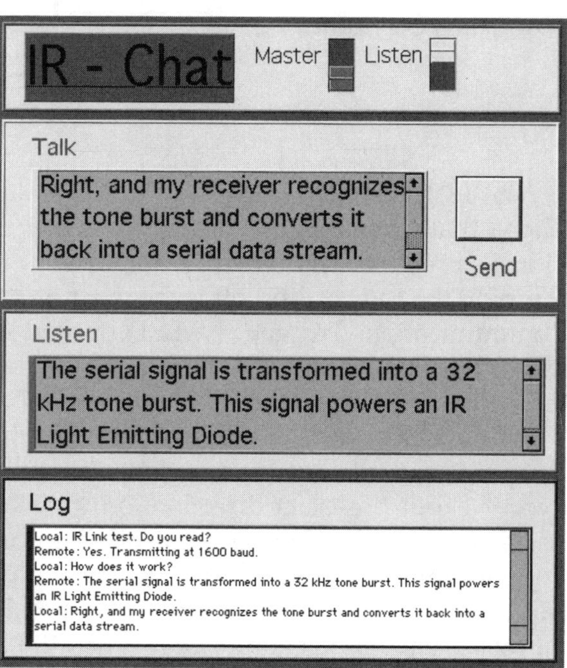

LabVIEW Front Panel—Infrared Optical Chat Line

With the advent of efficient light emitting diodes (LEDs) in the long wavelength region, infrared IR communications became a reality. Today, IR remote controllers are found in all types of consumer electronic products: TV, VCR, or stereo. In this chapter, the fundamental components of IR communications systems are studied and used to investigate IR sensing, free space IR communications, and an IR optical transponder.

The IR transceiver featured in this chapter consists of an IR LED as the transmitter and a commercial IR receiver package. Together with a few other electronic chips, the transceiver is interfaced to the computer's serial port.

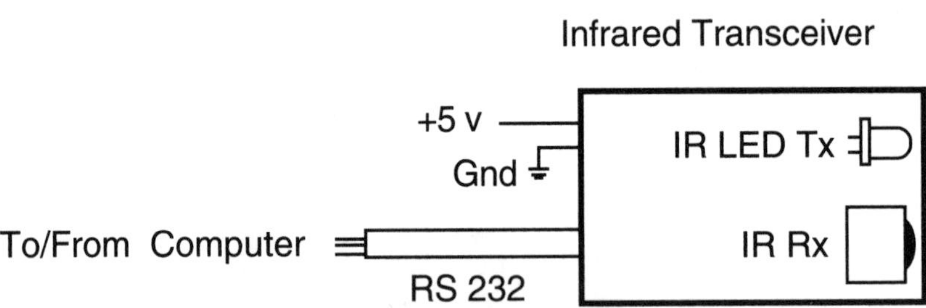

The IR transmitter consists of a 555 timer IC and a IR LED. Recall that when a light emitting diode is forward biased, radiation is produced. When it is not driven, no light is emitted. Turning the LED on and off is the most primitive form of optical communication. Turning the LED on and off in a pattern produces a communication system. The light radiates from the LED in a cone of light having a half angle of about 30 degrees. This wide coverage and partial directionality makes the LED ideal for point-and-click consumer controllers. One does not have to worry about the exact direction, only that it be pointed in the vicinity of the device to be controlled. In order for the receiver to discriminate between valid IR data pulses and other light sources, the LED is modulated. When the LED is on, a toneburst of pulses at 32 kHz is produced. When the LED is off, no signal is sent. This modulation provides a distinct signature for the LED transmitter.

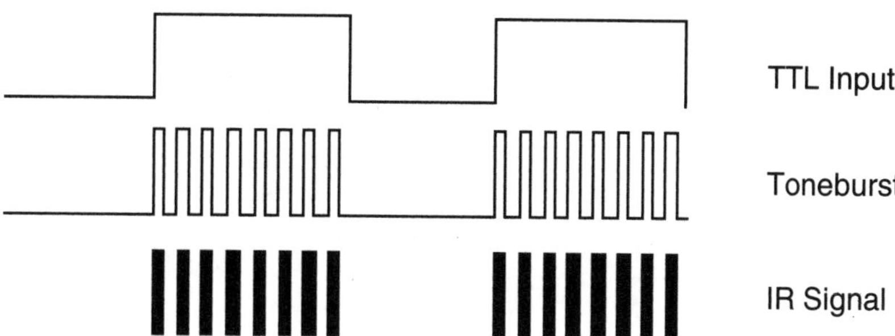

Incandescent light bulbs have a 60 Hz modulation on top of a bright steady background, and fluorescent light fixtures produce a 120 Hz modulation. These frequencies are a long way from the 32 kHz IR frequency, hence interference is minimal to nonexistent. In the communication jargon, the tone-burst modulation scheme is called "non return to zero."

The IR transmitter circuit shown on next page uses only three ICs, an IR LED, and a few resistors and capacitors. The 555 timer chip is used as a resettable astable oscillator. Two resistors—R2, R3—and the capacitor C2 are chosen to set the oscillator frequency, f = 1.44/(R3+2*R2)C2 to 32 kHz, and the duty cycle, DC = R2/(R3+2*R2), to about 49 percent. The tone burst is produced by raising the reset input pin (RST) high. Provided that the 555 timer is powered by +5 volts, the (RST) line is TTL compatible and other common digital ICs can be used to interface with the RS232 serial voltage levels. A line driver, the MC1489 IC, translates the RS232 signal levels into TTL levels. A 7400 Quad NAND gate IC reinverts the input signal that was inverted by the MC1489 IC.

The receiver circuit exploits a special IC designed for IR remote devices and is available at most electronic component retailers.

One such device, a 32 kHz IR receiver module (Digi-Key LT1033) is a simple 3-terminal package. One lead is connected to the +5 volt power supply, one lead is connected to ground, and the other lead is a TTL-compatible output. Do not be fooled by the simplicity of the package. Inside this package is a photodiode, a current-to-voltage amplifier, a limiter, a bandpass filter, a demodulator, an integrator, and a comparator.

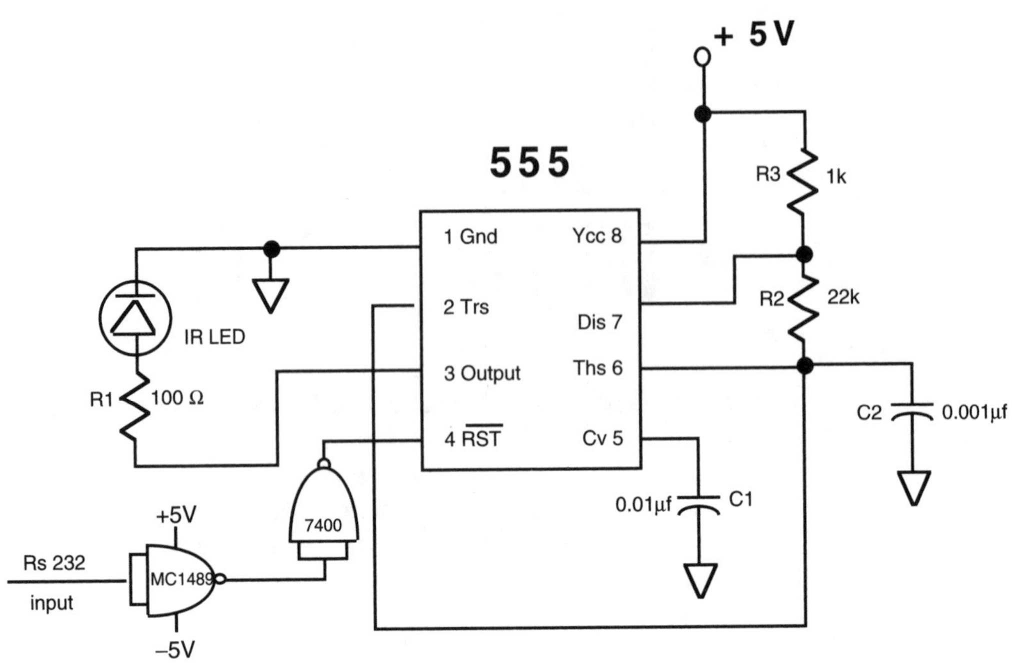
+ 5V
555
R3
1k
1 Gnd
Ycc 8
R2
22k
2 Trs
Dis 7
IR LED
3 Output
Ths 6
R1
100 Ω
C2
0.001µf
4 RST
Cv 5
0.01µf
C1
+5V
7400
Rs 232
MC1489
input
–5V

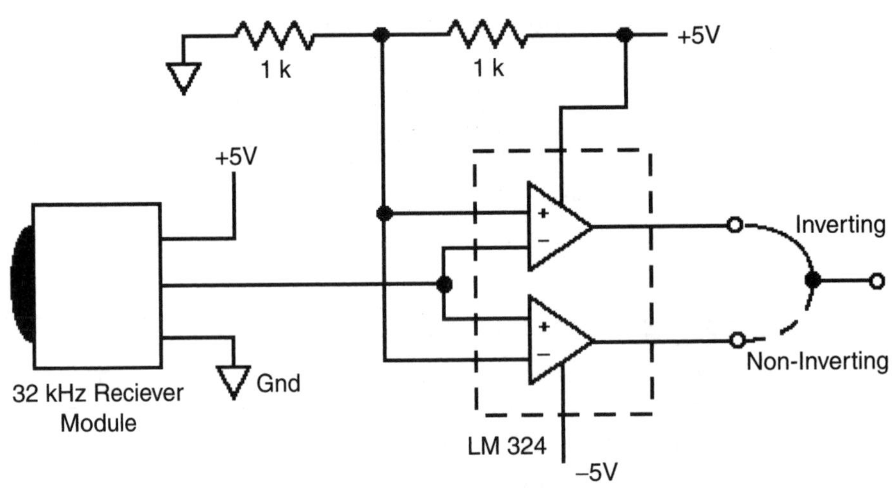
1 k
1 k
+5V
+5V
Inverting
Non-Inverting
32 kHz Reciever
Module
Gnd
LM 324
–5V

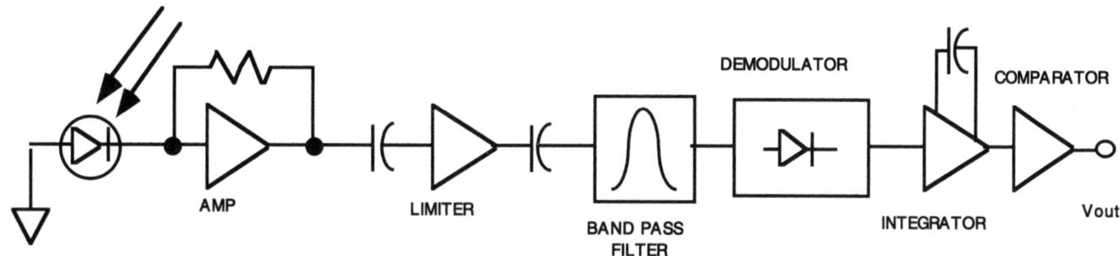

A blackened plastic lens in front of the photodiode attenuates visible light, but is transparent to infrared light. IR light impinging on the photodiode generates a photo-current that is converted into a voltage by a current-to-voltage OpAmp circuit. The limiter stage ensures that even when the transmitter is nose to nose with the receiver, the detector signal does not saturate. The bandpass filter is tuned to the 32 kHz signature frequency of the IR transmitter. The demodulator and integrator convert the tone burst into a TTL high level and the absence of signal into a low TTL level. The output of the IR module then goes to a pair of comparators, the LM324 dual OpAmp, to produce the RS232 signal levels. The output can be inverted or noninverted depending on the communication protocol (positive true or negative true logic).

IR Proximity Sensor

A proximity sensor detects the presence or not presence of an object. In the IR proximity sensor, an infrared beam is transmitted in the direction of interest. When an object is close enough, a fraction of the transmitted beam is returned to the receiver. When the reflected signal reaches a threshold level, the sensor responds with an acquisition message. In this experiment, the transceiver module is used as a proximity sensor and is placed in front of a door, a window, or other object whose location is important. LabVIEW generates a search signal on the serial port output pin that is translated by the transceiver into a unique sequence of pulses. When a target is acquired, the module sends a message back to the computer via the serial port.

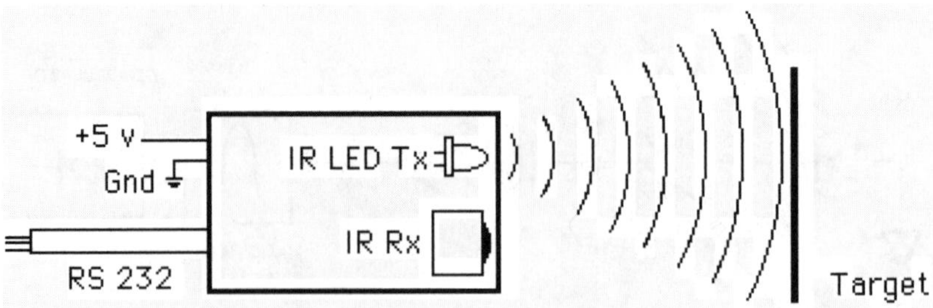

Recall the RS232 serial waveform discussed in Chapter 5. The data byte is preceded by a start bit (low) followed by the data bits (least significant to most significant bit) and terminated by a stop bit (high). If the data byte sent is the binary byte (01010101), $55, or "U" then a pulse-train consisting of 10 alternate 0s and 1s is produced on the serial port and dutifully converted into IR pulses by the transceiver module. Repeated transmission of the ASCII character "U" yields a squarewave for the search pattern. When the receiver detects a signal, it is coded and sent to the serial input port. When LabVIEW recognizes the received signal as an ASCII "U", the object is detected. Low level signals or bad reflections would appear as some other ASCII character and be ignored by the program.

Due to the speed of light being so large, the return signal is instantaneous at the IR LED ranges. A simple radar type of signal processing is also possible, that is, transmit-listen, transmit-listen, transmit-listen, and so on until a target is acquired.

Exercise: Study carefully the IR Sensing.vi in the chapter library.

IR Optical Chat Line

Most point-to-point networks and servers have some form of chat line. Chat allows two or more users to communicate using the keyboard as the input device and the video monitor as the output device. When implemented, two windows open up on the monitor. One is for outgoing messages and the other for incoming messages. Once a message is composed, it is sent to the other computer waiting in the listen mode. When the message is received, the second computer can now talk by composing a message and sending it. The cycle can be repeat until one of the users "hangs up." The IR optical chat line uses two IR transceiver modules and two computers to form a bidirectional free space optical data link: "Look, ma, no wires." The link has an effective range of about 50 feet and transmits at data rates up to 1800 baud.

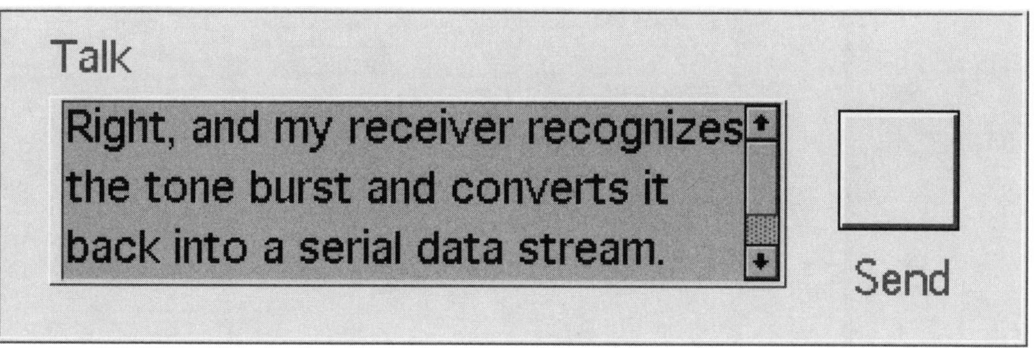

If you do not have IR modules, then an electrical chat line between two computers is possible using a serial cable line with the transmitter and receiver lines on one computer connected to the receiver and transmitter lines on the other computer. Such a connection is often called a null modem cable.

The first step in communications is to establish a transmission/reception protocol for the chat line. One computer will be the master and the other the slave. An indicator on the front panel signals the user which mode is active. Initially the users must decide who is the master and who is the slave. This is set on the front panel Master/Slave switch.

Once established, the protocol ensures proper operation for all future communications. When the master sends a message, a Master/Slave flag is appended to the data message. When the slave receives the message, it is displayed, and if the Master/Slave flag is present, then the slave becomes the master. The previous master switches to slave mode as soon as the Master/Slave flag is sent. An LED indicator on the front panel tells the user which computer is master (Talker) and which is the slave (Listener).

All messages are ASCII strings sent to the serial port and converted by the IR transceiver module into IR encoded pulse streams. The heart of the chat program is a case structure where the <|True|> state is the listen mode and the <|False|> state is the talk mode.

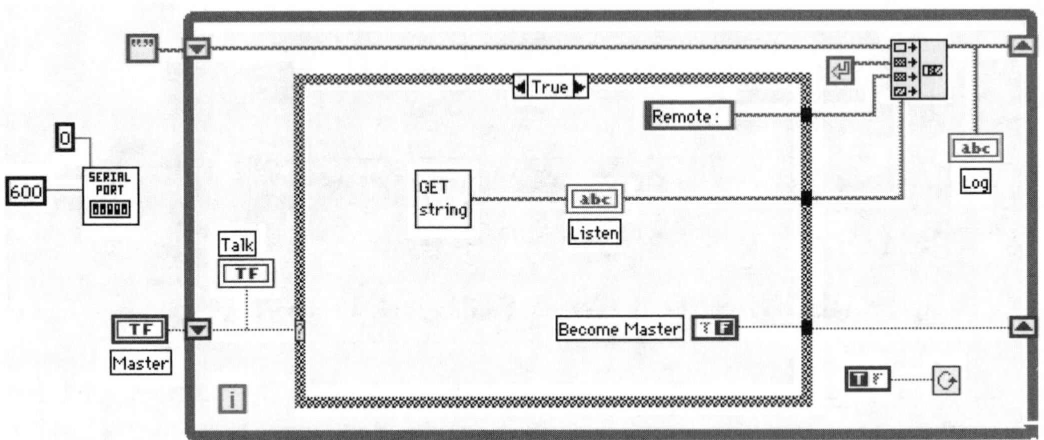

When the slave is chosen as listener, then the <|True|> case is active. Get String.vi listens to the serial port collecting data until the Master/Slave flag is received. The message is passed to the front panel message pad labeled Listen and forwarded with the header {Remote: } to the message Log.

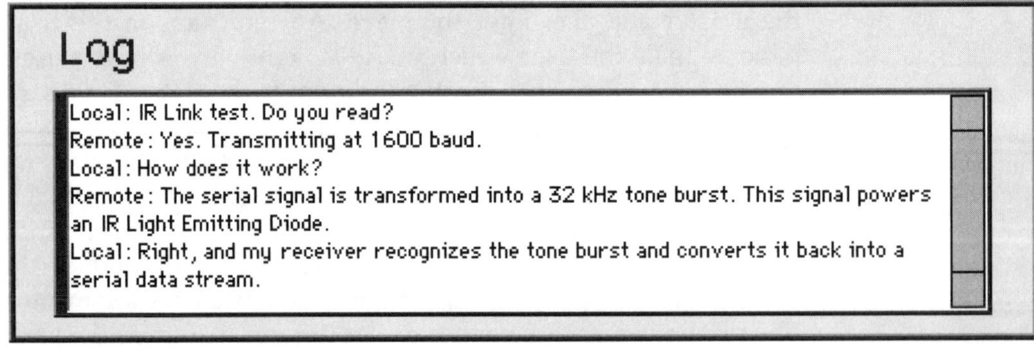

A Boolean constant [False] is sent to a front panel indicator via the shift register in the While . . . Loop. On the next iteration, this condition switches the front panel indicator from Listen to Talk and the <|False|> case state becomes active.

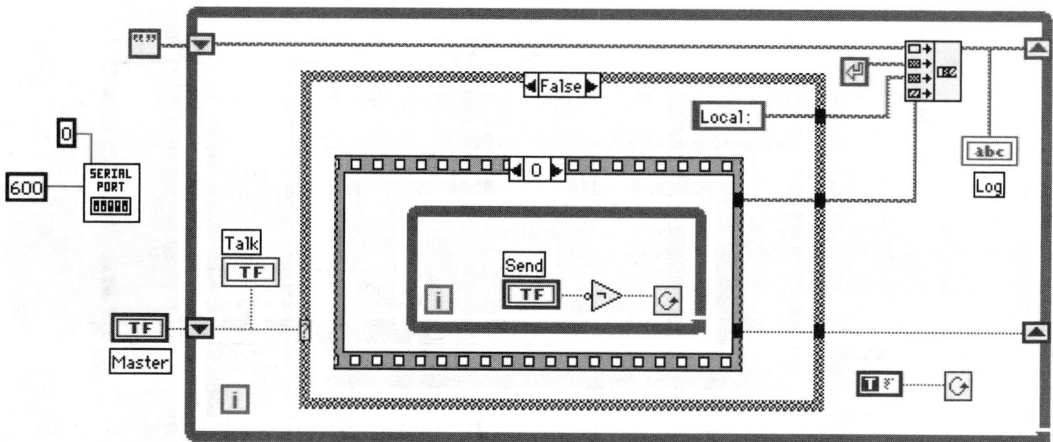

In the talk <|False|> mode, a message can be composed and edited in the Talk message pad. When completed, the [Send] button is clicked. This initiates a four-frame sequence.

Frame 0: Waits patiently until the [Send] button is pressed.

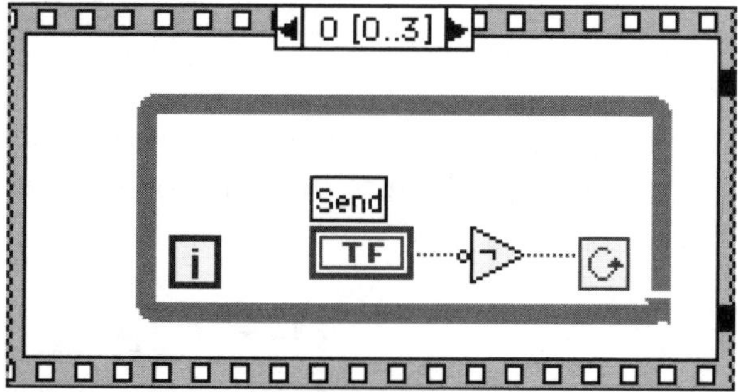

Frame 1: Takes the string message from the Talk message pad, passes it to the Send String.vi and onto the message log. The case statement outside this sequence adds the

ASCII string {Local: } to the message before passing it along to the message log. A Boolean True constant signals the front panel to become slave on the next iteration of the outside While . . . Loop.

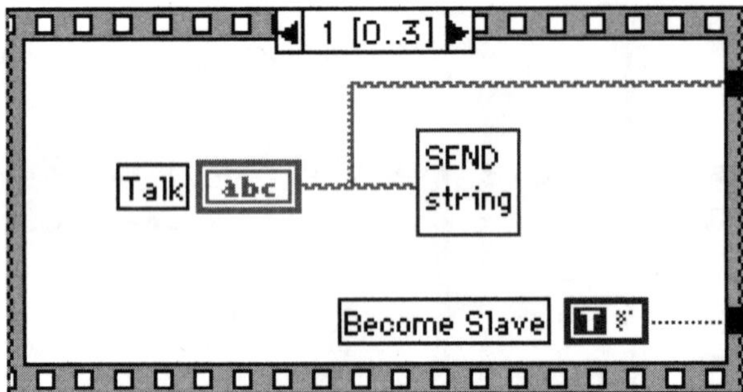

Frame 2: Waits 20 milliseconds to allow the message to pass to the serial port where it is sent bit by bit.

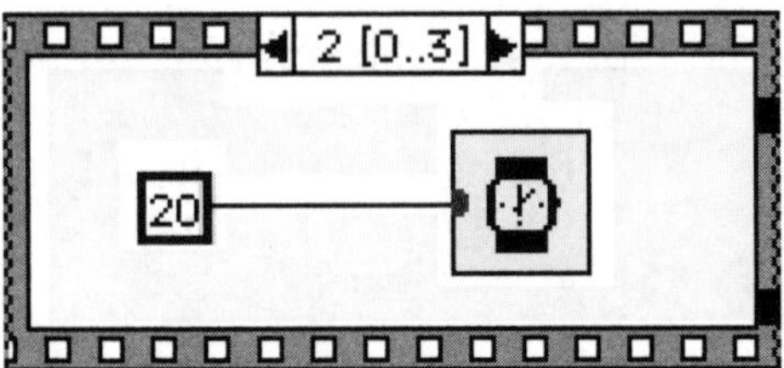

Frame 3: Clears the buffer in case any stray reflections trigger the receiver with some error characters.

The Case statement is now completed and processing passes to updating the message log. The message log is interesting since it uses an ASCII shift register to build the log. Initially, the message log contains only a Null message. The string concatenate function on each iteration creates a new log equal to

old log + <CR> + Header (Local: /Remote:)
+ new message

The Chat Serial VIs

Send String.vi takes the string from the {Talk} message pad and frames the message with a pair of ASCII characters "<" and ">." The header character "<" defines the start of a message. The trailer character ">" not only defines the end of the message but it is also the master/slave flag. When the receiver finds this character, it will switch from listener to talker.

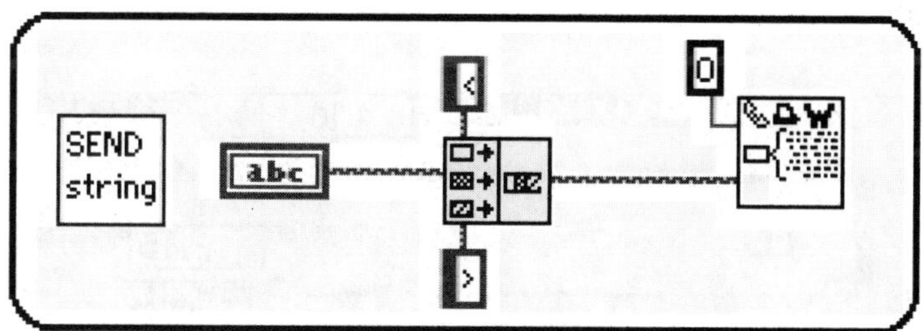

Get String.vi listens (next page) to the serial port and looks for a valid header character "<." Frame ϕ loops forever until the header character is found. When found, the While . . . Loop terminates and the sequence passes to the next frame. Note that not only is the header character detected but it is also stripped off from the incoming message.

Frame 1 listens to the serial receiver port and collects characters in a string shift register. When the trailer character ">" is

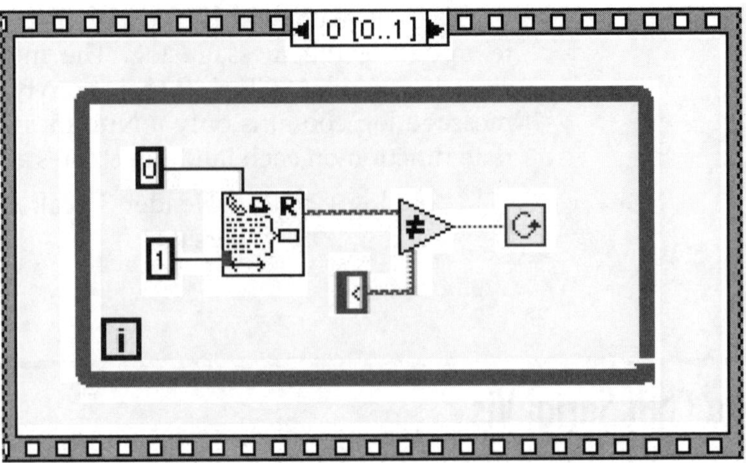

found, the case statement becomes False so that the end of message character ">" is *not* appended to the message. The While . . . Loop is terminated, the message is passed to the String indicator terminal, and the sequence is closed. Control passes to the <|True|> case in the main loop, where the message is passed to the front panel and the listener becomes the talker.

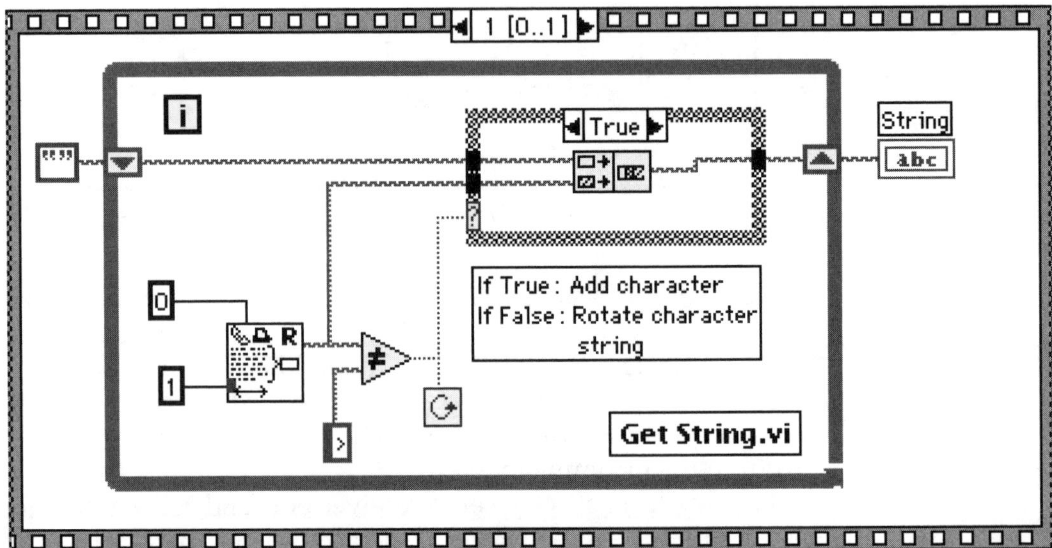

Clear Buffer.vi gets any characters that may have been received as stray reflections during the transmission. These characters are passed to a dummy buffer and not used.

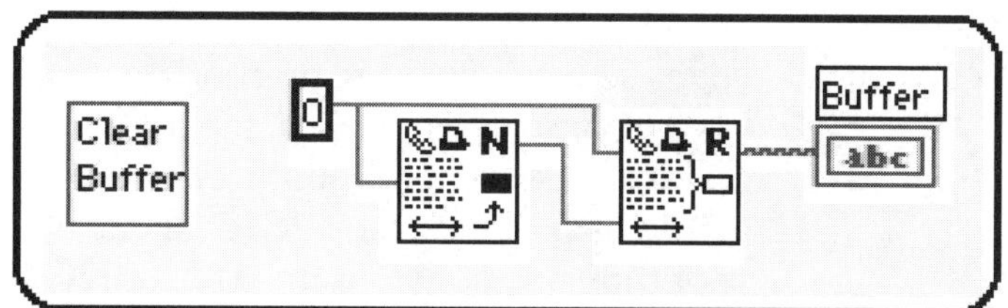

Load the program Chat.vi into two separate computers. If using the IR transceivers, connect one to the serial port on each computer. Place the modules in a line about 6 feet apart. Set the front panel switch on one computer to [Master] and on the other computer to [Slave]. Run both programs. The master speaks first, after which the protocol ensures the order of speaking line is automatic. Try different baud rates, wider separation, misalignment, and even bouncing the signal off a wall. If no IR modules are available, use a null modem cable between the serial port on the two computers.

Friend or Foe?

On a moonless night, a darkly painted helicopter winds its way along no man's land. It is flying at low altitude, with no lights showing, with no discriminating features. A lone soldier alerted by the soft whoop, whoop, whoop of the helicopter blades lifts a Stinger missile pack to his shoulder. A few switches are flipped and the gunsight comes into view. In the dark little can be seen, but the heat seeker in the nose of the missile peers forward and watches for any tell-tale heat signature. A soft growl is heard in the soldier's earphone alerting him that the missile has spotted a heat source. As he moves the launch tube to track the helicopter, an intense grooooooowl is heard. The missile has found the target and locked on. One more squeeze of the trigger and the night sky would be filled with light, first as the missile jumps from the launch tube, then streaks across the sky, and finally ends in a flash as the four-pound warhead explodes.

Instead, a short IR optical pulse is sent to interrogate the target, a precursor to the launch. The pulse is in the form of a universal "W" for "Who are you?" followed by the friendly forces code. The receiver, if present on the helicopter, strips off the "W" character and uses the following bytes (friendly forces code) as a seed condition for a pseudo-random number generator. It also clocks the PRNG ahead a number of counts corresponding to today's date. The PRNG output forms part of the coded reply. A laser on board the helicopter flashes on and sends its reply, the universal answer character "A" followed by the coded reply. Meanwhile, inside the missile's electronics package, a CPU has already calculated the expected response and awaits the return signal from the optical transponder.

A small LED inside the gunsight turns green, a friendly, and the growl stops. The soldier lowers the launch tube and lifts his hand in an unseen silent wave to friends aboard the helicopter.

Friend or Foe Program

The friend or foe program takes the form of a seven-frame sequence. On the front panel, the interrogate command "W" together with the current friendly force code "173" is entered.

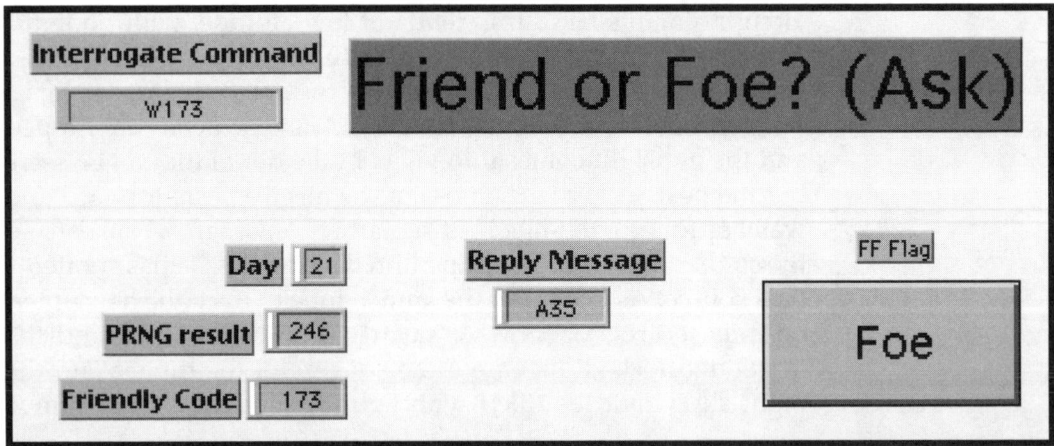

When the program is run, the system clock calculates today's day "21" and reports the expected return code "246" The reply "A35" is displayed in the [Reply Message] indicator. The first character "A", is stripped off and the following characters "35" are compared with the expected reply "246". If different, the Friend Foe flag is set to [Foe] otherwise set to [Friend].

On the block diagram Frame 0 is used to initialize the baud rate on the serial port and to clear any characters in the serial buffers. In practice, the 32 kHz modulation limits the transmission rates to 1800 baud. Can you think why this limit?

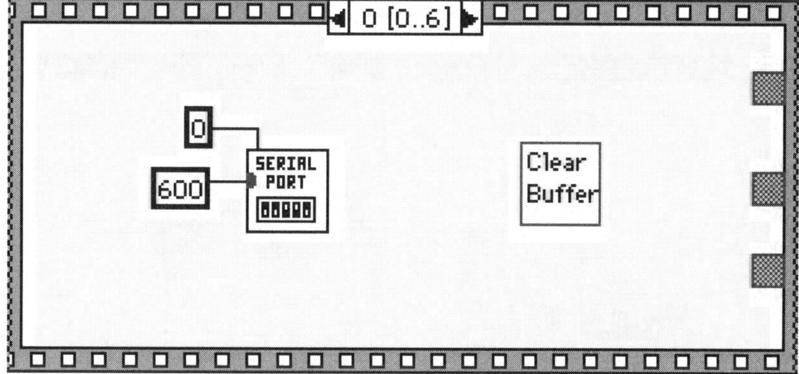

Frame 2 sends the command sequence consisting of the universal interrogate character "W" followed by the friendly forces seed code. For this demonstration the seed is taken as a three-digit number having an ASCII representation from 000 to 255.

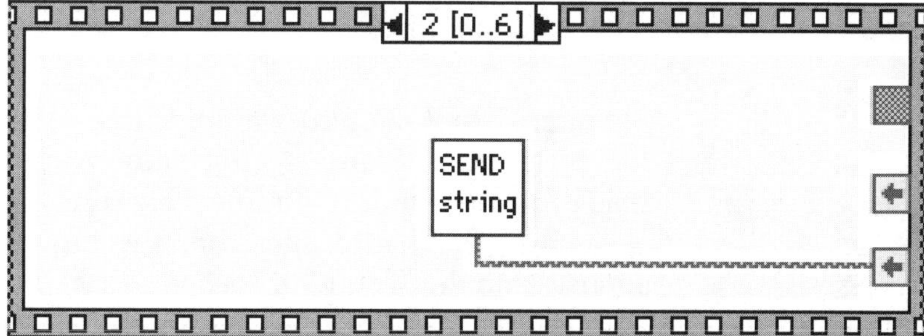

Meanwhile (actually before frame 2 in frame 1), LabVIEW has calculated the expected response message. The first character in the transmitted message "W173" is stripped off using a string subset function. The ASCII number in the sub-string is passed as an input to PRNG.vi and echoed to the front panel. Today's date is retrieved from the system using LabVIEW's **Get Date/Time** function. The data is formatted as day/month/year. LabVIEW's **Split String** function splits off the date "21" as a substring and it is passed as a second input to the PRNG. The day is the number before the first "/" character in the Get/Time Date function output. PRNG calculates the expected response and passes it to the front panel and on to "verify the response" in Frame 6.

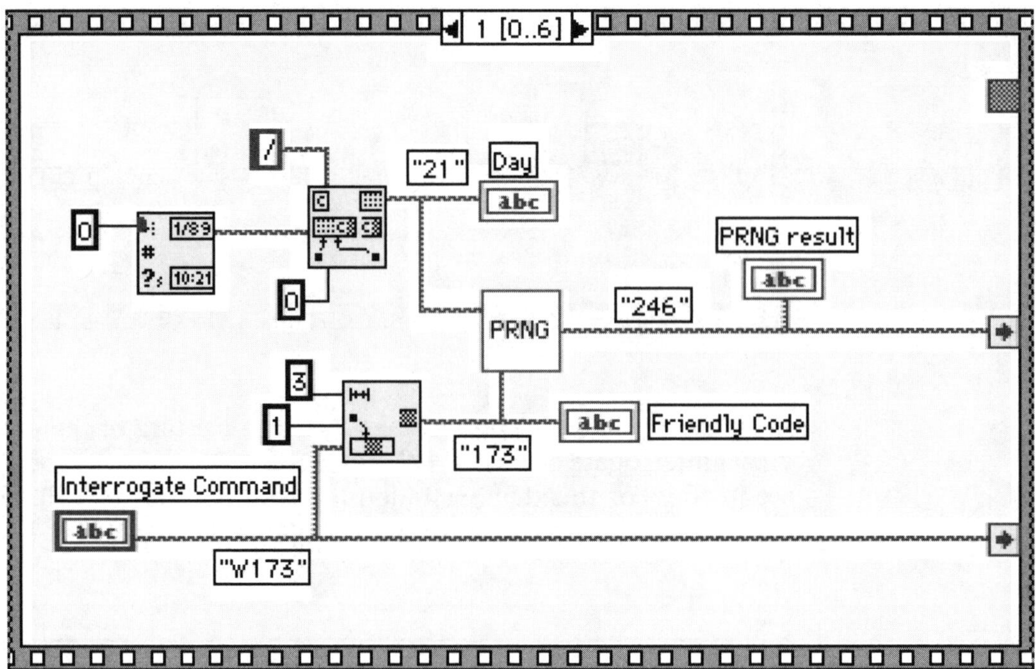

Frame 3 (not shown) waits for the message to pass through the serial port. Frame 4 (not shown) clears the input buffer ready for the response. Frame 5 executes the Get String.vi to receive the ASCII message from the target and passes the response onto the "friend or foe" verify frame.

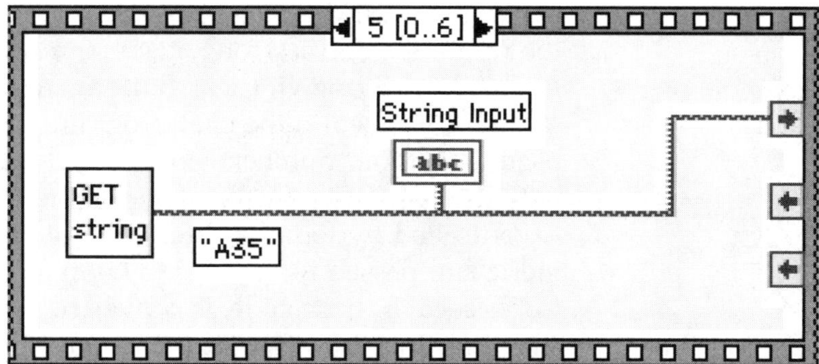

Frame 6 strips off the first character in the response string to see if it is the universal answer "A" character. The substring after the answer character is then passed to the compare function where the received response is compared with the expected response. A wrong response code is flashed to the front panel as a foe.

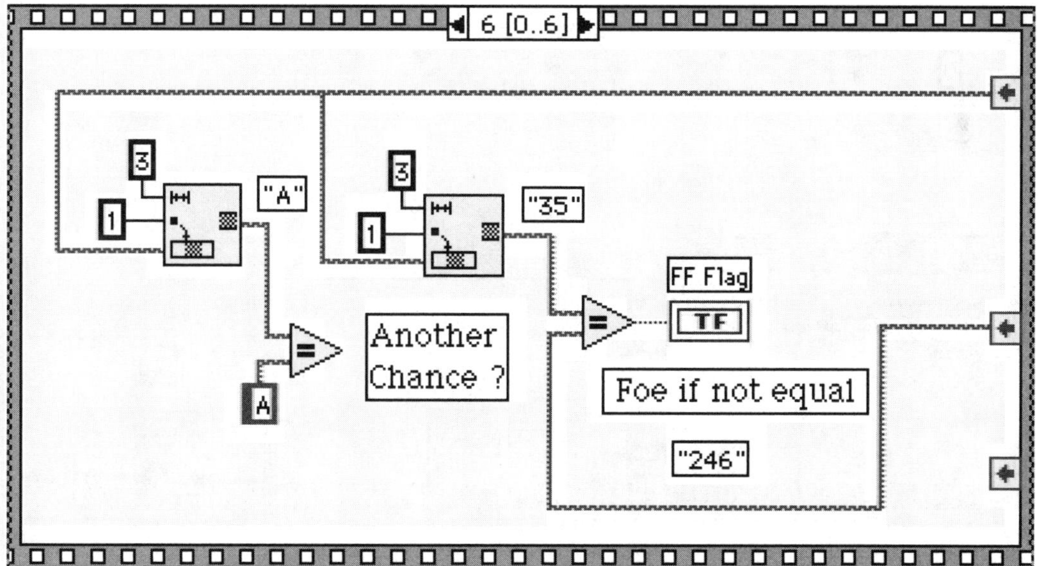

If the first character received is not the answer character "A", it does not necessarily mean that the target is a foe; only that the signal levels may not be high enough for a clear response. Perhaps you might want to give the target another chance. But if the response code is wrong, keep your head down or fire.

The Send String.vi, Clear Buffer.vi, and Get String.vi are the same VIs as used in the chat program. PRNG.vi is based on the pseudo-random number generator discussed in Chapter 3 but with a twist. The number of cycles the PRNG counter executes is controlled by today's date. The date is converted into a numeric and passed to the For . . . Loop terminal [N]. The friendly forces seed, a string code, is converted into its binary equivalent using an analog to digital VI called A/D.vi. For example, the string seed "254" is converted to the binary code (11111110). This 8-bit binary number becomes the initial value of the PRNG. After clocking forward the PRNG counter by today's date, the friendly forces code is formed on the 8-bit shift register. Its decimal equivalent is converted by the DAC.vi into a numeric and then into a string with the **Format & Append** function.

The Friend or Foe answer program running on another computer is similar in structure to the calculations done in Frame 2, only now the character "A" is tacked onto the front of the response.

Exercise: Study carefully FF Ask.vi in the chapter library.

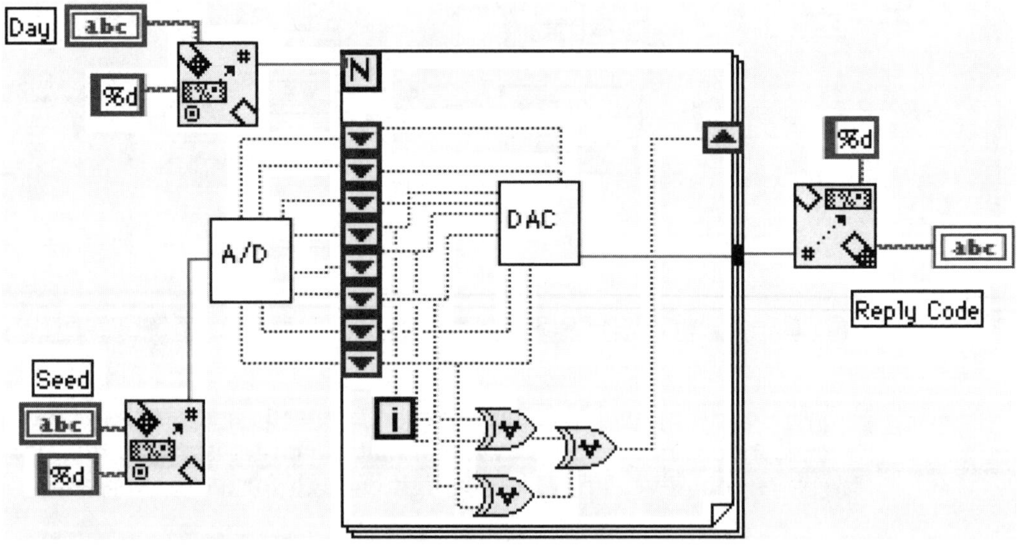

How to Use the Program

Load the programs FF Ask.vi and FF Answer.vi into two separate computers. If using the IR transceivers, connect one to each serial port on the computers. Place the modules in line about 6 feet apart.

If you are using a null modem cable, then the operation is straightforward and you can concentrate on making the algorithm better.

If you do not have access to the IR modules and/or two computers, you can still exercise the "Friend/Foe?" program using a simulator version found in chapter library called Friend/FoeSim.vi. It requires only one computer and no hardware.

LabVIEW Challenge: A Robust Optical Transponder

Build some redundancy and a time-out feature into Friend or Foe to make the program more robust.

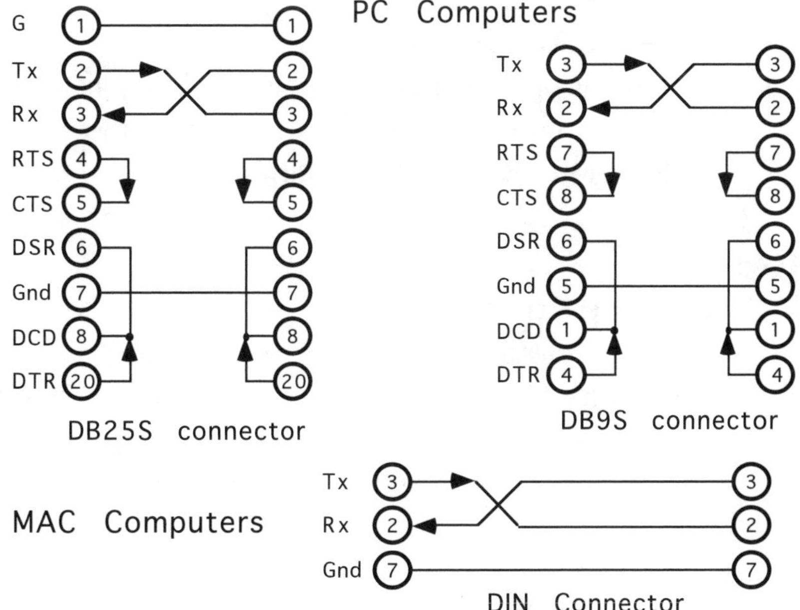

Overview

Barometers have been used for centuries as a predictor of oncoming weather. In this chapter, a LabVIEW program is designed to measure the current atmospheric pressure, keep a record of the past pressure changes, and make predictions of future weather. In simulation, a pressure VI simulates the arrival of stormy weather followed by a longer clearing trend. With the electronic barometer circuit discussed in the text, the barometric pressure can be measured and input into the LabVIEW program over one of the computer ports and the barometer project can analyze real data and make weather predictions.

Goals

- Study silicon pressure sensors
- Build an electronic barometer
- Design a simple barometer simulation
- Design an algorithm for detecting a rise in the barometric pressure
- Design an algorithm for predicting the arrival of stormy weather
- Design a weather predicting barometer

Key Terms

- Silicon pressure sensors
- Window comparators
- Range selector
- Data smoothing
- Mixing numeric and Booleans
- User-friendly front panels

The Barometer

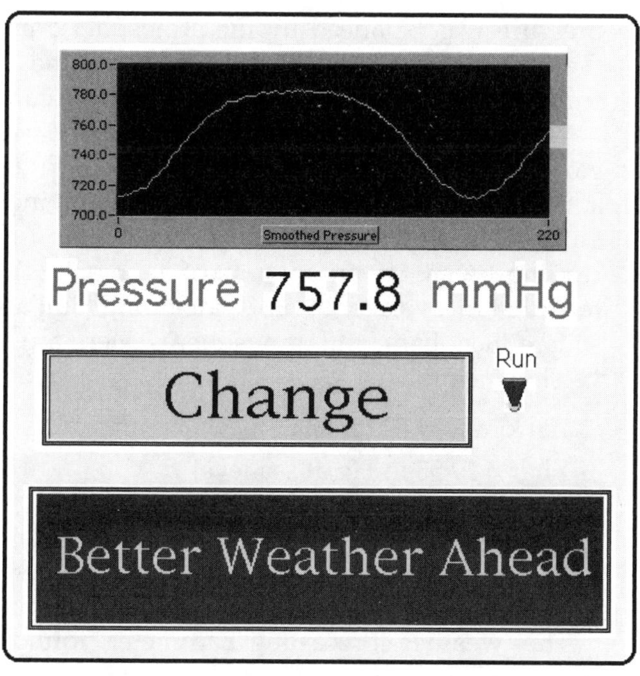

LabVIEW Front Panel: Forecasting Barometer

The barometer is an instrument that measures atmospheric pressure; it is used chiefly in forecasting weather and in measuring altitude.

History of Barometers

Evangelista Torricelli, a student of Galileo, in 1643 took a long glass tube closed on one end and filled it with mercury. Placing his finger over one end and inverting the tube, he placed it into a bowl of mercury. Upon releasing his finger, the level of mercury in the tube fell until the air pressure acting on the surface of the bowl provided a force strong enough to hold up the column of liquid mercury within the tube. The pressure in the closed end of the tube was near a vacuum and provided a reference point. The pressure of the mercury column at the surface of the bowl was equal to the density of mercury times the acceleration of gravity times the length of the mercury column. This pressure was balanced by the atmospheric pressure acting on the surface of the bowl of mercury. When the atmospheric pressure changed, so must the height of the mercury column.

The height of the mercury column above the mercury surface gave a direct measurement of atmospheric pressure. Early measurements gave the pressure reading in inches of mercury. Later when scientists began using metric units the pressure readings were given in mm of mercury. Today pressure is recorded in a variety of units, including kiloPascals, pounds per square inch, bars, and atmospheres. Here are some useful conversion factors:

100 kPa = 29.529 in Hg

1 bar = 29.53 in Hg at sea level

14.7 psi = 29.92 in Hg

1 atm = 760 mm Hg at 0°C.

In this project, pressure will be measured in mm Hg.

The 'weather forecasting barometer' found in many homes uses a sealed bellows containing a reference pressure, usually vacuum. A change in the atmospheric pressure causes a displacement in the bellows extension. Connected to the bellows face is a mechanical spring assembly and a pointer. The angular position of the pointer reflects the bellows extension and hence the atmospheric pressure.

Modern-Day Integrated Silicon Pressure Sensors

Modern pressure sensors—such as the Motorola MPX family of pressure sensors—use a reference pressure cavity where one side is enclosed with a flexible silicon membrane. This membrane bends in response to an external pressure. Microminiature strain gauges mounted on the membrane convert the bending motion into a resistance change. Four strain gauges are used in a bridge configuration to null out temperature effects to first order. In addition, laser trimmed thermistors around the edge of the sensor aid in the temperature compensation. The completed sensor is a four-terminal device requiring only DC voltage excitation. The sensor output is connected to an analog in-

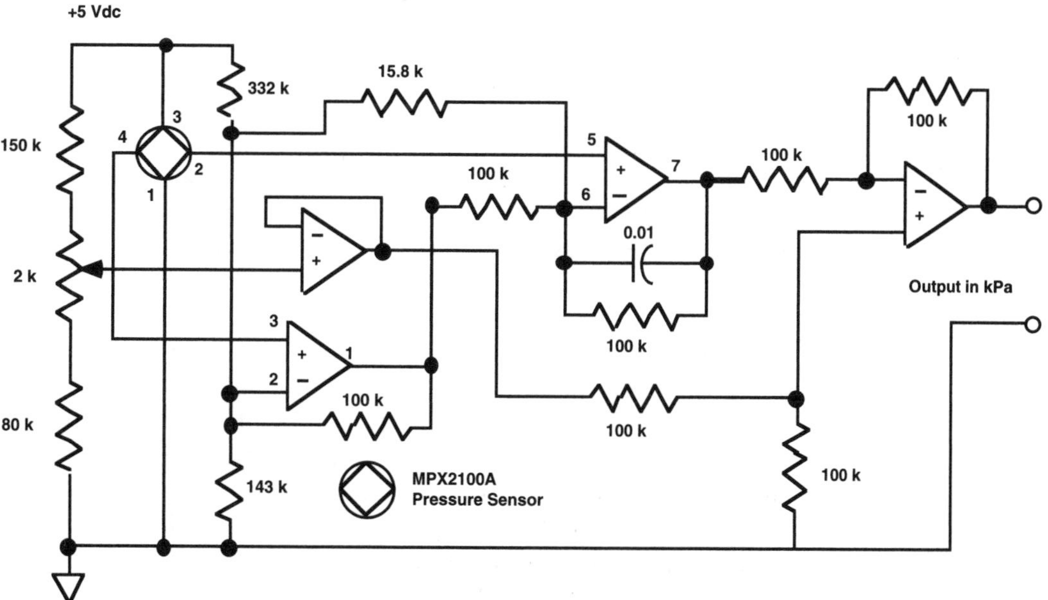

terface, which converts the sensor output into a voltage level suitable for computer interfacing.

The Motorola pressure sensor MPX200 is ideal for an electronic barometer. It has an output of 400 mV at 1 atmosphere (740 mm Hg). The device is inexpensive, requires low excitation voltage (5 to 15 V) and is temperature compensated over a wide temperature range (–40 to 125°C).

The circuit on the previous page produces an output voltage

$$V(mV) = 41.89 + 1.175\ P(mm\ Hg).$$

Simulating a Forecasting Barometer

In about the eighteenth century people began to keep careful pressure records and subsequently discovered that the barometer readings track weather disturbances. They found that a record of past hourly readings could give an indication of future weather. With this observation, the weather forecasting barometer was born. Even today with all the satellite imagery, vast data banks, and supercomputers, weather forecasting is an inexact science. However, the lonely barometer can still make many useful predictions.

As a storm rolls through, the barometric pressure falls. A deep low indicates a severe storm. After the storm passes, the pressure rises and better weather is ahead. These changing conditions—a falling or a rising pressure—indicate a change in the weather either for the worse or the better. On the upper end of the scale, a high barometer reading indicates very dry conditions. In this simulation, it is assumed that most of the time, the weather is good and the pressure will be high.

Pressure.vi simulates the pressure changes that would accompany the passing of a severe storm through a region of otherwise good weather. Each time you call this VI, a different pressure will show up on a conventional-looking linear display. This display mocks a mercury barometer where the pressure is read by aligning the mercury meniscus with a linear scale. A digital display has been added to make reading the mercury level easier.

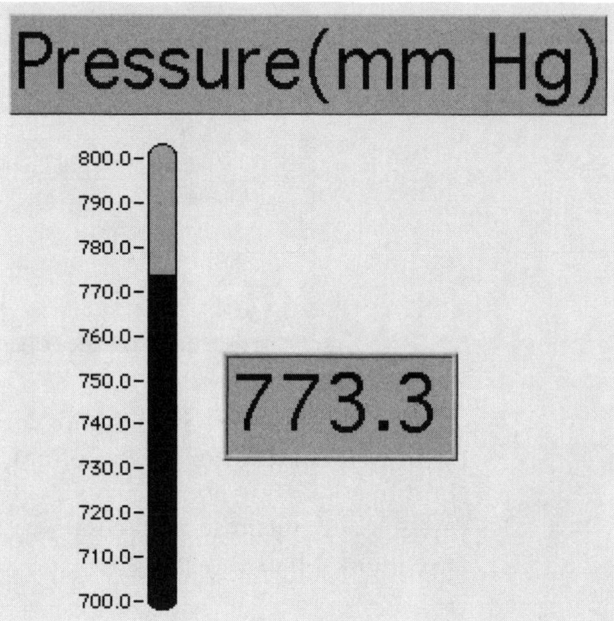

Pressure.vi is intended to simulate pressure readings taken on an hourly interval. In keeping with spirit of a real simulation, some randomness has been added to the pressure output. Take a look at the diagram panel to see how this is accomplished. Pressure.vi is also a sub-VI with only one link, the current pressure. Each time this sub-VI is called it reports the current hourly pressure.

Calling All Stations

One advantage of an electronic barometer over a conventional barometer is that you can chart the pressure readings in a graphical format and see the storm brewing. To observe the pressure record as the storm passes through our region, make a new VI called Pressure Record.vi. We will need one display, a waveform chart, and a Boolean switch on the front panel. On the diagram panel add Pressure.vi and wire it up to the waveform chart inside a While . . . Loop. After wiring up the Boolean switch, run the program to observe the pressure log.

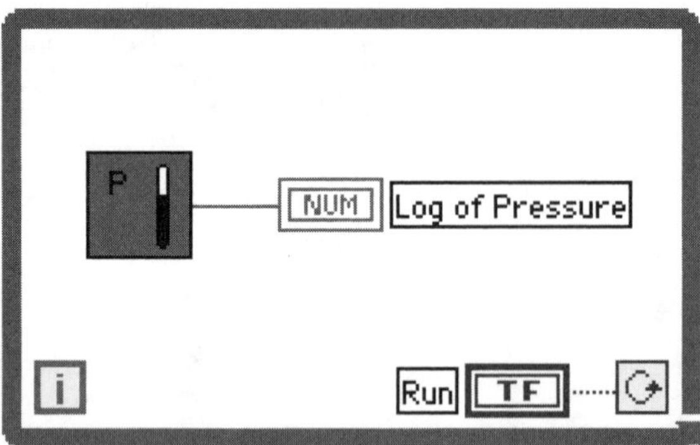

Although pressure is high most of the time, a fall in barometric pressure indicates an oncoming storm. A severe storm is short lived, lasting about 10 hours. Then the pressure rises, indicating good weather is on its way. The cycle repeats with a seven-and-a-half-day period.

Barometer Simulator Version 1.0

Version 1.0 of the electronic barometer simulates all the features found on conventional mechanical barometers. The electronic barometer will read the barometric pressure, record that pressure, plot an electronic log, and indicate the current weather.

Mechanical barometers tend to display the current weather by dividing the pressure range into five regions: "Very Dry," "Fair," "Change," "Rain," and "Stormy." While each manufacturer uses slightly different ranges, Version 1.0 will use the following:

780 –> 800 mm Hg barometer reads "Very Dry"

760 –> 780 mm Hg barometer reads "Fair"

745 –> 760 mm Hg barometer reads "Change"

730 –> 745 mm Hg barometer reads "Rain"

700 –> 730 mm Hg barometer reads "Stormy"

To add the range feature, the pressure range is divided up into these five regions and appropriate LED indictors are asserted whenever the current pressure falls within a particular region. The first task is to build a new sub-VI called HiLo, which will set a Boolean flag whenever the pressure falls between a high and low limit. In the analog electronic world such a function would be called a window comparator. Whenever the voltage level falls within a voltage window, the output high is set high. The HiLo.vi design follows.

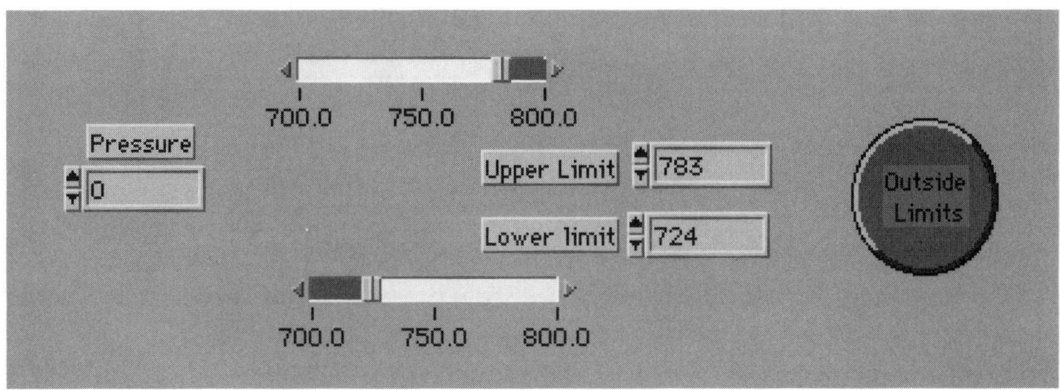

Use two horizontal slide controls to set the high and the low level limit. A simple digital control will be our link to Pressure.vi and a Boolean LED indicator will set a flag whenever the current pressure is within the HiLo window. The diagram panel uses two compare functions and a Boolean AND function to simulate a window comparator.

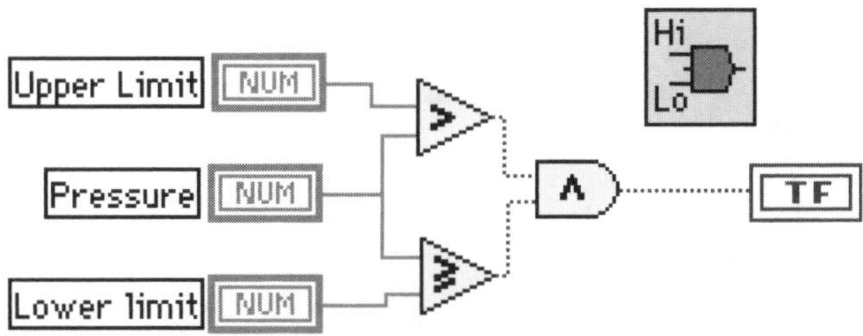

Four sub-VI links from the **Show Connector/Show Icon Patterns** in the **Icon** menu are chosen to link pressure, high limit, low limit, and flag output to other VIs. Save this sub-program as HiLo.vi.

The current weather is displayed with a new VI panel called Pressure Range.vi. It can be used as a stand-alone weather announcer or linked to the other VIs to build a more sophisticated barometer. Each weather range uses a color-coded LED display to indicate when the pressure is within a particular range. Red is used for "Very Dry," Orange for "Fair," Yellow for "Change," Blue for "Rain," and Brown for "Stormy." Use the operating tool to select the current state of a Boolean indicator and use the coloring tool to color the LED displays. In the off state, color the LEDs gray; in the on state, color the LED the appropriate range color. On the left side of each LED indicator, add a string display. You can select a large font size and color the text background to match the color of this pressure range. When a range is selected both the colored LED and name of the range will be displayed. An additional string indicator, [Message Board] is used to pass the current weather on to other VIs in the form of a string message. For now, the pressure input is chosen as a vertical slide indicator to mimic the mercury barometer.

Seven connector links are used with this sub-VI: the input pressure, the message board, and the five Boolean LED outputs.

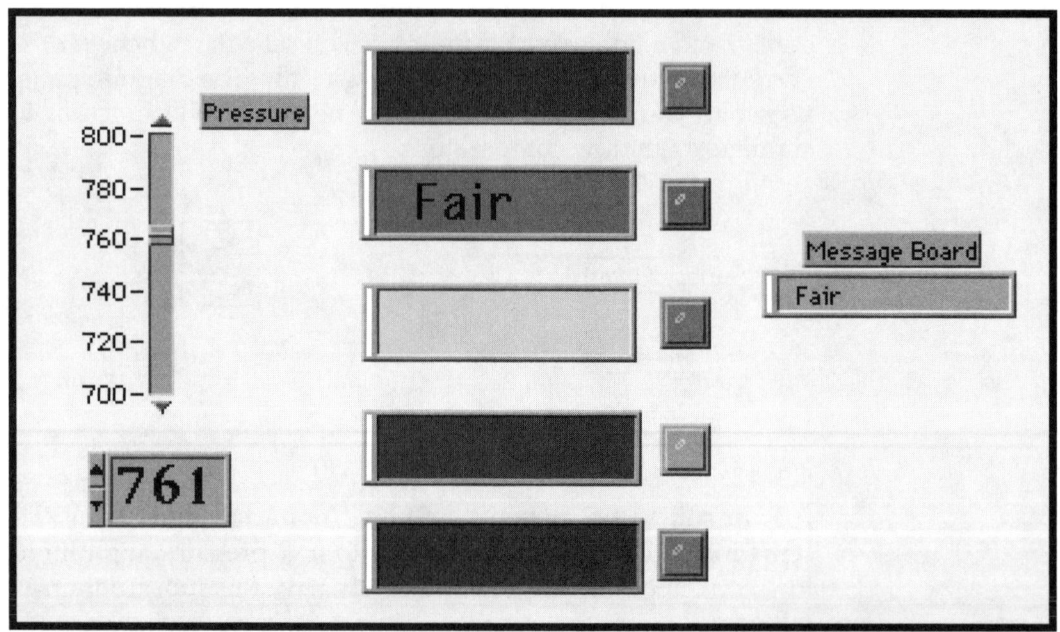

The diagram panel uses five HiLo.vi window comparators. Each HiLo sets a Boolean flag, which in turn selects either the <|True|> case, a string range message, or the <|False|> case, a null string. All messages, real or null, are concatenated together and sent to the message board. On the front panel, the appropriate LED display turns on, the range is announced, and the string message is displayed.

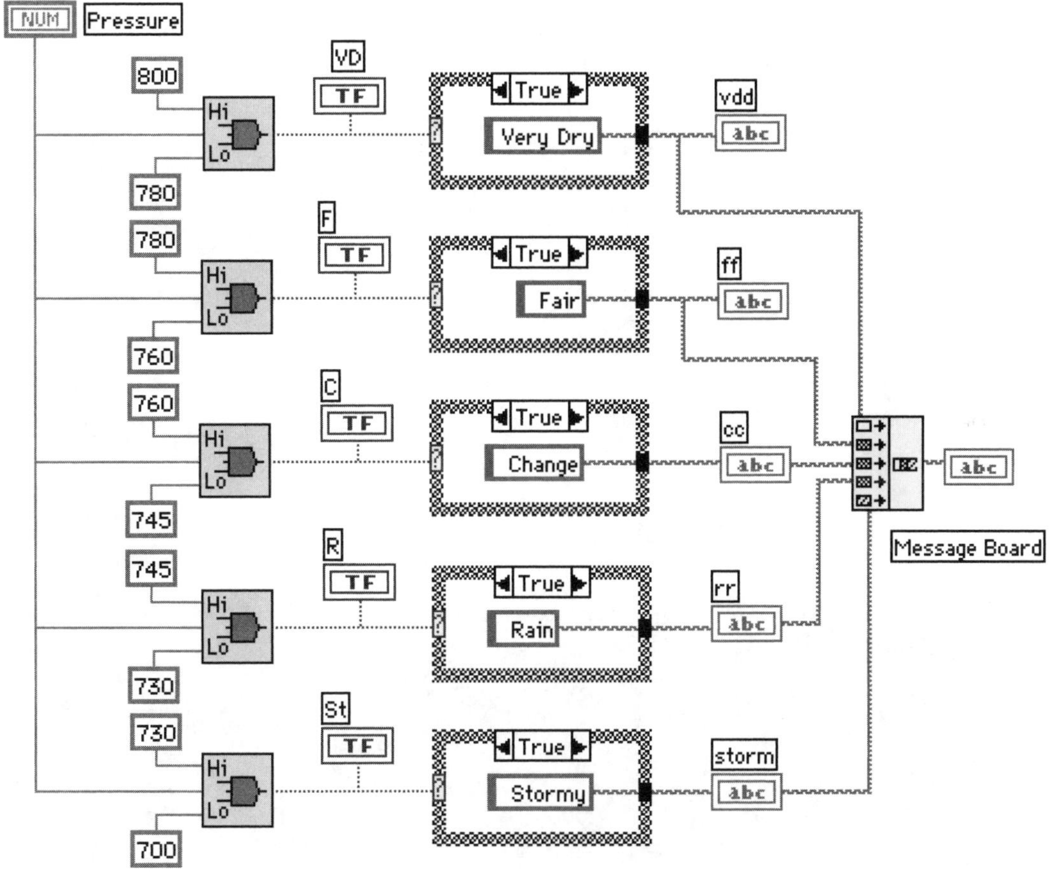

To test your VI, run the program continuously and observe the front panel as the hand (operating) tool changes the pressure on the vertical slide wire. Save this program as Range.vi.

Exercise: LabVIEW provides a "window comparator" icon [**In Range?**] found in the **Functions>Comparison** subpalette. It executes the same function as the HiLo.vi but is more compact. *Rewrite Range.vi using this function.*

We are now in a position to put several VIs together to create the Electronic Barometer Version 1.0. The simplest way is to modify the Pressure Range.vi. Replace the pressure control with Pressure.vi, the source of pressure readings. Add a rotary dial and make it an indicator. Wire it up to the Pressure.vi to provide an interesting visual effect when the program is running. Good software design calls for placing the program inside a While . . . Loop and adding a run button to the front panel.

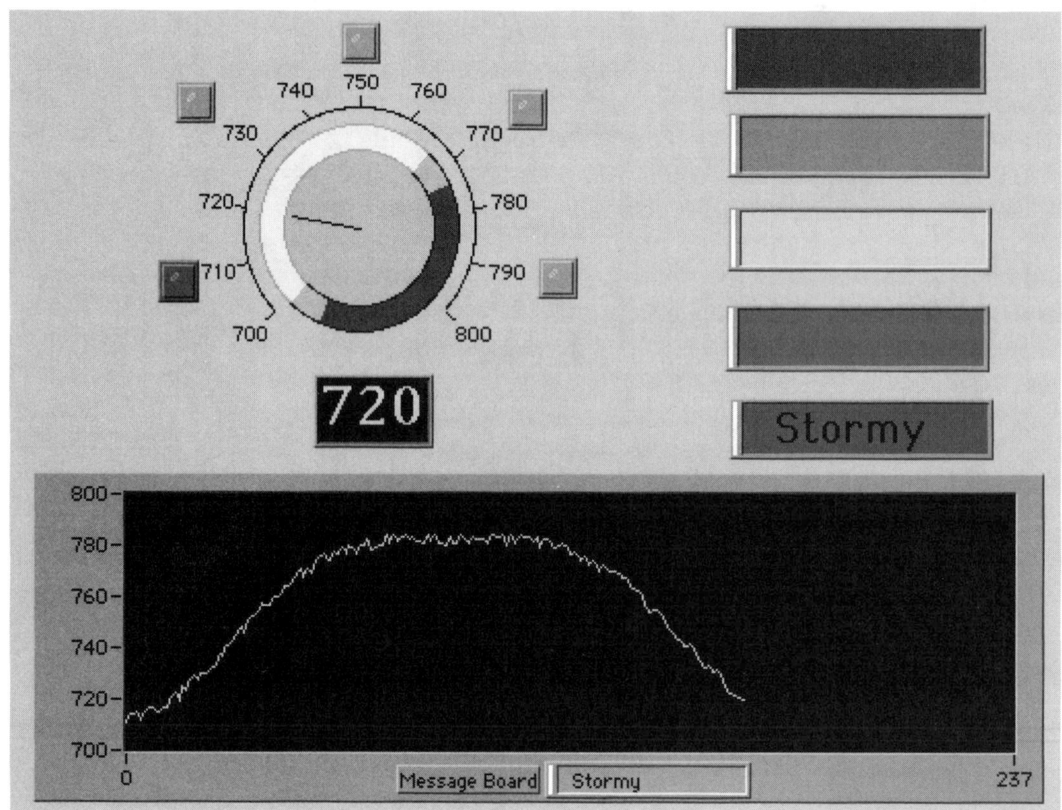

On the front panel modify the font size of the pressure digital indicator to be more readable. The Boolean range indicators can be placed around the dial in appropriate locations. Add a chart display for logging the pressure measurements and set the Y-axis range from 700 to 800. Place the [Message Board] at a convenient location near the chart. Run Version 1.0 to verify that it has all the features of a conventional barometer and save it.

Weather Forecasting Barometer

It is always difficult to make accurate weather forecasts, and even more difficult when having only one parameter, the pressure. However, the rate of change of the barometric pressure is often used to make forecasts. From our pressure record, it is clear that the variability in the hourly pressure reading will make a simple rate calculation inaccurate. Since the pressure varies slowly over the period of a few hours, a three-point average will produce a smoothed pressure curve. The slope of the averaged pressure curve will be the parameter used in the forecasting routines. Look first at a simple averaging VI consisting of two additions and one division function.

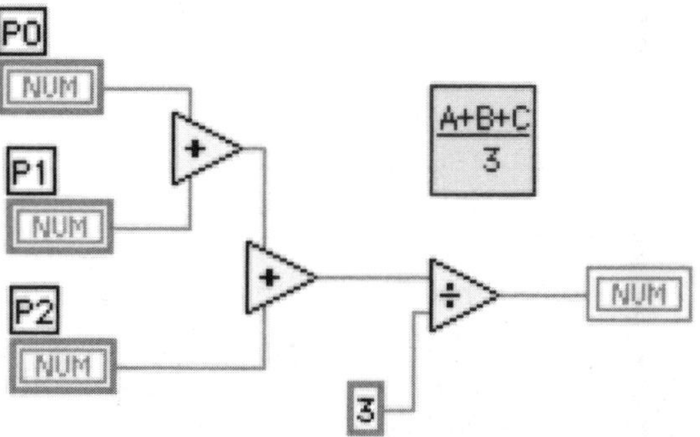

This VI has three digital controls for the inputs and one digital indicator for the output. Save it as Av3.vi. To observe the smoothing of a data set, create a new VI that uses Av3.vi inside a While . . . Loop with a three-element shift register to form the three-hour pressure average. Note that the shift registers are initialized to the mid-range pressure value. Hence, the average pressure will not be correct until we have at least three valid pressure readings. Add the Pressure.vi inside the loop and place two charts on the front panel. One is wired to the hourly pressure and the other is wired to the smoothed pressure curve. Running the program will quickly demonstrate the value of averaging a noisy data set. Save this program as Smoothing.vi.

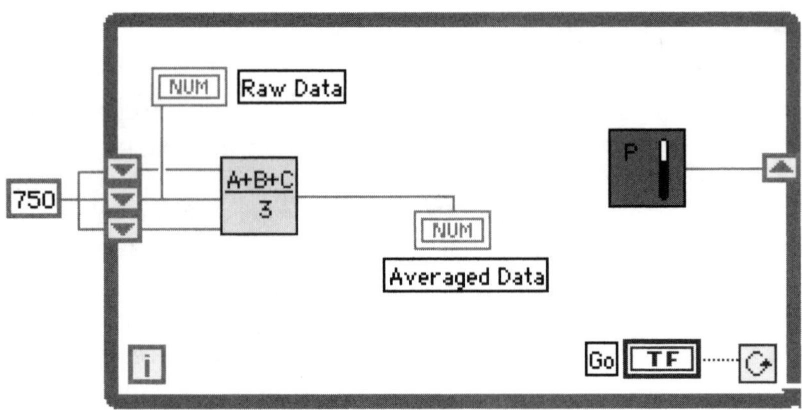

Barometer: Rising or Falling?

The rate of pressure change tells how fast the weather is changing, and its sign tells whether the pressure is rising or falling. The output of the Av3.vi gives the average pressure over the last three hours. Let's call the initial value $\overline{P0}$. At the end of the next hour, a new average will be formed, $\overline{P1}$. The weather predictor is the slope of the smoothed data set and can be written as $(\overline{P1}-\overline{P0})/1$ hour. If this parameter is negative, poor weather is on the way, while if the parameter is positive, then better weather can be expected. A simple comparison function is used to determine if the slope is greater or equal to zero and its Boolean output provides a sign flag. If the flag is true, then the barometer is rising or if flag is false then the barometer in falling.

While the sign of the slope gives the trend, the magnitude of the slope indicates a condition. Analysis of the barometric

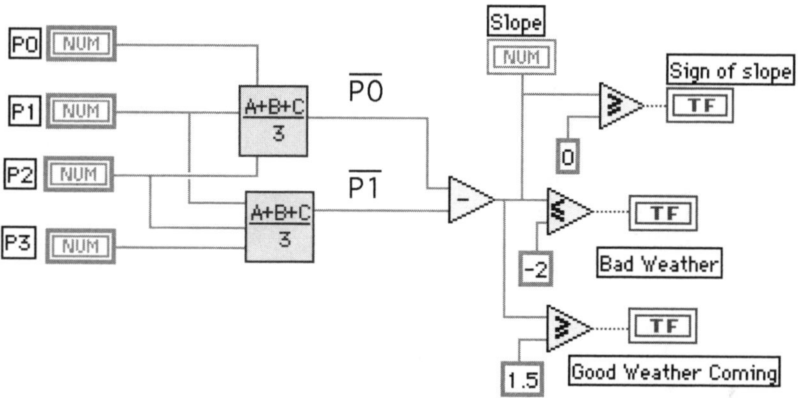

record shows that a negative pressure change greater than 2 mm Hg per hour is an indicator of bad weather ahead. On the other hand, storms seem to linger at times, hence a slower rate of +1.5 mm Hg per hour indicates good weather is on its way. Bad and good weather flags are added in the Slope.vi with the addition of two comparison functions.

Stormy Skies Ahead

In order to strengthen our prediction of an approaching severe storm, several other conditions must be met. If the smoothed pressure is falling, AND the current pressure is within the "change" region, AND the rate of pressure change exceeds 2 mm per hour, then it is time to announce a severe storm warning. This is accomplished by ANDing the sign flag from the Slope.vi with the "Change" flag from the Pressure Range.vi and the "Bad Weather" flag also from Slope.vi.

When the above criteria are met, a storm flag is set and the "Storm Warning" message is asserted. A shift register element is used to indicate that the storm flag is set. It will remain set until the pressure reaches the "stormy" region. Since the storm warning is only useful before the storm, it needs not be announced once the storm is upon us. Take a look at Storm Warning.vi and observe how the announcer operates.

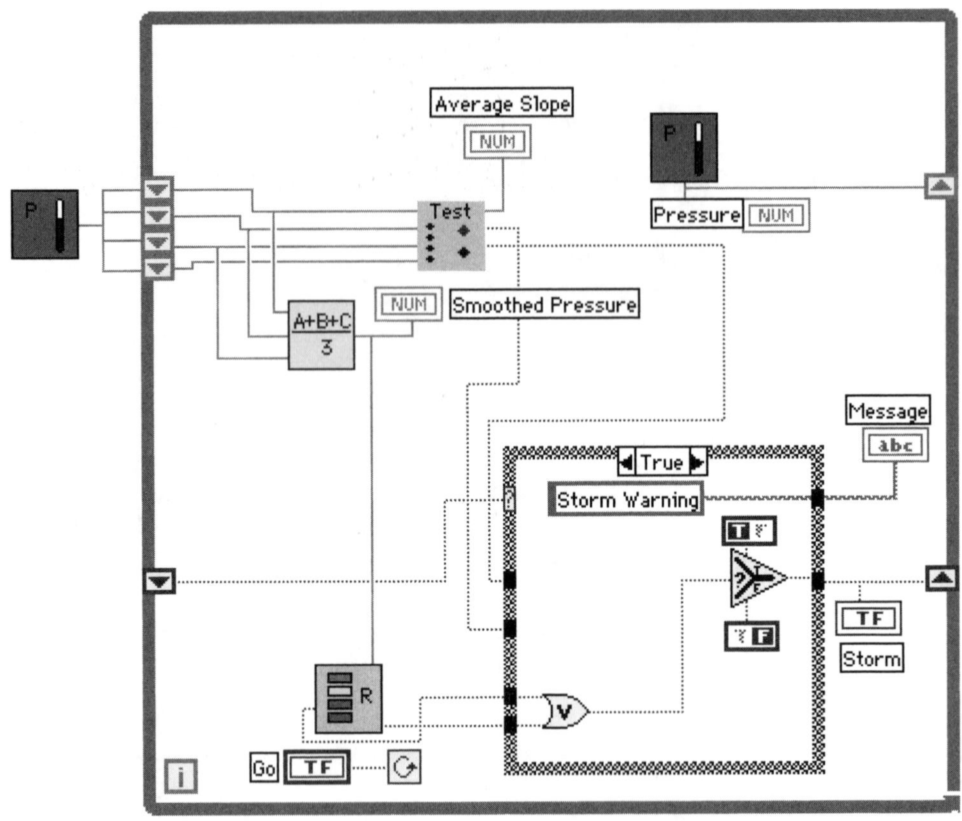

Better Weather Ahead

After the storm passes and the pressure begins to rise, it would be useful to announce "Better Weather Ahead." For this announcer, the condition is that the current pressure is within the "Change" region, AND the smoothed pressure change is positive, AND its magnitude is greater than 1.5 mm Hg per hour. Each of these conditions is available from earlier VIs.

Better Weather.vi uses these three flags to determine when to announce the good news. As in Stormy Weather.vi, a case structure is used to terminate the announcer. In this case, the announcer flag is cleared when the pressure reaches the "Fair" pressure region. Test out the operation of the Better Weather.vi.

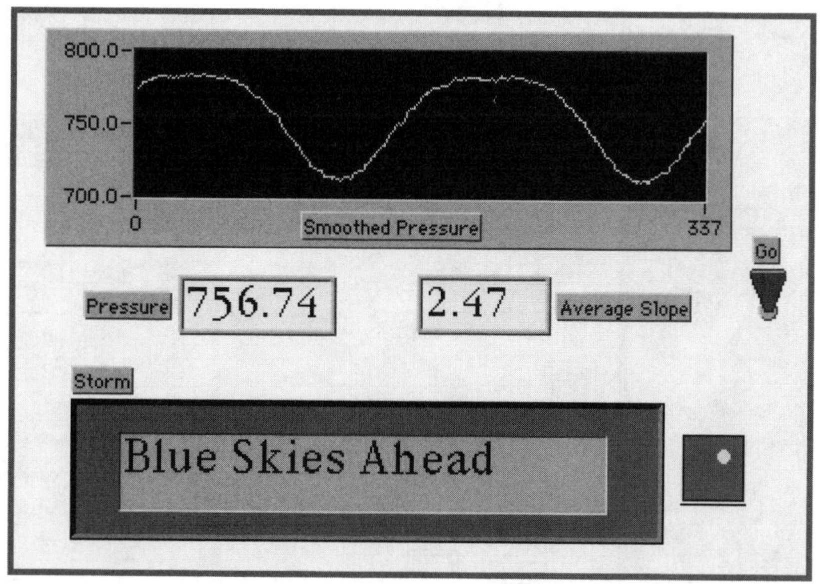

The Electronic Barometer Version 2.0

The electronic barometer Version 2.0 enhances Version 1.0 with the weather predicting logic. All the required sub-VIs have been assembled into a program called Barometer2.0.vi. You should look carefully at the diagram panel to see how the sub-VIs are integrated into a compete instrument. Note the extensive use of shift registers to bring memory of previous calculations forward to current calculations.

Once a program has been written, tested, and debugged, the real challenge is to make the front panel as user friendly as possible. LabVIEW provides a rich range of icons, colors, and attributes to achieve this goal. The information that the electronic barometer brings to the user is the current pressure, current weather condition (very dry... stormy), a pressure log of the last few days, and special forecasting indicators to herald the onset of a severe storm or of better weather ahead. The front panel allows the LabVIEW designer to place his or her own unique style to the presentation of information. The electronic barometer Version 2.0 is one example of a barometer front

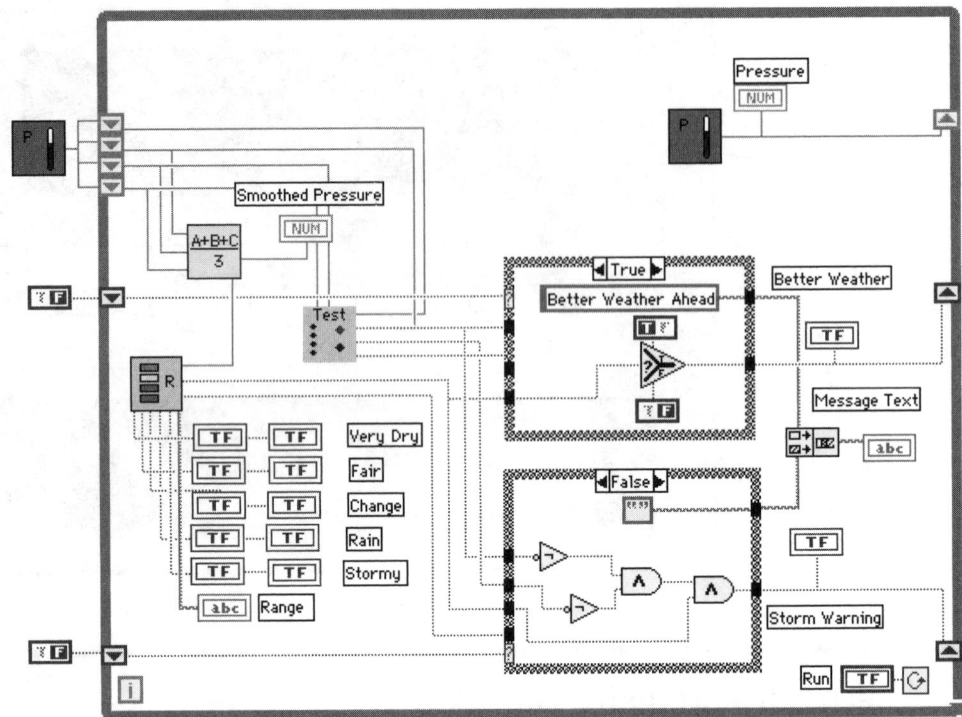

panel. Several other adaptations can be found in Versions 2.1 and 2.2. Carefully study each version to learn how the medium portrays the message.

LabVIEW Challenge: Design a Better Barometer Front Panel

Build a better front panel for the barometer.

LabVIEW Challenge: Add a Reference Needle to the Barometer Display

Some mechanical barometers have a second needle that can be rotated by an observer to the current pressure. Then, in future hours, the observer can see how fast or slow the barometer is rising or falling. LabVIEW provides a gauge numeric control that can have two or more needle outputs. *Add a second needle to your barometer and a push button to set the needle to the current pressure.*

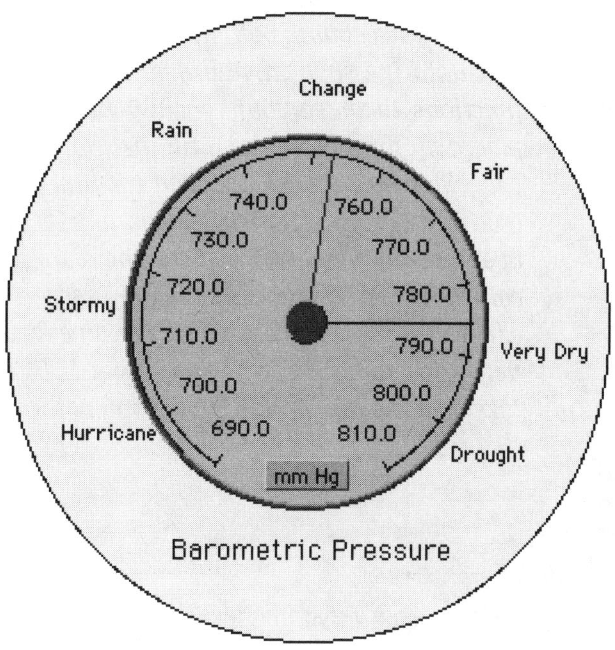

How to Use This Project

If a pressure sensor and interface is available, you can replace the Pressure.vi with a data acquisition VI to sample the real atmospheric pressure and give the electronic barometer a practical test. Components can be purchased from an electronics parts distributor. Using the circuit shown earlier, an electronic barometer sensor can be put to together for less than $40. If neither option is available, then some effort could be expended into designing a better atmospheric pressure simulator, Pressure.vi. Better yet, data collected from your local weather station can be used to build a database of interesting storms that have passed through your region.

Overview

Video conferencing brings image monitoring to the internet. It is a two-way street with both the client and the server having video, audio, and data message capabilities. LabVIEW TCP/IP icons add control functions to video conferencing. In this chapter, a video camera is placed on top of a pan and tilt mechanical mechanism. Simple electrical relays operate the pan and tilt functions. A serial-to-parallel microcontroller receives string commands from a LabVIEW program over a serial port line to control the relays. The camera and its pan/tilt controller are configured as a Web server. Remote users called clients can access the local camera and receive its video images over the Internet. If the remote user is running a LabVIEW client program, it can take over the pan and tilt operation of the video camera.

Goals

- Design a virtual joystick VI
- Build a string message processor for an external microcontroller
- Design a simulator for remote operation of a pan/tilt device
- Learn the Internet TCP/IP protocols
- Design client and server algorithms
- Integrate video conferencing with a LabVIEW TCP/IP process

Key Terms

- Video conferencing
- IP addresses
- Nested case structures
- TCP/IP open and close icons
- TCP/IP read and write icons

Video Surveillance

11

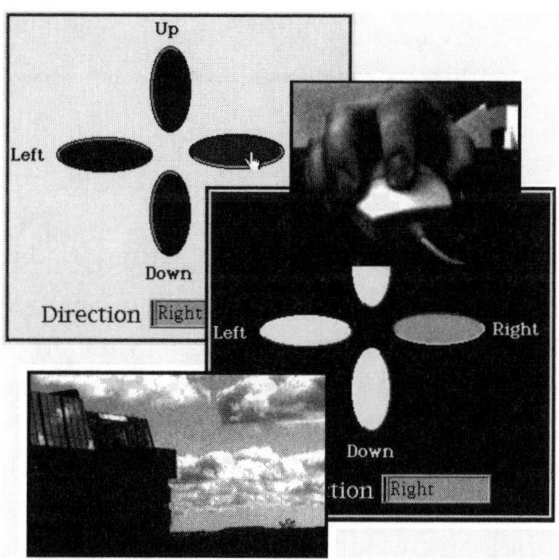

Client/Server Remote Video Surveillance: The client (upper two frames) controls the orientation of the video camera at the server site by pressing one of the virtual joystick buttons. Here the Right button is selected by placing the operating tool over the button and clicking on the mouse. A command is issued over the internet to pan the video camera to the right. The received command is executed and displayed on the server's front panel as the Right indicator, turned on. The camera rotates and the image captured from the server's location is sent back to the client to appear as the bottom frame.

Have you ever wondered, "What is happening at a sensitive experiment in the lab?" "Is the dog still in the backyard?" or "What's the weather's like at the cottage?" If so, then video surveillance presented in this chapter provides an answer. LabVIEW is used to control the orientation of a video camera by sending commands to a pan/tilt controller over the Internet. The image from the camera at a remote site is sent also via the Internet back to the local site. The user can see the video in real time and send commands to orient the camera from a LabVIEW front panel.

LabVIEW version 3.0 and higher allows the user the ability to interact with an experiment from a distance. Once the connection is established, processes can be run, devices controlled, and the results observed as if you were sitting at the terminal next to the source. In this chapter TCP/IP internet protocol is featured to demonstrate remote sensing and control over a remote process. Once the server is running, then it can be accessed by any client in the world wherever an internet port exists.

Pan/Tilt Driver

The pan/tilt output device uses two DC motors to execute the pan and tilt operations. The polarity of power supply defines the direction of rotation, clockwise or counterclockwise. Two motors and two rotation directions provide four functions (up, down, left, and right). In the pan/tilt driver, the four directions are controlled by four mechanical switches. To interface the pan/tilt device with a computer, the four switches are replaced with four relays, which are in turn controlled by four bits on a parallel output data port. The serial-to-parallel interface introduced in Chapter 5 will be used to provide the computer link. If not available, a DAQ card works just as well. Output bits b0 to b3 are mapped as follows:

Up <—> b0
Down <—> b1
Left <—> b2
Right <—> b0

Recall the serial-to-parallel interface uses an ASCII command string to set the port bits. The command message is composed of four ASCII characters:

AP"X"<CR>

where A is the local address of the microcontroller (taken as 3), P is the microcontroller byte command, "X" is the ASCII encoded port bit assignments, and <CR> is the microcontroller's execute command. The byte command outputs the 7-bit binary code of the ASCII character "X" to bits 0–6 on the parallel port. To make the codes transparent, the ASCII numeric character code is mapped directly to the bit code. For example, the ASCII code "1" or hexadecimal $31 is mapped to the port as (011 0001). Since the upper four bits are not used by the relay card interface, the code "1" sets bit 0 on the parallel port to the high state. In a similar way the other ASCII codes mean:

"1" sets bit 0 high which switches the	Up relay	(On)
"2" sets bit 1 high which switches the	Down relay	(On)
"4" sets bit 2 high which switches the	Left relay	(On)
"8" sets bit 3 high which switches the	Right relay	(On)

The standby condition, which ensures all the relays are (Off) is the ASCII character "0". The complete command message suite is

3P1<CR> tilts the camera	Up
3P2<CR> tilts the camera	Down
3P4<CR> rotates the camera	Left
3P8<CR> rotates the camera	Right

3P0<CR> leaves the camera at the current position.

Virtual Joystick

A virtual joystick on the front panel is formed with four push button controls arranged in a logical up, down, left, and right pattern. Each time a control is selected by placing the cursor on the button and clicking the mouse, the appropriate command message is formed and sent to the serial port. An ASCII message is also sent to the front panel message pad to indicate which function is currently selected.

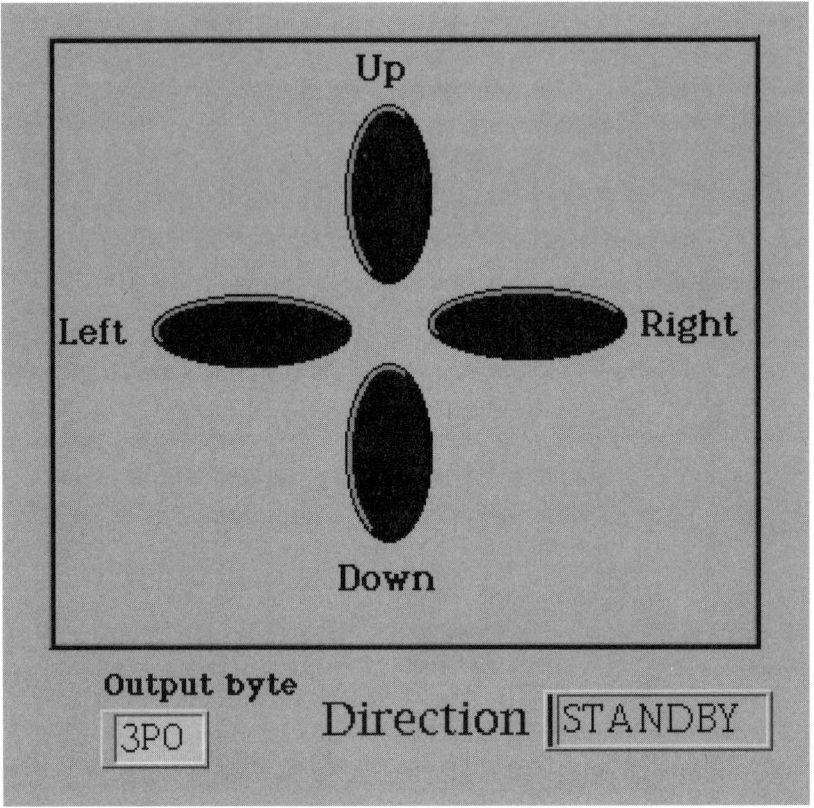

To achieve the correct operation of the switches their **Mechanical Action** must be set to **Switch Until Released.** Pop up on the Boolean control and select the proper action from the subpalette. To clearly distinguish between the selected (clicked) and non-selected state of the front panel buttons choose the on and off colors with good contrast so that the two states can easily be discerned.

The diagram panel for the virtual joystick is composed of a set of nested case structures. This construction is chosen as an efficient means to cycle through all the switch commands. If no switch is pressed, all the cases are false and the Standby code is sent. Each time a button is pushed, a <|True|> state in one of four case structures is selected and the appropriate command message is sent. When the switch is deselected, the case structure returns to scanning the switch positions.

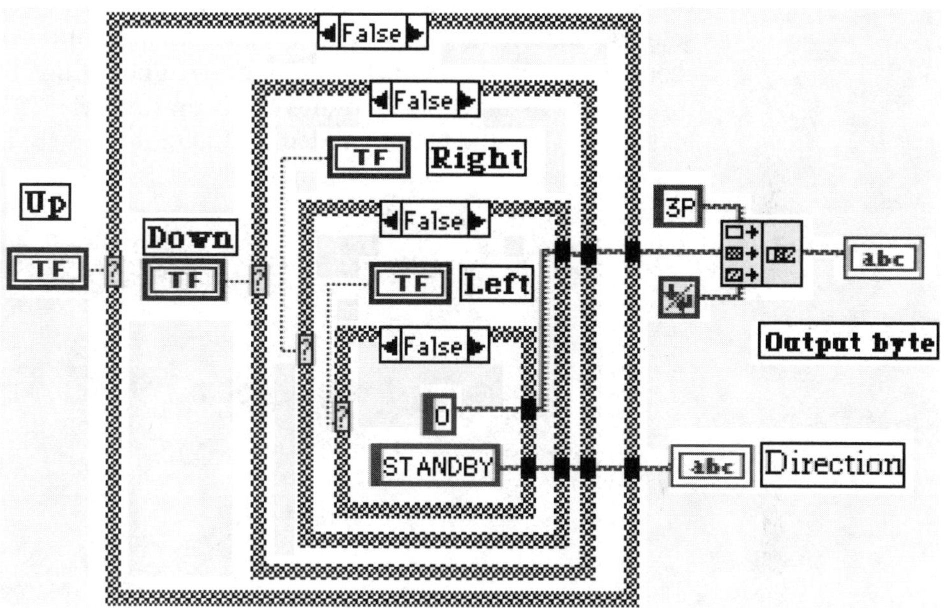

This structure can be written in conventional programming code as:

```
IF the Up button is pressed send up code                            3P1<CR>
         Else if the Down button is selected send down code         3P2<CR>
         Else if the Left button is selected send left code         3P4<CR>
         Else if the Right button is selected send right code       3P8<CR>
Otherwise no button has been pressed, send Standby                  3P0<CR>.
```

Load the program Joystick.vi from the program library for this chapter. Study the diagram page and observe the VI in action. This VI is also a sub-VI whose inputs are the four Boolean buttons, and outputs are the ASCII command message and a direction status message.

Relay Simulator

Whenever a direction is selected, one of four ASCII messages is sent to the serial port. The microcontroller sets the port bits as received from the command message. If a bit (0–3) is selected, the addressed relay is turned on and the camera rotates in the selected direction. Relays.vi simulates the operation of the microcontroller and relay interface. It determines which of the four directions has been selected and turns on that relay. This action is displayed on the front panel as a LED indicator.

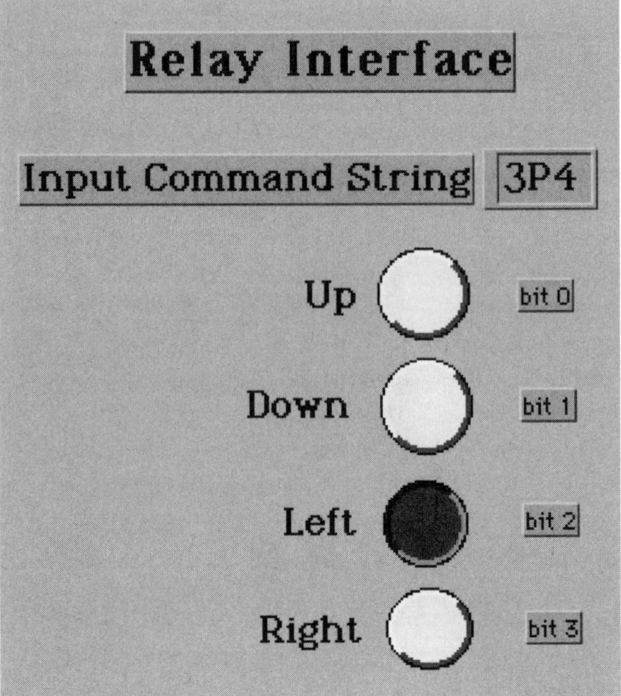

The first two characters [3P] of all messages are the same and can be stripped off using the **String Subset** function. Recall that the string offset defines the start of the string subset and the string length defines how many characters are in the substring. The third character defines the relay code. The bottom four binary bits of the 7-bit ASCII code are passed to the Port and then onto the relays. A binary (1) turns the relay on and a binary (0) turns a relay off. The last character in the message <CR> is not needed in the simulation but is essential for the microcontroller, because it is the execute command to update the parallel port bits. In the simulation, four Boolean constants inside a <|True|> state of each the case structure set the LED relay displays. These outputs signal which condition—Up, Down, Left, Right, or Standby—has been selected. When the update byte command is received, LabVIEW's compare function is used to determine which switch, if any, has been pushed. Note the polymorphic nature of this function. It works just as well with ASCII characters as it does with numerics. Again, the nested case structures make for efficient coding. Whenever a direction is deselected (unclicked), the virtual joystick responds with the standby message 3P0<CR>. In simulation, the inner <|False|> state is selected and all four relays are set to OFF.

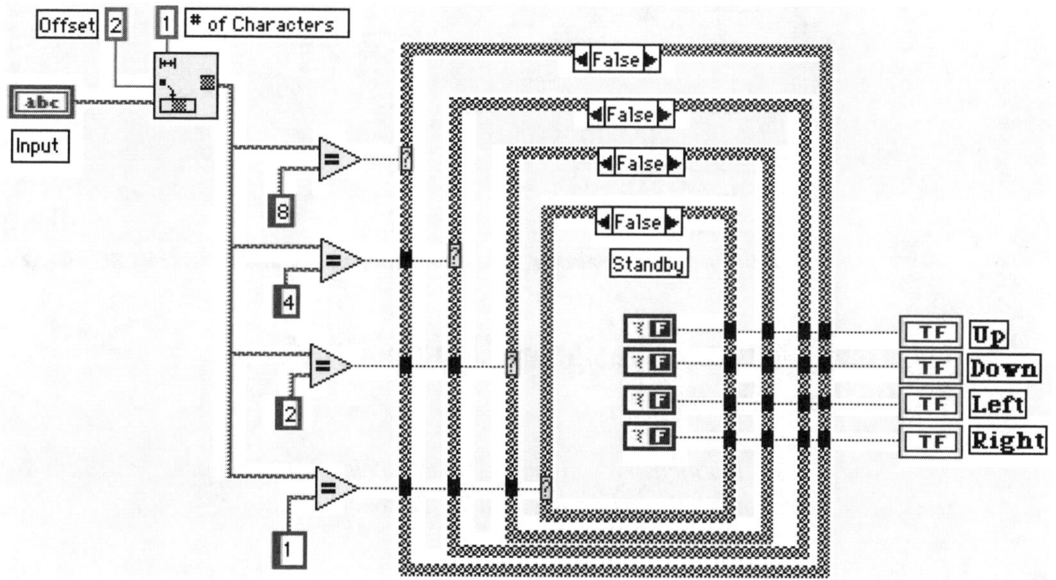

Exercise: Link these two programs Joy Stick.vi and Relays.vi together to observe the operation of the virtual joystick and relays working together.

Since the microcontroller port is latched, it is not necessary to send a command message more than once. Multiple commands achieve nothing and slow down the response time. A command is sent only when the current command is a different from the last command. This requirement is met by bringing memory into the joystick VI. A shift register added to a While . . . Loop, a compare function, and a case structure provide this functionality. The current command is compared with the last command sent. If they are the same, no action is taken. However, if they are different, the current command is passed to both the output port and to the shift register where it becomes the last command sent. Again note the polymorphic nature of the shift register.

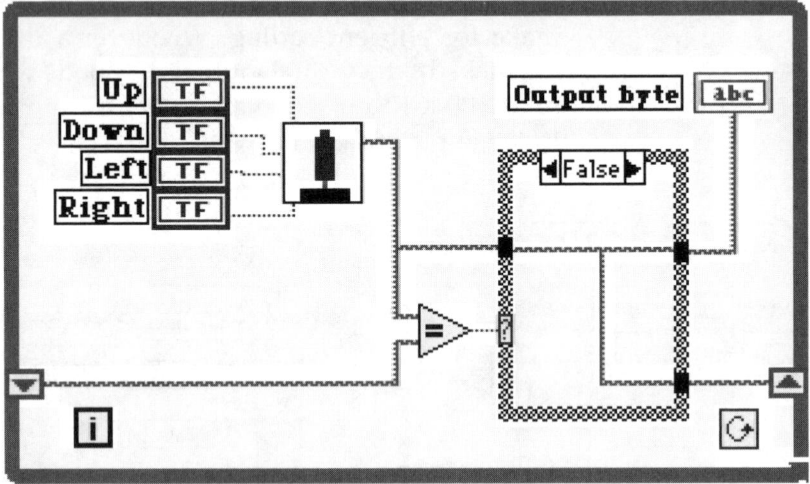

■ LabVIEW Challenge: Design an Upgate Structure for the Pan/Tilt Driver

Add this memory construct to the previous joystick program to ensure that only transitions are sent to the relay simulator.

Pan and Tilt Simulator

To add realism to the front panel display, a knob-shaped indicator is programmed to act like the pan and tilt operation. The pan/tilt device can pan from –180 to +180 degrees, while the tilt action is limited to +/–30 degrees from horizontal. The pan operation requires a up/down counter with limits set at +/–180. On the front panel choose a knob indicator. Set the limits to –180 and +180. With the circular cursor drag the two limits of the indicator together. Each time the counter is called, the angle display is to be rotated by one degree. If the ASCII command calls for Right, then the counter is incremented by one. If the ASCII command calls for Left, the counter is decremented by one. This is accomplished with a data switch whose update values are +1 or –1. In a similar manner, the tilt operation can be programmed. Note in each case, the limits are sticky—that is, when a limit is reached, it remains there until an update brings the output below the limit.

Exercise: Load the Pan/Tilt Sim.vi from the chapter library. Run the program to observe the action. Carefully study the diagram panel to see how the simulation is implemented.

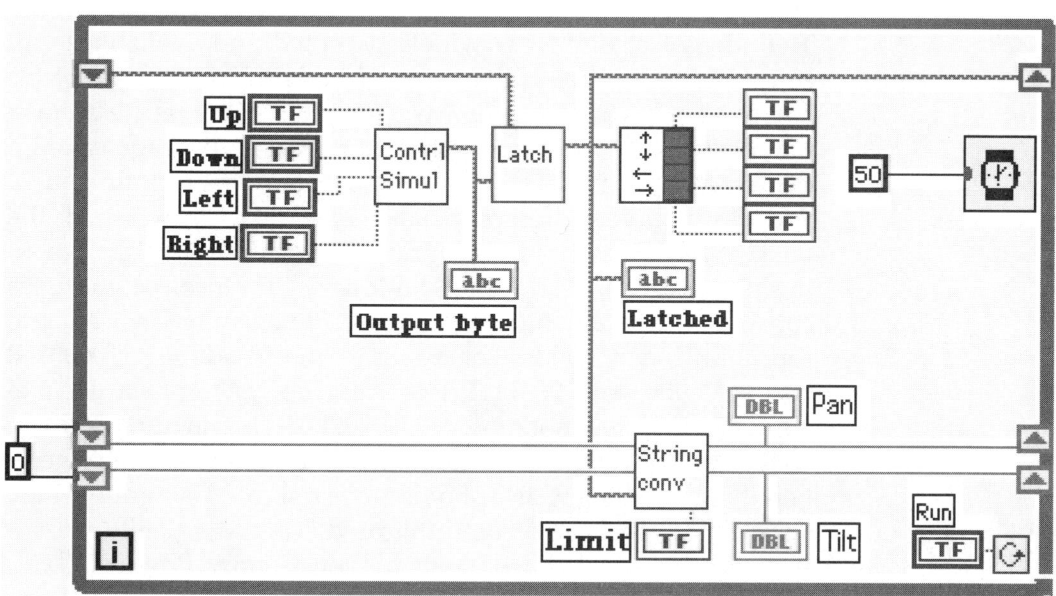

The action begins when one of the push buttons (Up, Down, Left, or Right) is pressed. The sub VI [**Contr**oller **Simul**ator] detects a button press and creates a string message. This message passes to the sub-VI [**Latch**] which compares this command with the previous command. If the same, no action is taken. If different, it is passed onto Relays.vi and the state of that relay is echoed on the front panel LED displays. The message is also used to update the shift register with the new command. In addition, the sub-VI [**String Conv**ert] converts the message into an numeric command (increment or decrement) for the pan and tilt rotary indicators. Embedded inside this VI is Sticky Limits.vi, which detects and creates the sticky limits for these displays.

The Web

The internet, or the Web, as it is commonly known, consists of a world-wide communications network. Observing the standard communications protocol allows the user access to thousands or even millions of data ports. Several network protocols have emerged as acceptable standards, but in this chapter only TCP/IP is featured. It is widely accepted and available to all computers. The Web is like a world-wide telephone system for computers. Each data port has its own IP address (telephone number) and to reach an address, an IP data package called a datagram is formed and sent. These datagrams contain source and destination addresses, header, and data. Once issued, the packet winds its way across various networks eventually reaching the destination address. The Internet Protocol (IP) address consists of a 32-bit binary number conveniently divided into four 8-bit binary fields. For example, the IP address (10000100 00001101 00000010 00011110) can also be written as a decimal equivalent dotted notation (132.13.2.30). This in turn is often converted by a domain name server into a more easily remembered address such as (photon.phys.dal.ca). Datagrams may have errors, get lost, or be duplicated. To add reliability, a supervisory protocol called Transmission Control Protocol (TCP) manages the datagram flow, retry, and error messages. TCP ensures that datagrams are delivered in sequence, with no error, loss, or duplication.

LabVIEW passes ASCII messages to TCP, which together with supervisor control bits, addresses, and data, forms the IP datagrams. The process is reversed at the destination: that is, datagrams —> messages. If an error occurs, TCP does not acknowledge to the sender receipt of a datagram and it is resent by the source. Larger data files are often broken up into smaller sized data blocks for transmission. TCP correctly reconstructs the file from the parts received at the destination site.

TCP allows multiple and simultaneous connections. Each unique process carries a local TCP port address. To access that process, LabVIEW requires both the IP address and the port address for that process. The local port address is a 16-bit binary number ranging from 0 to 65535. All that is needed is a computer connected to the Internet and a LabVIEW process running with the TCP/IP VIs.

As an example, consider sending a command from the virtual joystick. The address of the camera site is 129.173.21.107. The local process that controls the orientation of the camera is 6343. These parameters are added to the front panel of the pan/tilt client.

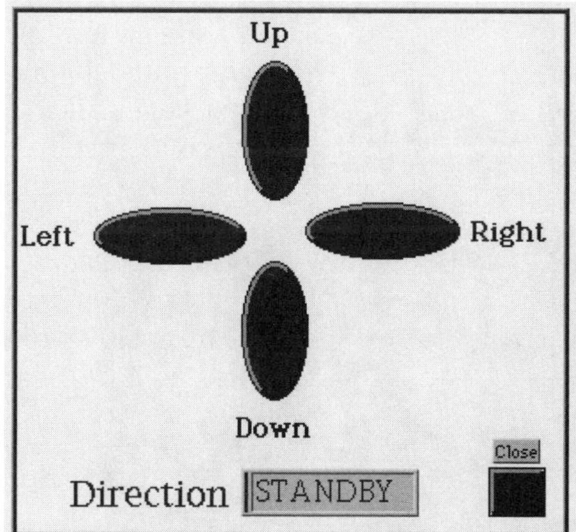

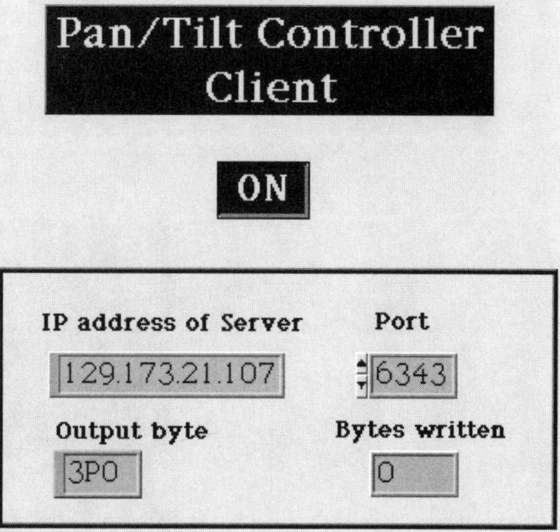

The only additional parameter not found in the previous joystick example is a Close button on the front panel. When pressed, it will send a command to the server that will terminate the TCP/IP link between the sender and receiver.

Client/Server Model

LabVIEW uses the client/server model for network applications. One set of VIs (the client) requests services from another set of VIs (the server). In this model the server must be running before the client makes a connection. In this paradigm, the server waits for a connection request. Once the connection (IP address and port address) is received from the client, the server enters the wait for command phase. Once the command is received, LabVIEW executes the command and generates a response if required. The program loops here, awaiting new commands or the close command. Then the connection is closed and the server returns to the listener mode waiting for another connection.

The client paradigm is even simpler. The client opens up a connection by sending the IP address and port number followed be an ASCII command string. Once the connection is established on the server, the command is executed and the server re-

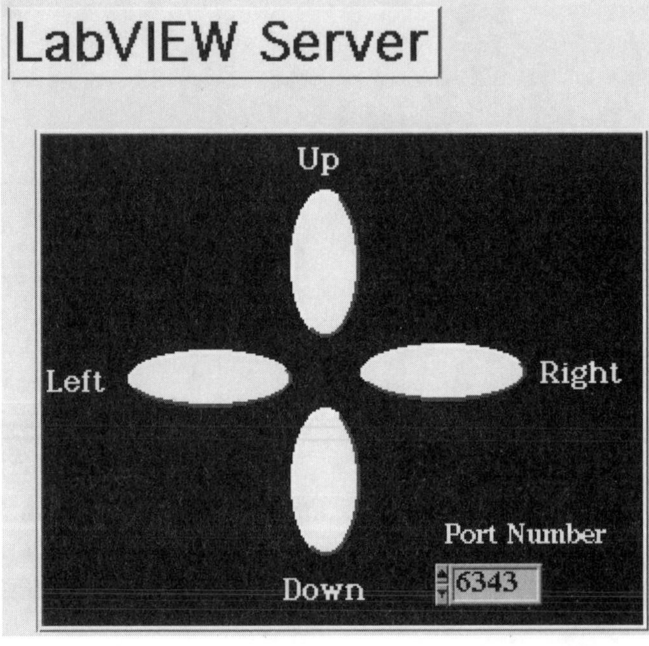

sponds with a reply to the client. The client receives the response, processes it, and reports any results to the front panel. New messages may be sent and results processed until the LabVIEW client has completed all the tasks. At the end of the dialog, the connection is closed.

To demonstrate the operation of the client/server model, we will use the virtual joystick as input for the client and the Relay.vi on the server as an output. The client/server interaction is a sequential operation, hence a sequence structure is appropriate. Only a three-frame sequence is required for both the client and server programs.

Relay Server

The front panel relay LEDs have been replaced with indicators that resemble the push buttons on the client's front panel. The only new piece of information presented is the port number (process identification number) that uniquely defines this process. When the client pushes a joystick button, both the client and server button/indicator are expected to respond.

On the diagram window in frame 0, the server listens for a connection using the **TCP Listen.vi** found in the **Functions>Communications>TCP** subpalette. The only variable is the port address. The IP address is inherent to the computer and already established when the computer was connected to the internet.

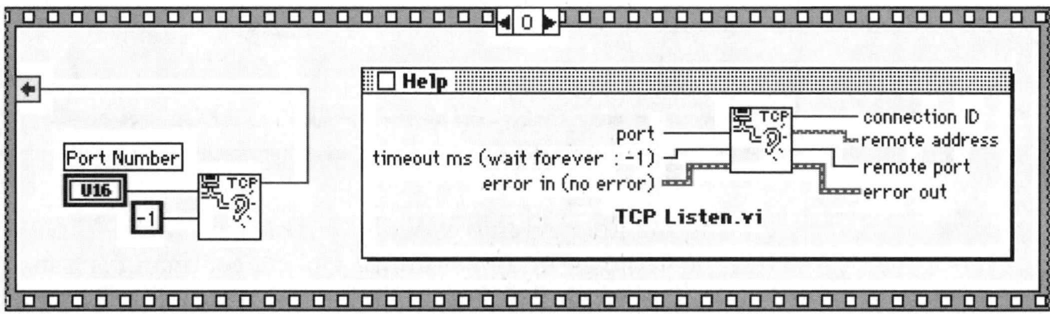

The input [–1] sets the time-out to listen forever or until an operator closes the server. Frame 1 is entered once the connection is established. The TCP Read.vi waits for an ASCII command string (bytes to read) to appear at the Ethernet Port and then passes the string message onto the Relay.vi for processing.

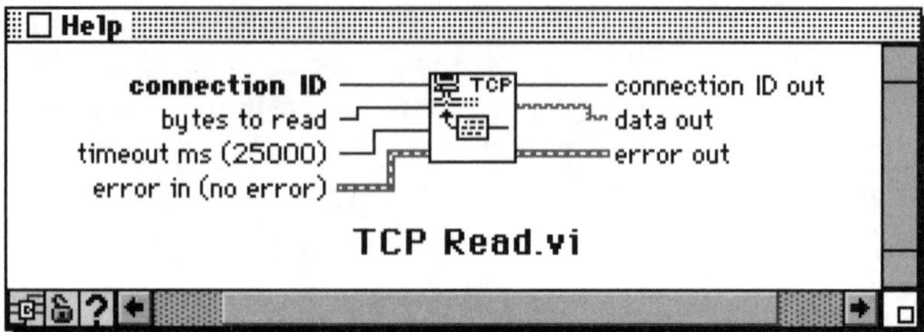

Any errors are reported at this time to the client using the TCP Error.vi. It is also possible to define a Timeout so that if no messages appear in a fixed period of time, the process closes gracefully. Again, once the message appears, the processing in Relay.vi proceeds.

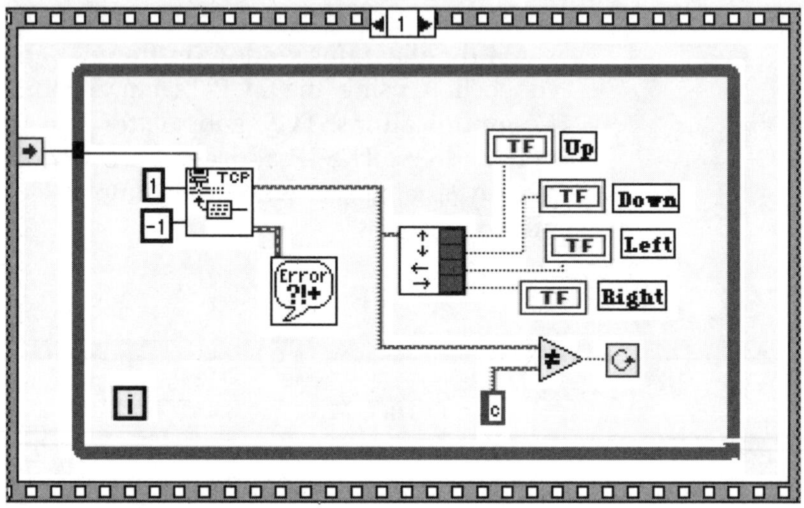

Placing the Relay.vi into a While . . . Loop allows the process to be repeated over and over until a closing command is issued by the client. In this example, the single ASCII character "c"

generates the close command. When received, the While . . . Loop is terminated and the action moves onto the next frame. Frame 2 terminates the connection with the TCP Close Connection.vi.

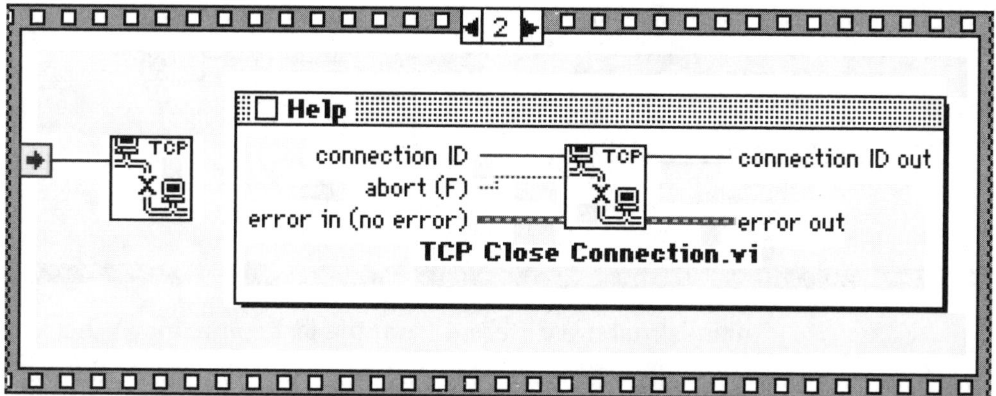

If the server is to be run continuously, then the frame sequences are placed inside another a While . . . Loop (not shown) and the server returns to the listen mode after a connection is broken. In this case, only the server supervisor can stop the server.

Joystick Client

The client also uses a three-frame sequence. On the diagram window, frame 0 forms the datagram using the IP address and the port number. TCP Open Connection.vi sends the connection request over the internet.

Here again the time-out is set to [–1], no timeout. Once the connection is made, a connection ID is passed on to the next frame. Recall, in general, that multiple TCP/IP sessions can occur and the ID number keeps the messages routed to the correct process.

Frame 1 takes the current message from the Joystick.vi and passes it along to a front panel display [Output Byte] and to TCP Write.vi. The case structure on the right side of the diagram is part of the memory function used only to send com-

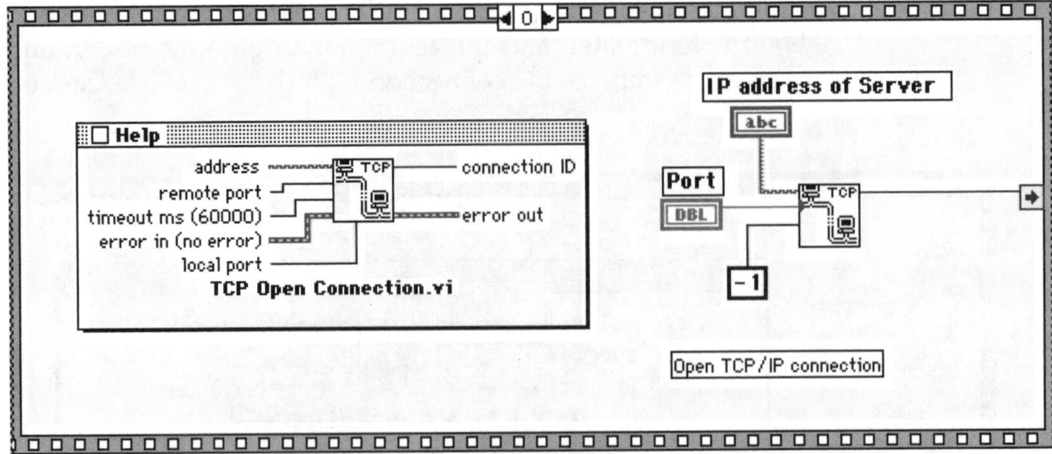

mands that are different from the last command. When the joystick commands are the same, the <|True|> case is empty and no message is transmitted, only recirculated in the shift register. When a new command is different, the case state is <|False|> and the command is sent over the network and the shift register is updated.

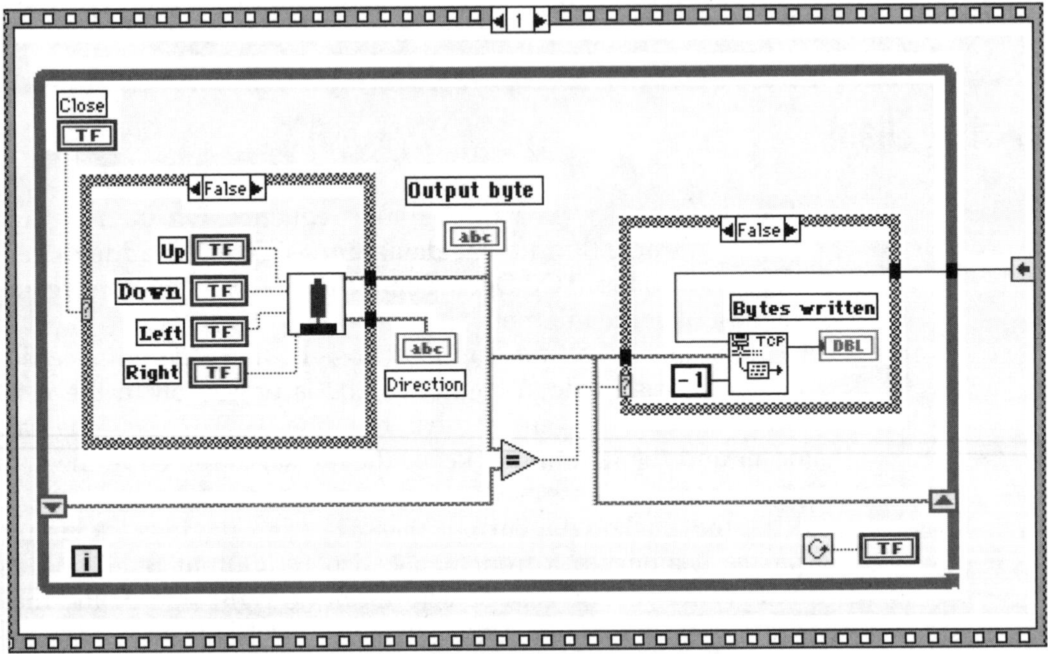

The Case structure on the left has been added so that once the link is established, it can be run continuously until a Boolean switch on the client's front panel is pressed. This action generates the close "c" command. The <|True|> state of the case structure will be selected and the request to close the data links sent to the server. On receipt of the close command, the server completes the current frame and moves on to the final frame to terminate the process. The client loops in frame 1 until the front panel [close] button is pressed. Then frame 2 closes the local client with the TCP Close.vi.

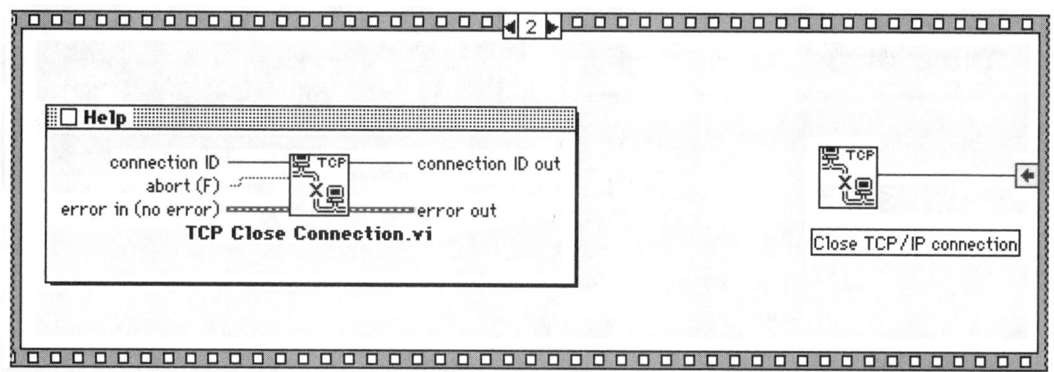

■ LabVIEW Challenge: Client/Server Operation of the Pan/Tilt Driver

TCP/IP is a read and write process that can be executed locally on one computer. *Design the client and server VIs for operating the relays as a server with the virtual joystick as a client.* Load each program and run first the server then the client. If you have two computers nearby, run the server on one, the client on the other.

In this action shot, the client running on the left has the Up button pressed. The client issues the command over TCP/IP at the IP address 129.173.21.107 to initiate process on port number 6343. The server running on the right receives the command and turns the Up relay on. This is signaled on the front panel by the Up LED display turned on. The relay LEDs have been placed and fashioned in a shape similar to the virtual controls in

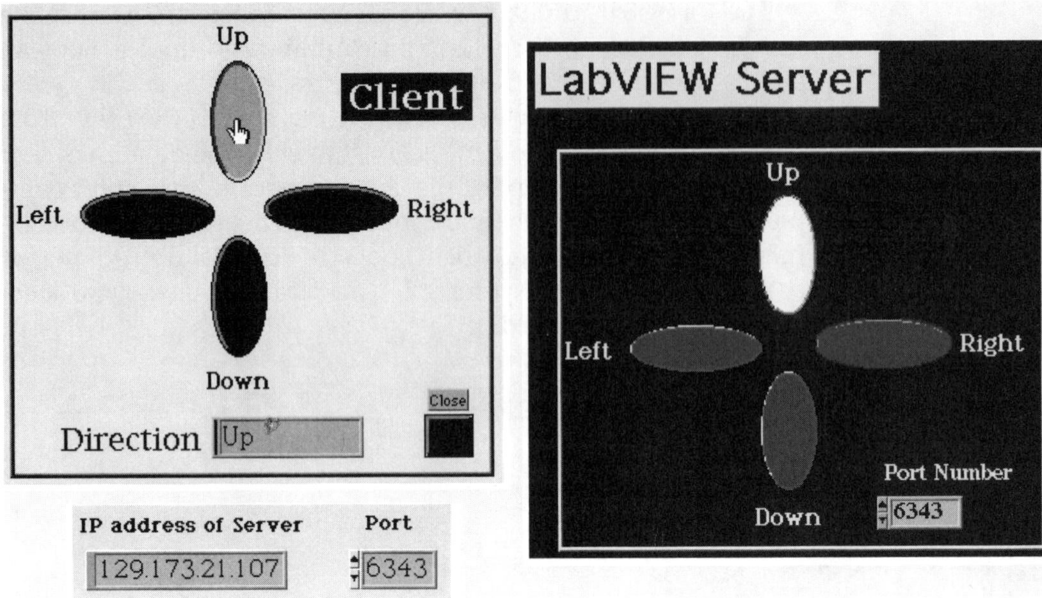

order to demonstrate the symmetry of the client/server operations.

Video Communications

In uncompressed form, video and audio files require copious amounts of memory, and real-time transmission is impracticable over the internet at the current bandwidths. To reduce the file size, data is compressed using a codec (COmpressor/DEComposer). In 1992 Cornell University, in partnerships with other universities and organizations, commenced a project to produce compression algorithms to send video and audio over the internet. The solution to the bandwidth problem comes in the form of clever codecs and incremental video transmission. The initial frame is sent in small video blocks and the receiver gradually builds up the frame over a period of a few seconds. Future transmissions send only changes in the video image from the last sampling period. If you are watching a scene, then only changes such as leaves blowing, flags flying, or people walking are transmitted. Video processing on the client's computer integrates these changes into the current image. The updates are sent at whatever bandwidth the network can allot. Even frame rates less than 1 frame per second are often adequate for video conferencing and video monitoring.

Numerous video conferencing packages are currently available, but Cornell University's CU-SeeMe is featured in this chapter. To run the internet video conferencing software, a reasonably fast CPU, lots of RAM, 10 megabytes of hard disk space, and an internet connection are required. What makes video monitoring possible is that the video processes can be run concurrently with the LabVIEW control program.

Video Conferencing

Video conferencing is the ability to send video and sound over a network. In some cases, keyboard chat windows and graphic workspaces can also be sent. Conventional video conferencing is point-to-point communications between two users. However, group video conferencing is also possible using a computer dedicated as a reflector site. In this case, the server logs onto the reflector and sends its images. Other clients can now log onto the reflector site to view the image and send their images to the server. Since only one user at a time can effectively control the orientation of the camera, point-to-point conference is best for this project.

Video Monitoring

To complete the video monitoring and control project, first run the video conferencing software, then the LabVIEW server program. The client establishes a video link at the server's IP address using the video conferencing software. Once the video image is observed, the client program is run. Here is a potpourri of images captured using video monitoring.

Overview

Some sensors and transducers require careful calibration procedures before they can be used. LabVIEW's program structures allow the software designer the ability to build the calibration or processing procedures into the data flow in such a manner that only by following the experimental protocol can a measurement be carried to completion. In addition, the great variety of front panel designs and patterns can produce both a functional and ergometric design.

Goals

- Translate experimental protocol into a graphical program
- Build a transmissivity sensor
- Design a Help feature using pop-up dialog boxes
- Use local and global variables
- Design a user friendly front panel

Key Terms

- Beer's Law
- Pop-up dialog boxes
- Local and global variables
- Clusters

Beer's Law: Determination of the Impurity Concentration in a Liquid

12

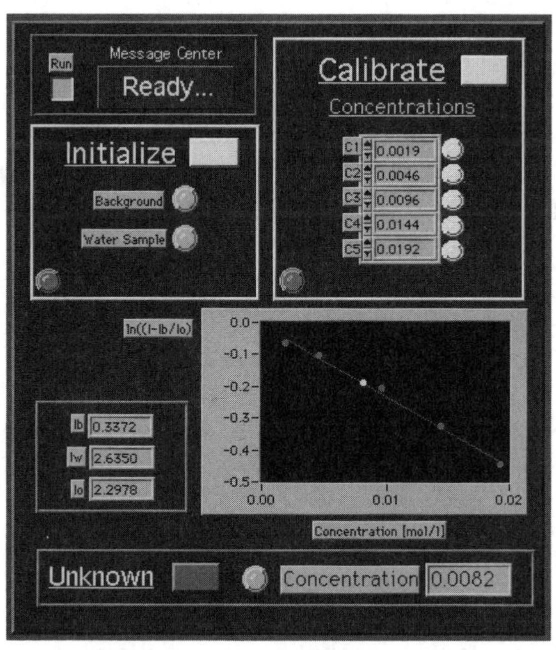

LabVIEW Front Panel: Beer's Law

Sensors can be made to detect specific physical processes with a high degree of sensitivity. However, the operating environment, material degradation, and aging all conspire to reduce sensitivity. In some cases, these factors can change the nature of the sensor to the point where measurements are in question. When this occurs, calibration of the sensor is an essential process prior to its use. Proper and careful calibration procedures can ensure the sensor will produce accurate and reproducible results. The procedure may be tedious, time consuming, and perhaps boring, thus it is easy to get a little careless in repeating the same procedure over and over. Great care and a certain discipline are required to overcome human error. After calibration of the sensor, the unknown sample is measured and the physical parameter is recorded as a single number. There is a constant temptation to believe this number as it is presented. However, good calibration techniques carried out precisely as prescribed in the measurement protocol are required to ensure the number will be meaningful.

In this chapter, transmissivity is used to determine the concentration of particles suspended in a liquid. This is the type of experiment that requires careful calibration in order to place high confidence in the final result. Some LabVIEW constructs are featured that help produce a user friendly front panel that is both easy to read and follow. As well, the program design inflicts a certain discipline on the user to ensure the calibration procedure follows the measurement protocol precisely.

Beer's Law

As light passes through an absorbing media, the intensity or radiant power falls off with the exponential of the path length. The thicker the sample, the more light is lost. The transmittance T is given by the ratio of the transmitted light, I, to the incident light, I_0.

$$T = I/I_0$$

Beer's law states that the transmittance for a liquid is given by

$$T = \exp{-(kLC)}$$

where k is the extinction coefficient, L is the thickness of the sample, and C is the concentration of scattering centers within the liquid. For the chemists, k is also called the molar absorptivity, provided L is measured in centimeters and C is in moles per liter.

On a plot of ln(T) versus C, a straight line implies that Beer's law is followed. The slope is –kL, and if the path length is known, the extinction coefficient can be determined. It can then be used together with the path length to determine an unknown concentration.

In a trace analysis experiment to determine the concentration of a specific element, a source of light is chosen to be selective to the material in solution. This ensures a high contrast between the coloration induced by the impurity and the host liquid. A series of known concentrations are measured and fitted to Beer's law to determine the extinction coefficient. When an unknown sample is measured, its transmittance can then be used to calculate the unknown concentration.

Experiment Design

A photodiode is used to measure the transmission through a colored solution. If the coloring of the liquid is a result of impurities in suspension, then Beer's law can be used to determine the concentration of these impurities. The optical transmission is defined as the intensity $I(L)$ measured through a path length L divided by the intensity $I(0)$ at zero path length. In practice, the path length is fixed and $I(0)$ refers to the intensity with no impurities present in the solution.

An opaque block is fitted with three channels. A test tube containing the liquid is placed in the vertical channel. A light-tight cap is placed over the top of the test tube to prevent ambient light striking the sample and causing a background error. One side channel allows light from a LED source to flood the solution contained in the sample chamber. A third channel aligned with the source channel and the sample chamber houses a photodiode. The photodiode current is given by:

$$I = I_0 \exp(-\mathrm{kLC}) + I_{\mathrm{Offset}}$$

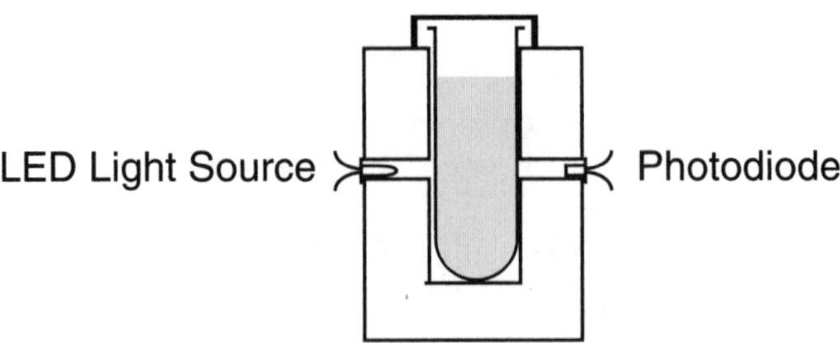

Transmissivity Sensor

I_{Offset} is the dark current generated by the photodiode with no light present. In general, a current-to-voltage OpAmp circuit is used to boost the photocurrents (μa) up to the analog-to-digital voltage levels (0–10 V). The above equation can be rewritten as:

$$V = V_0 \exp(-kLC) + V_{offset}$$

In general, the OpAmp may itself display some offset. These two signals—the dark current and the amplifier offset—are included in V_{offset}.

The constant (V_0 + V_{Offset}) can be found from the intensity when the test tube is filled with a transparent liquid ($C = 0$). The offset voltage is found from the intensity with the LED source off ($I_0 = 0$).

A verification of Beer's law for the experimental setup would be a straight line on a semi-log plot of $\ln(V - V_{Offset})/V_0$ versus the concentration C of scattering centers. The slope gives the extinction coefficient, provided L, the inside diameter of test tube, is known.

Experimental Protocol

In order to accurately determine the concentration of impurities within an unknown liquid sample, the following measurement protocol is to be followed.

1. Turn on the electronics. Let the apparatus warm up for 10 minutes.

2. Place a clean test tube filled with transparent liquid into the sample chamber. Place the dark cap over the top of the sample tube and measure the dark current, V_d.
3. Turn on the LED and measure the reference signal level, V_{00}.
4. Calculate $V_{Offset} = V_d$ and $V_0 = V_{00} - V_d$.
5. Insert a calibrated sample with a known concentration C_i into the sample chamber. Cover the top and measure the voltage, V_i.
6. Calculate the value $\ln(V_i - V_d)/V_0$.
7. Repeat steps 5 and 6 for the other calibration samples.
8. Plot a graph of $\ln(V_i - V_d)/V_0$ versus C_i. Beer's Law is valid only if a straight line is observed. Calculate the extinction coefficient from the slope (–kL).
9. Insert the unknown sample into the sample chamber, cover and measure the signal level, V_x.
10. Calculate the impurity concentration $C_x = (-1/\text{kL}) \ln[(V_x - V_d)/V_0]$.
11. Repeat steps 9 and 10 for other unknowns.
12. Turn off the power.

This is a rather long and somewhat complex procedure to measure an unknown concentration. However, without a measurement protocol, one may, or perhaps should, question the validity of the final number for the unknown concentration. Several observations can be made from the protocol:

1. There are three distinct procedures, initialization (steps 2–4), calibration (steps 5–8), and calculation of the unknown concentration (steps 9–10).
2. A certain discipline or order of events must be followed at certain stages. For example, the dark current must be measured before the LED is turned on.
3. A Beer's Law plot is essential to verify that the calibrated samples follow this law.
4. Only after initialization and calibration can the unknown sample be measured.

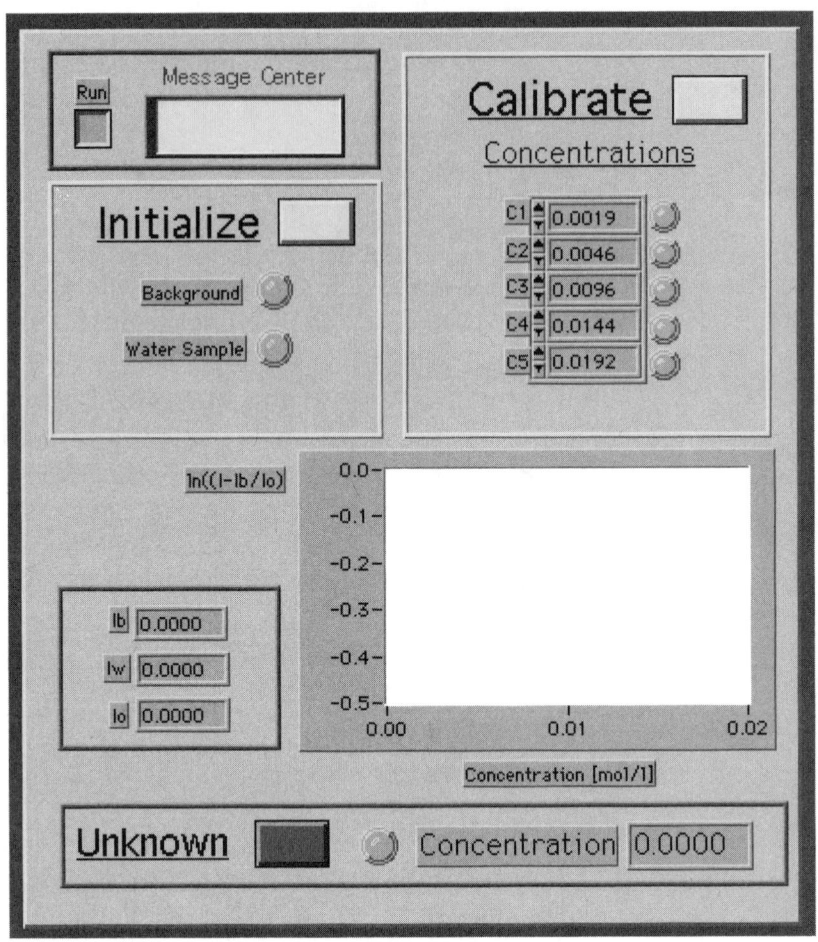

User Friendly Front Panel

The prime purpose of a front panel is to clearly display the status of processes and the results of any measurements and/or calculations. The front panel should aid the operator in completion of the tasks at hand and in this application provide a certain discipline on the operator to ensure the measurement protocol is followed. The front panel is conveniently divided and highlighted with colored borders into the four major activities: starting the program, initialization, calibration, and calculation.

Included in the run area is a message pad called Message Center that tells the operator the current program status. When the program is first run, {Initialize} comes up on the [Message Center], meaning that the program is waiting for the initialization to start.

When the initialize button is pressed, a yellow LED comes on telling the operator that the background is to be measured. In addition, a dialog box pops up with specific instructions for the operator, in this case, "Insert the test tube filled with distilled water into the chamber and cover. Press OK to proceed." After placing the calibrated sample into the chamber and covering the top [OK] is pressed.

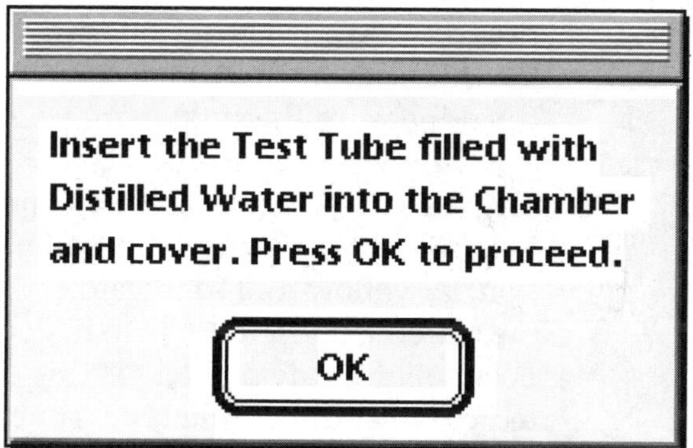

At this point a measurement of the dark current is taken, recorded, and displayed on the front panel in an indicator labeled [I_b] (background). A second dialog box pops up with a message for the operator to turn the LED source on. After turning on the LED, press [OK]. A second signal level is taken, recorded and presented on the front panel beside the label [I_w]. The corrected intensity I_o is now calculated and reported on the front panel. Upon completion of the initialization, the [Message Center] presents a new message {Calibrate}, meaning that the instrument is ready for the calibration procedure. Note that a green LED shows up in the lower left corner of the Initialization box. This comes on when the initialization process has started and remains on until the operator has completed the initialization.

This particular design for the Calibration phase allows the calibrated concentrations to be entered manually on the front panel controls prior to pressing the Calibrate button. When the [Calibrate] button is pressed, a series of calibrated samples may then be measured. The pop-up dialog box provides instructions for the operator to follow. Each time a concentration is measured, a yellow LED beside one of the controls is lit to signify which concentration is being measured. Although the concentrations are given in increasing order, this is not necessary. The only restriction being that concentration C1 corresponds to the measured sample [Test Tube #1], and so on. At the start of the calibrate cycle, a green light comes on in the lower left corner of the calibration box indicating that the calibration procedure is in progress. At the end of the fifth sample, the LED stays on and the [Message Center] reads {Ready}.

To measure the unknown, press the Unknown green switch and follow the instructions in the pop-up box. If you have pressed [OK], the procedure continues with a measurement of the unknown sample followed by a calculation of the concentration. A yellow LED indicator comes on during the unknown measurement cycle. When it goes off, the concentration and Beer's plot are displayed. At this point, the calibration curve, fitted curve and the unknown sample appear in a Beer's Law format on the front panel graph. A straight line is fitted to the five calibration points and shown as a solid line. The five calibration points are shown on the same graph as discrete points. The unknown sample is displayed as a single point. If the calibration points do not fall on a straight line, then something is wrong with the preparation of the calibration samples or the apparatus. The data is suspect and the concentration value is meaningless. If all points fall along the fitted solid line, then Beer's Law is followed and the concentration calculated is valid.

New unknowns can be measured at this time and each is reported in the same manner. However, the dialog box for the unknown does allow the operator to go back to the calibration stage without having to reinitialize the transmission cell. This can be done with the Cancel button in the two-button dialog box.

Software Design

The advantages of dataflow programming can clearly be seen in the program flow on the block diagram. Four Boolean switches are used to select the type of processing, run the program (Run), initialize the sensor (Init), calibrate the sensor (Calib), and measure the unknown concentration (Unknown). LabVIEW cycles through the inputs testing each for a task message. Run is a simple toggle action so that once started the While . . . Loop will

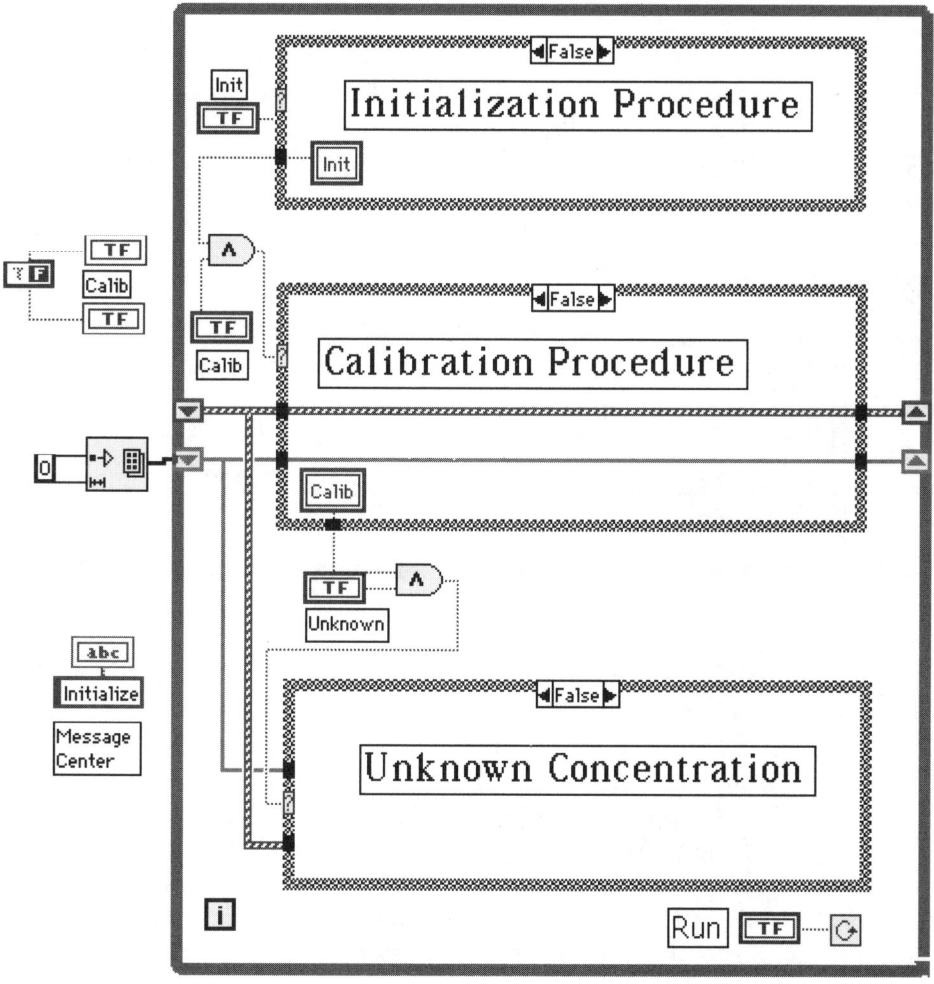

continue until the [Run] button is toggled off. The other task switches have their mechanical action set to **Switch Until Released** to ensure that when the switch is pressed, the task will be completed only once. When the task is finished, the switch can be used again to restart that process. The false state in each of the Case structures indicates no front panel activity.

Various parameters and displays are initialized by placing their icons outside the While . . . Loop. Note that two shift registers—one for arrays and the other for clusters—bring memory into the processing for two of the processes. Discipline in the measurement protocol is ensured by the use of two AND gates. Calibration can only occur if the initialization is completed and the [Calib] button has been pressed. The unknown concentration can only be calculated if the calibration has been completed and the [Unknown] button has been pressed. If any of the processes is attempted in the wrong order, no activity will take place. Note the use of global and local variables for {Initialize}, {Init} and {Calib}.

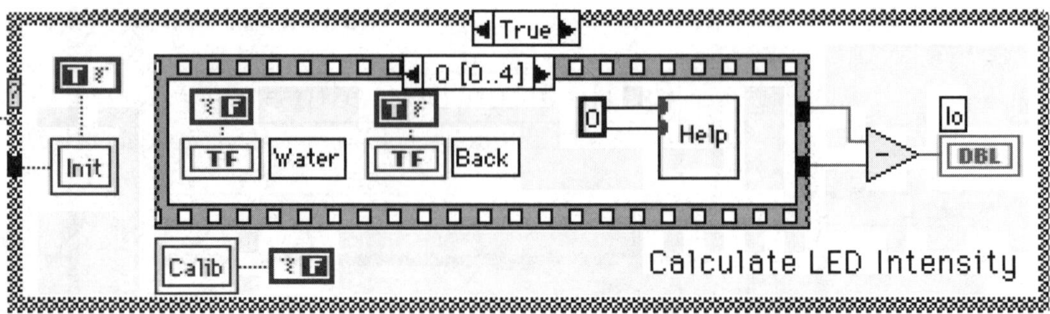

When the [Init] button is pressed, the <|True|> state of the initialize processes commences. Several local variables, {Init} and {Calib}, are set. In the first sequence, the background LED is lit and level 0, sub-case 0 of the Help.vi is entered.

Help is an interesting subroutine because it generates all the dialog boxes for each part of the procedure. It uses two inputs—one called Level defines which of the three main processes is active, 0- Initialization, 1- Calibration, and 2- Calculations. The second input, called Case, defines which of the subcases, if any, are to be active. In the case of Initialization, there are two subcases, one for the background measurement and the other for the intensity measurement. In the Calibration cycle, there are five subcases corresponding to the five calibration samples. There are also five Boolean outputs (c1–c5), which are used to control the activity LEDs in the calibration front panel box.

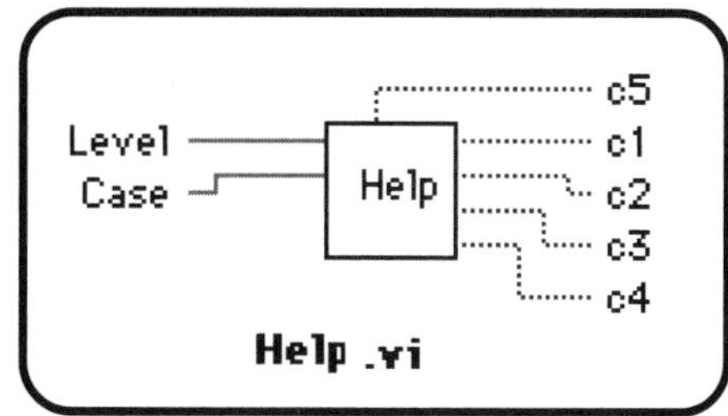

The calculation case has no subcases.

Inside the Help.vi are two forms of dialog boxes that pop up on the front panel when the processing reaches that icon. The contents of the dialog box can be set by the software designer. In this application, clear messages are presented to the operator as to the next step in the measurement protocol. Dialog boxes have either a one-button or two-button control. In the simple case of an alert box, a one-button dialog is sufficient to pause the processing until the operator is ready for the next step.

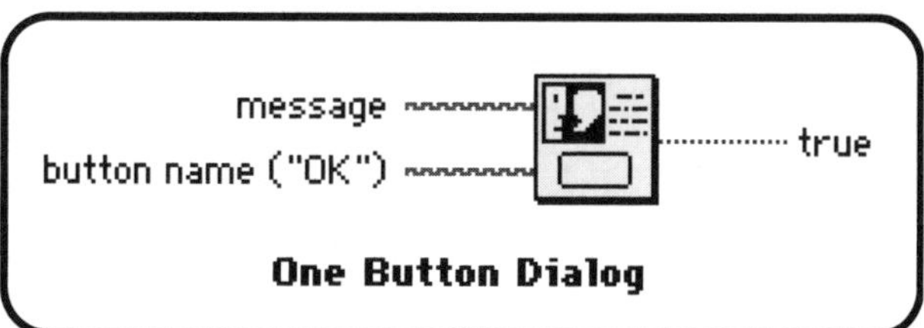

In the two-button dialog, processing can be continued as in the one-button dialog using the ("OK") input, or switched using the ("Cancel") input. Which path is taken is signaled by a Boolean output.

Frame 0 on the previous page prompts the operator with a dialog box containing the message "Insert the test tube filled with

message
T button name ("OK")
F button name ("Cancel")
T button?
Two Button Dialog

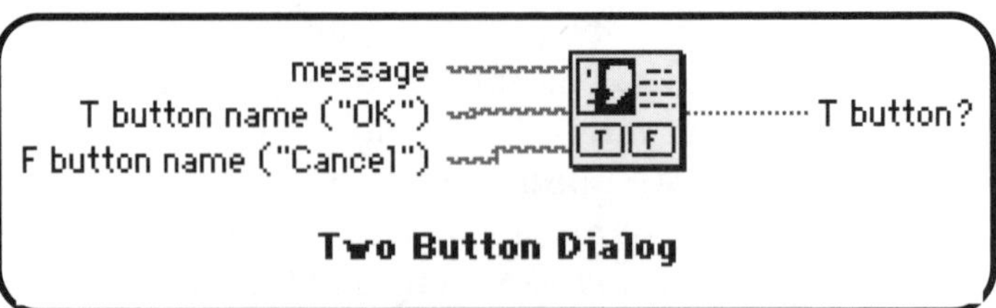

distilled water into the chamber and cover." The user completes this operation and clicks on "OK." Frame 1 uses [Read Init] to measure the background intensity, I_b. The Wait function in Frame 1 simulates the action of the measurement acquisition time and is not needed for real VIs. Frame 2 pops up another dialog box with a message to turn on the light source. When the operator completes this task and clicks on "OK," Frame 3 again

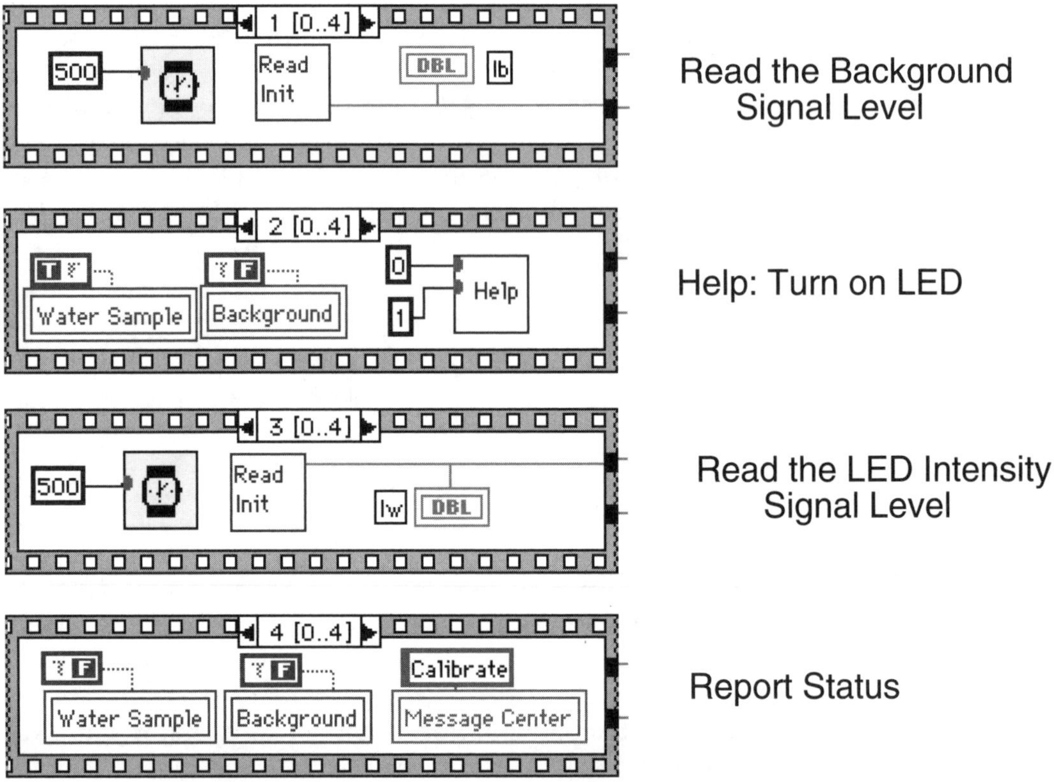

1 [0..4]
500
Read Init
DBL
Ib
Read the Background Signal Level
2 [0..4]
Water Sample
Background
0
1
Help
Help: Turn on LED
3 [0..4]
500
Read Init
Iw
DBL
Read the LED Intensity Signal Level
4 [0..4]
Water Sample
Background
Calibrate
Message Center
Report Status

uses [Read Init] to read the reference signal, I_w. In these sequences Read Init is the measurement call to the analog-to-digital converter. In simulation, the call reads from a data array, but in the real world it could be a DAQ card or a digital voltmeter read over the RS232 or IEEE 488 instrument bus. Finally, frame 4 resets the front panel LED used to indicate the reference measurement and reports the status to the Message Center.

A sequence structure was chosen here to force the measurement protocol on the operator.

There are two outputs, one for the background measurement, I_b, and the other for the reference measurement, I_w. After frame 4 is completed, the actual intensity I_0 is calculated from $(I_w - I_b)$.

Calibration Procedure

In the calibration phase, an array initially zeroed is filled with the measured signals from the five calibration samples. A For . . . Loop calls the measurement sequence five times. The Loop index is used to signal the Help.vi which sub-case message is appropriate for display and which LED indicator on the front panel is to be set on.

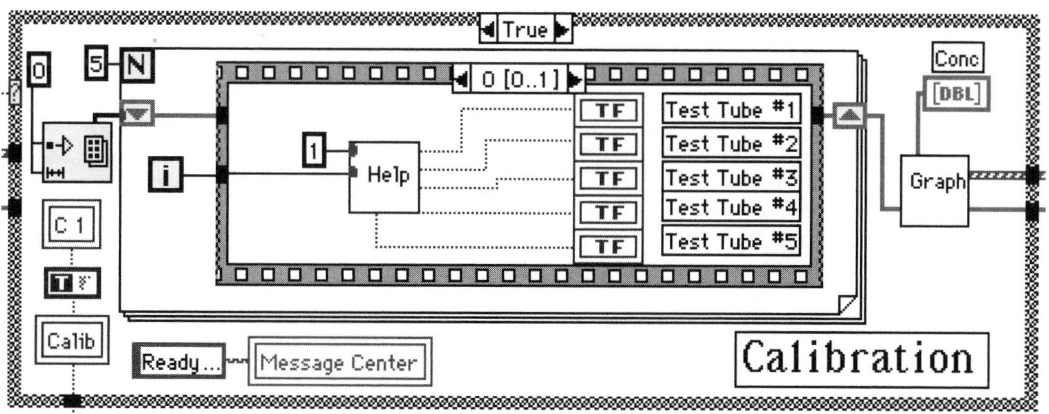

Frame 1 of the sequence does the real work. [Read Calib.] measures the calibration sample using the same data acquisition technique used in the Initialization phase. [Math] calculates the

Y–value $\ln[(I - I_b)/I_0]$ for the Beer's Law plot and appends it to the data array. Again the Wait function is only added to simulate the acquisition time.

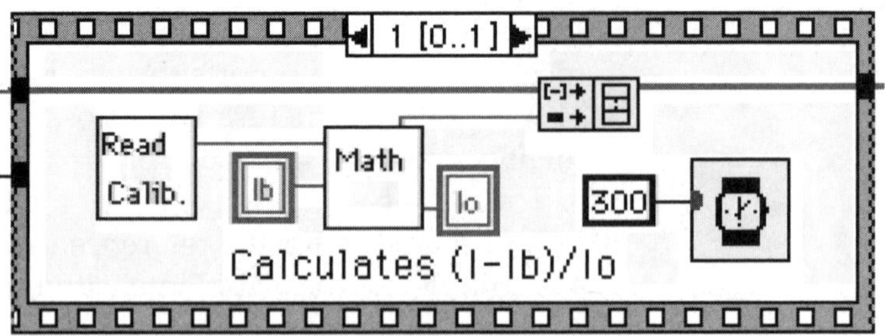

At the end of the five calibration measurement cycles, the data array is passed to [Graph] (diagram previous page) where the Y data array and the X data array (sample concentrations) are passed to the Curve Fit function.

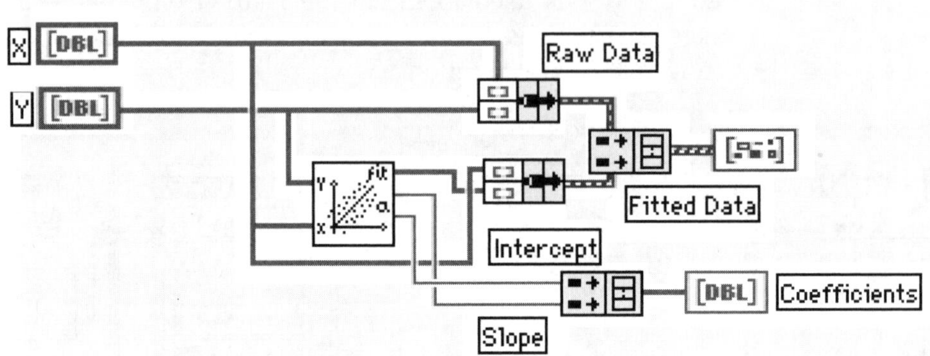

By default **[Curve Fit]** fits a straight line using a least squared fit algorithm to the Y and X 1D data arrays. The actual data sets (X_i, Y_i) are bundled together as one data set in a 2D cluster path. The fitted values (X_i, Y_i) are also bundled together as another 2D cluster path. These two cluster paths are in turn built together into a multipoint data cluster for the XY plot. When plotted on the front panel graph, the original data will show up as discrete points and the best fit to a straight line shows up as a solid line. The cluster output is then passed to the cluster shift register.

Calculation of the Unknown Concentration

When the calculation of the unknown concentration phase is entered, a two-button dialog box announces that the unknown sample is to be placed into the sample chamber and to press the [OK] button when ready. If not ready, the cancel button can be pressed to return the operation to the main program. This might be used if the operator was not ready with the unknown or realized a mistake had been made in entering the concentration values. Recalibration is in order.

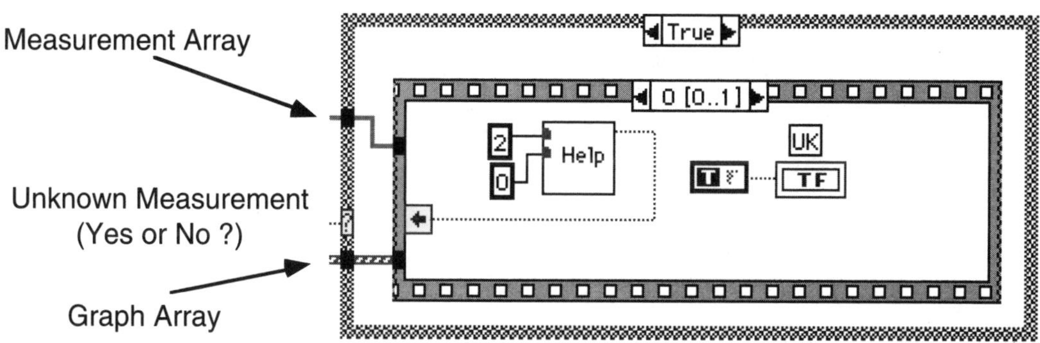

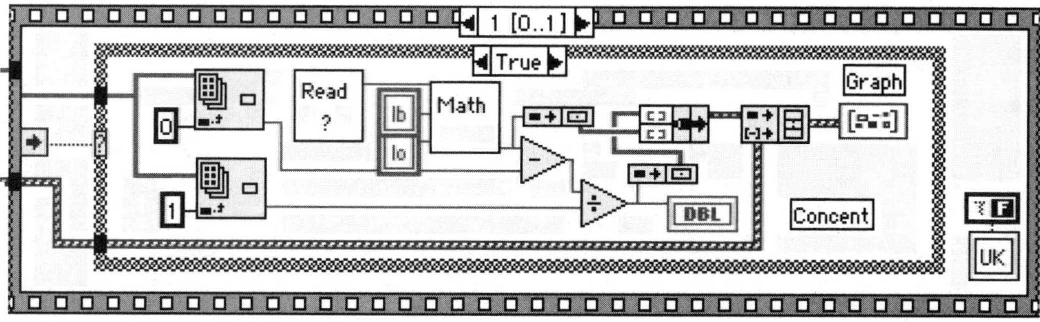

Again the real work occurs in Frame 1. Here [Read ?] is called to measure the unknown sample intensity. [Math] calculates the parameter value $\ln[(I? - I_b)/I_0]$, which is used with the fitted constants from the least square fit (intercept and slope) to calculate the concentration. The Y parameter and concentration are then bundled and added as a third data set to the cluster path. On passing to the Graph, the three data sets are plotted: the calibrated data, the fitted data, and the calculated unknown. The operator can easily tell from the graph whether Beer's Law is followed and if the calculated concentration is reasonable. The actual value of the measured concentration is given as a digital number on the front panel.

LabVIEW Challenge: Add Error Calculation to the Analysis

It would be interesting to calculate the error in determining the concentration and present it on the front panel along side the measured concentration. This can be estimated from the error in the fitted line to the calibration values. This calculation is left as a challenge for the reader.

Overview

A common form of probing a surface is to scan over the surface in a raster scan, recording the measurements on the fly. Each measurement has three parameters: the X position, the Y position, and the magnitude of the measurement. Two-dimensional arrays are a common data type used for these measurements, since the X and Y coordinates can be scaled to be the array indexes and its contents the measurement. LabVIEW has a wealth of array functions to manipulate data sets—one of the most interesting functions is the intensity graph. This chapter simulates flying over a surface and measuring the magnetic field at equally spaced intervals hunting for magnetic field anomalies.

Goals

- Design a raster scan algorithm
- Design a data acquisition VI using a DAQ card
- Design line and area scan programs using DAC outputs
- Use local and global variables
- Build a scaleable raster scan algorithm
- Use intensity graphs

Key Terms

- Raster scanning
- DAQ cards
- DAC outputs
- Analog inputs
- More global and local variables
- Intensity graphs

Hunt for Red October

13

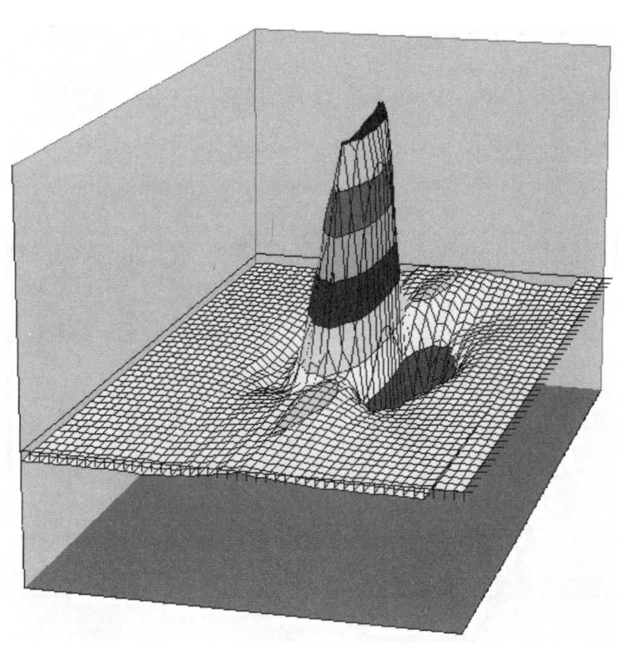

3D Rendering of a Magnetic Field Anomaly Scanned by a LabVIEW VI "Hunt for Red October"

Flying 500 feet above the icy cold waters of the North Atlantic is exhilarating at any time, but it is especially harrowing in early spring, when the skies are gray, the fog is gray, and the sea is gray. Trailing behind the lumbering Orion at the end of 800 feet of cable is MAD, the magnetic anomaly detector. It measures with great precision the precession of protons in a liquid rich with hydrogen atoms. The nuclei act as small magnetic dipoles and normally align with the earth's magnetic field. A coil wrapped around the cylinder of liquid generates a sudden magnetic field 100 times larger than the earth's magnetic field and in some other direction. In response to the impulse field, the protons align along the new direction. When the applied field is suddenly removed, the protons return to the original alignment by spiralling or precessing about the earth's field direction with a period of about 0.5 ms. It takes 1 to 3 seconds to return to the original orientation. The frequency of the precession is directly proportional to the earth's magnetic field and can be measured with high accuracy. Any large iron body gliding beneath the sea distorts the earth's magnetic field. By flying a grid pattern across the skies, a magnetic field map can be built up. Any distortions show up as ripples on the map, whose center leads right to the location of that large iron body.

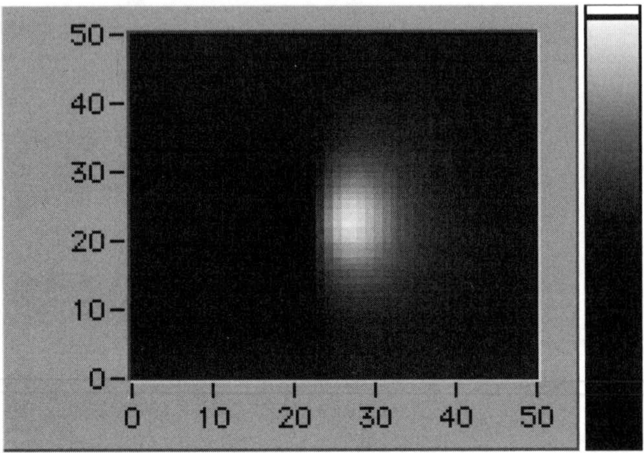

In this chapter, LabVIEW is used to map out the magnetic field distortions over a two-dimensional grid pattern. A Hall Effect sensor (Chapter 2) is used to simulate the MAD sensor, which is placed in the pen holder of an analog XY plotter. Two analog channels of a DAQ card are used to generate a raster scan pattern and simulate the aircraft flying over the ocean. In real life data is taken continuously and the exact location of each measurement is given by a GPS, Global Positioning System, signal. In simulation, the data will be collected at discrete points on the center of the measurement grid. Writing these points to a 2D array allows a magnetic field map to be built up as the aircraft covers the grid. Any anomalies will show up as colors on an intensity plot overlaid on the grid search pattern. The search area can then be focused onto a smaller area and rescanned with greater precision until the target is found.

The Search Pattern

For the benefit of discussion, the "ocean" is represented as an area 10 inches by 8 inches, fairly typical for a chart recorder. The absolute ocean coordinates are chosen to be 1000 by 800. If the "ocean" was divided into sub-areas 100 by 100, then there would be 80 elements in the "ocean" grid.

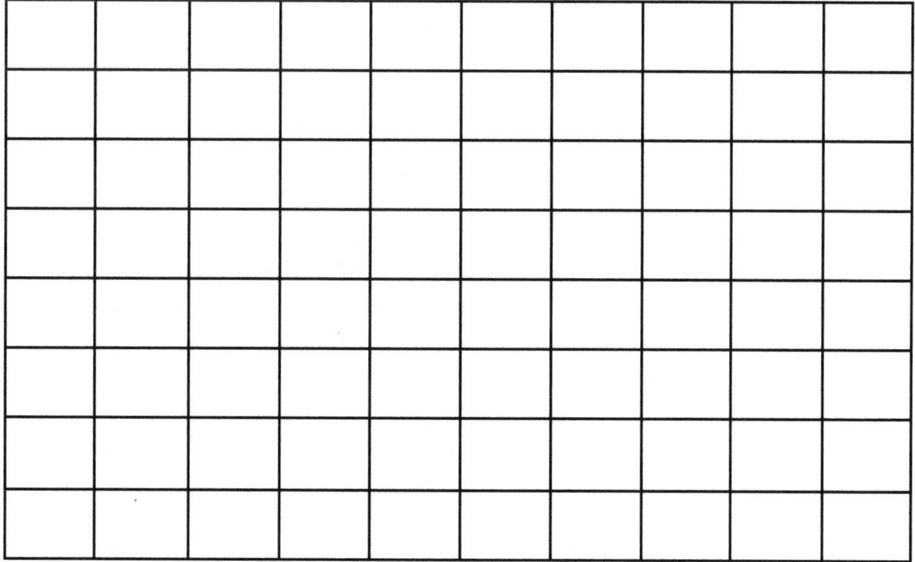

World coordinates for this exercise will be chosen as 1000 by 800. The coordinate (0,0) is chosen as the HOME position. In practice, measurements are made at the center of each grid subblock. The first measurement point would be offset by (x,y), where x and y are 1/2 of the subblock's length and width respectively. The center points of each subblock form the measurement grid. Storage of these data points will be conveniently placed in a 2D (tab delimited) spreadsheet file. Thus the data set can be quickly plotted in a variety of formats by popular 3D rendering software.

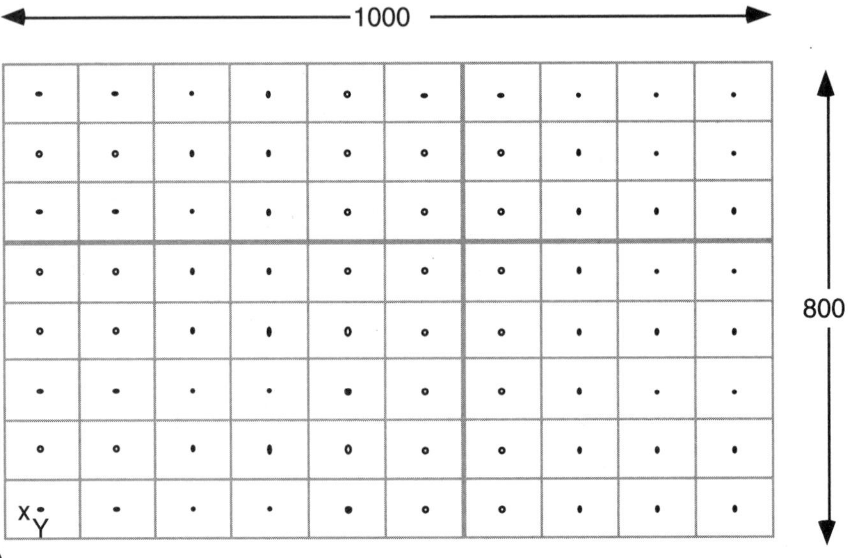

The raster flight plan is formed by flying a linear path along a specified heading x for a fixed period of time, then doing two sharp 90-degree bank turns, and returning along a new path offset in y but at a reverse x heading.

To generate this raster scan in a LabVIEW program, we rely on the high symmetry of the problem and LabVIEW's natural modularity in designing the sub-VI's. The linear scan is made up of a sub-VI to translate the aircraft at distance 2x. At the beginning of each module the magnetic field is sampled. Repeat-

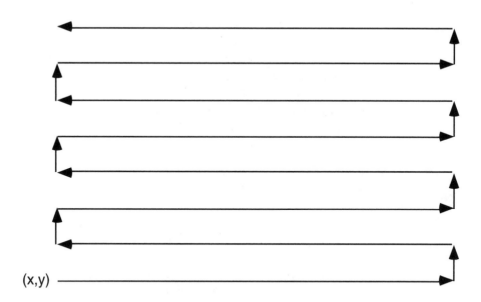

ing the x translation N times covers the first row of the measurement grid. At the end of the linear scan, a similar VI causes a translation 2y in the +y direction. As in the x translation, the measurement of magnetic field occurs at the beginning of the

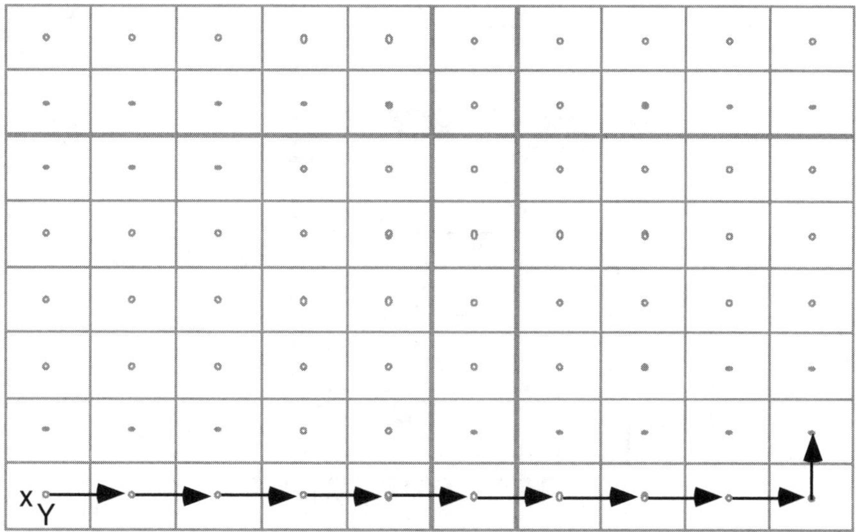

module. Now, we fly back, by repeating the x scan but in the opposite direction. Repeating the +y translation completes a cycle. These routines can be called over and over until the measurement grid is covered.

The natural VIs would consist of a Measurement.vi (○) with a Translate.vi X(→) and a similar translation for the Y direction Translate.vi(↑).

Data AcQuisition (DAQ) Card

Data acquisition, or DAQ, is simply the process of measuring a real-world signal and bringing that information inside the computer for further processing, analysis, storage, or presentation. National Instruments provides a great range of DAQ cards that plug directly into the computer bus, be it PC/XT, AT, EISA,

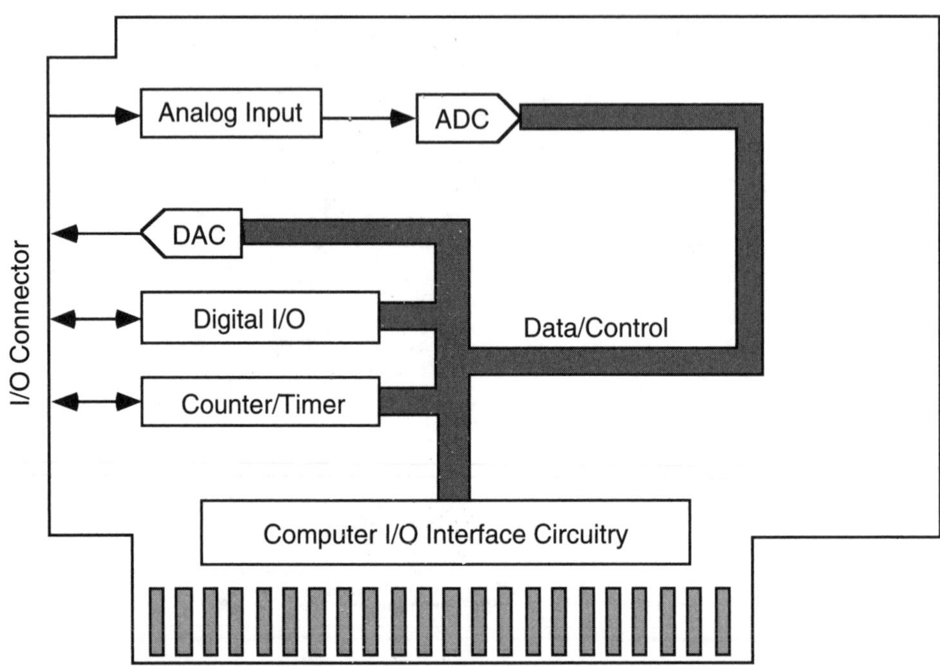

PCMCIA, NuBus, Micro Channel, or Sbus. With these cards not only are data acquisition functions available, but also the real world can be stimulated with digital or analog output signals. In a typical general purpose DAQ card, one might have available analog-to-digital ADC inputs, digital-to-analog DAC outputs, digital logic levels for both input and output, and some counter/timer functions. With such a variety, many common data acquisition needs can be met with just one interface card.

In this application, three analog channels are used: one to read the MAD detector, and two to drive the x and y axes of the analog chart recorder. In addition, if you wish to simulate the low flying passes executed when the aircraft skims across the ocean surface, a digital signal for pen up/down could also be used.

Increment Translation VI

All DAQ functions in this application are executed independently of each other, thus the basic DAQ VIs can be used. A translation is accomplished by outputting a voltage from one of the DAC channels connected to the x-axis or y-axis input terminal of a chart recorder. The driver VI found under the **Functions>Data Acquisition>Analog Output** subpalette called **AO Update Channel.vi** completes the output task. Double-clicking on this VI reveals the following driver.

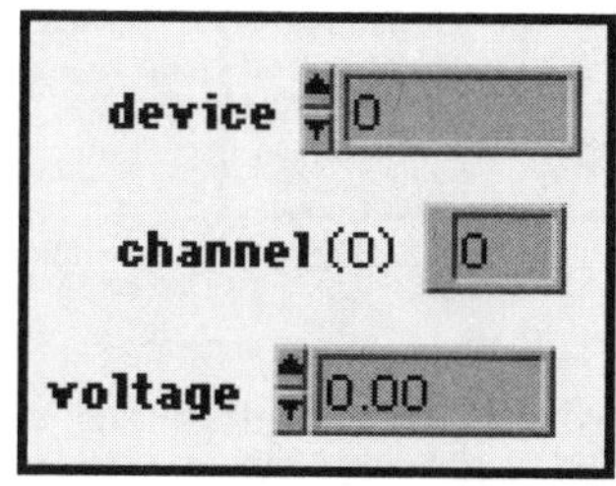

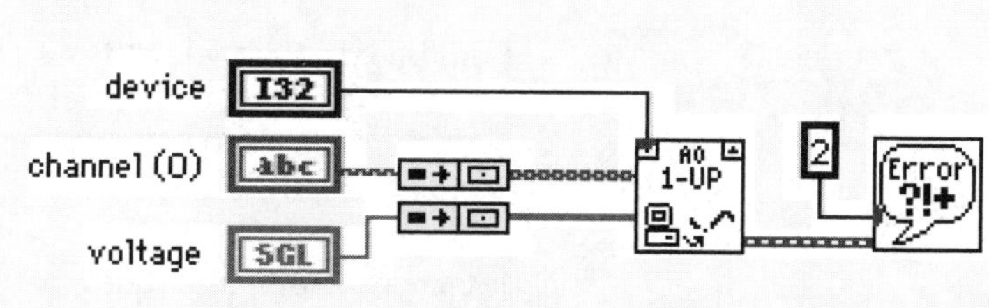

Note how the single element inputs are converted into arrays using the **Build Array** icon to link with the VI [A0 1-UP].

When called, the above VI transfers a numeric into a voltage and places it on a specific output pin on the DAQ card connector. The numeric (voltage) must be within the allowed limits of the digital-to-analog converter. For example, if the chosen DAQ card supports 0–10V outputs, then the range is 0 to 10 volts. Any value outside of this range will be stuck to the closest limit. Other higher level DAQ VIs allow the limits to be set within the VI. In this application the range of values runs from 0–8V on the y-axis to 0–10V on the x-axis. Two other inputs are necessary, the device number (the slot where the DAQ card is placed on the computer bus) and the channel number. In this program channel 0 corresponds to the x-axis and channel 1 corresponds to y-axis. For this project, the AO Update Channel.vi has been renamed Inc Translate.vi. Each time the VI is called, a new translation results in the X or Y direction. Since the numeric input will be an increment value so too will the translation ΔT be incremental.

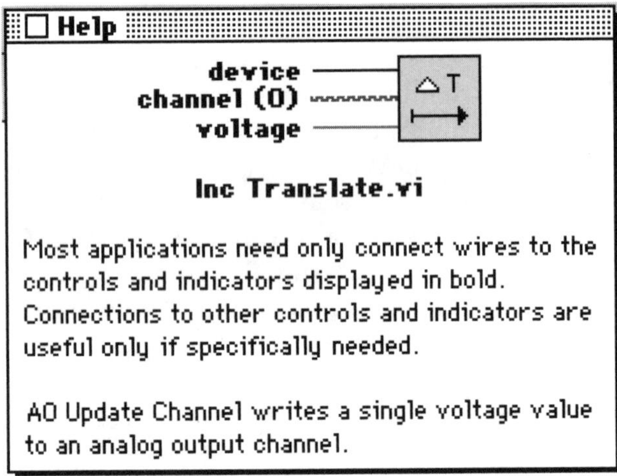

The device number and channel number can all be defined within the VI as its default value (recall the "Make Current Values Default" operation). Only one parameter (new ΔT) needs to be passed to the VI. The port value (DAC output) is latched so that the DAC output remains at its current value until a new value is sent to the port. This is why the DAC VI is correctly called AO Update Channel.

Get(V).vi

The analog input VI is almost as simple as the DAC output.

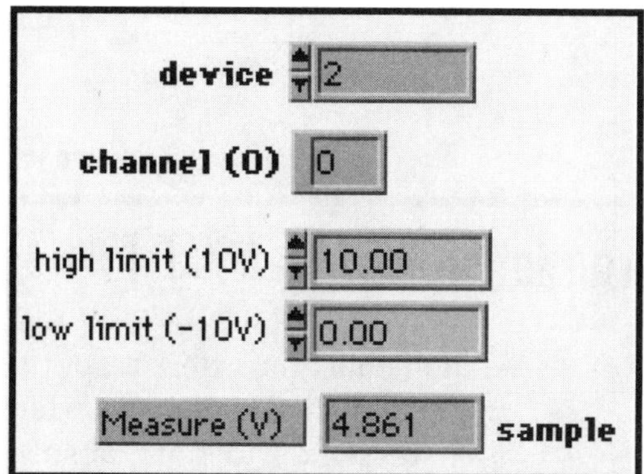

In addition to the usual inputs, device and channel number, the maximum and minimum analog levels, high and low limits are required. If you have an input signal that falls within the default limits, then no connection is necessary and the default values of (10 V) and (–10V) may be appropriate. However, most DAQ cards have additional gain settings in the analog input circuitry that can be changed by setting the limits.

$$\text{Gain} = \text{Board Input Range} / (|\text{High Limit}| - |\text{Low Limit}|)$$

For example, most DAQ cards have a input range of 20 volts. In this application the magnetic field sensor generates a signal of 3 to 7 volts. The correct gain setting would be:

$$\text{Gain} = 20 \ / \ (|7|-|3|) = 5$$

The DAQ card then rounds of the calculated gain to the nearest preset gain of 0.5, 1, 2, 5, 10, 20, 100, depending on the particular DAQ card.

The above VI, entitled **AI Sample Channel.vi**, is found in the **Functions/Data Acquisition>Analog Input** subpalette. In fact, this VI calls a lower level VI, analog input [**AI 1-Scan**], which allows scanning of several input channels and is the reason for the cluster input. The VI also has some error reporting capability.

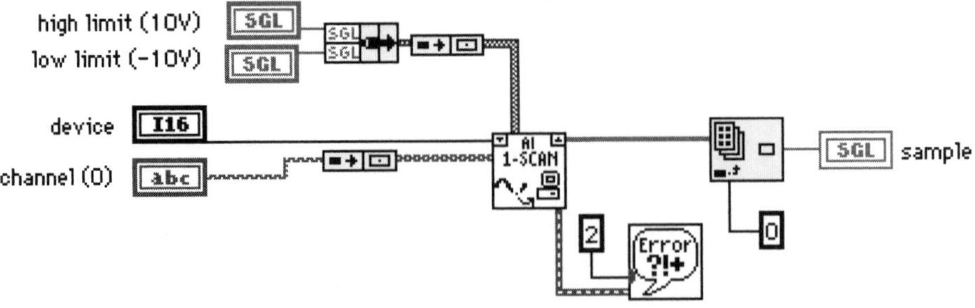

Absolute Translation VIs [X–>X'], [Y–>Y']

Analog chart recorders respond to absolute voltage coordinates, unlike the digital XY recorder discussed in Chapter 6, which often responds to both absolute and incremental commands. The relationship between the incremental and absolute coordinates is

new position = old position + increment

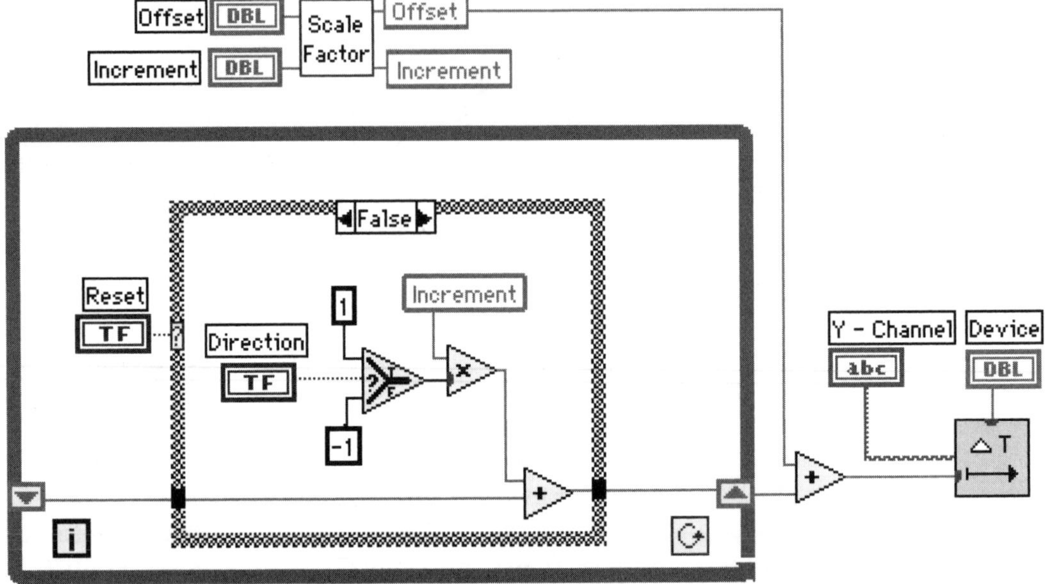

This relationship requires memory, hence a shift register will be used inside a While . . . Loop to remember the previous position. The new position, <|False|> case, is calculated by adding or subtracting the increment (+1 or −1) with the Data Select function. The <|True|> case is the reset function, which loads the initial offset position.

The new position is passed outside the loop where an additional offset can be added so that a smaller scanned area can be set from front panel controls and subsequently scanned with greater precision. The signal is then passed to the DAC output via the translate VI [**ΔT**].

The elementary operation of a magnetic field measurement [Get(V)] followed by a translation [Y→Y′] is accomplished with a two-frame sequence, called Sample.vi.

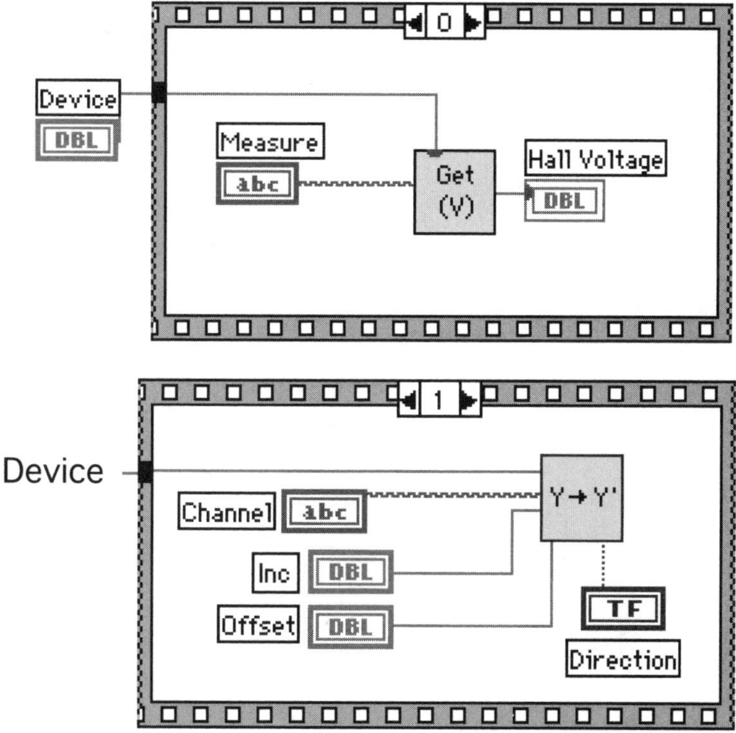

Frame 0 gets the Hall voltage measured on the [Measure] channel. Frame 1 takes the increment value [Inc] adds it to the offset [Offset], and outputs the voltage in world coordinates to the X or Y direction, [Channel] input.

Line Scan

The creation of a single line scan called Line Scan.vi results from repeating the Sample.vi ($n - 1$) times in the X direction. A subtle point here is that the last point on the line is just a measure with no x translate. However, to preserve programming symmetry, the nth measurement is combined with a Y increment translate in a VI called [ΔY + meas.]. Note how the nth data point in Frame 1 is added into the measurement array.

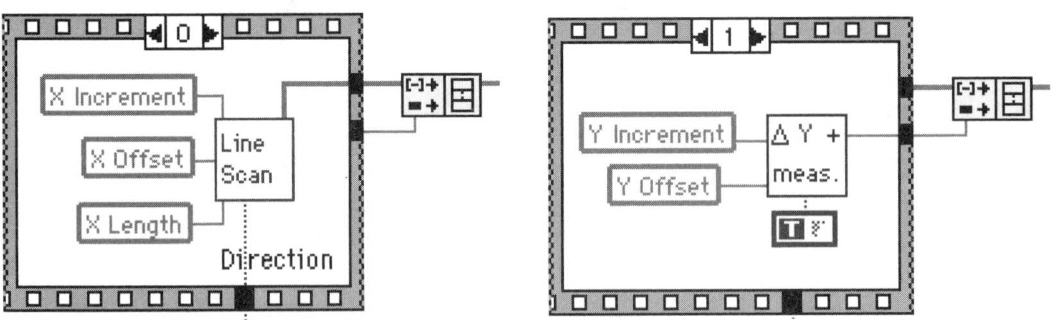

In other words, the line scan consists of flying along a (t or –) X heading, and then banking to the Y heading, measuring as we go.

Area Scan

An area scan is formed by systematically flying over all the area of interest. The natural method would be to follow a line scan and then repeat the line scan for N rows. In the following example, the computer program simulates the raster flight plan. A line scan is completed followed by a sharp bank and a return flight along the opposite direction offset by one row. Note how the data order for even rows data is reversed so that data can be passed to an array in the standard 2D array format, so that the array matches the physical area.

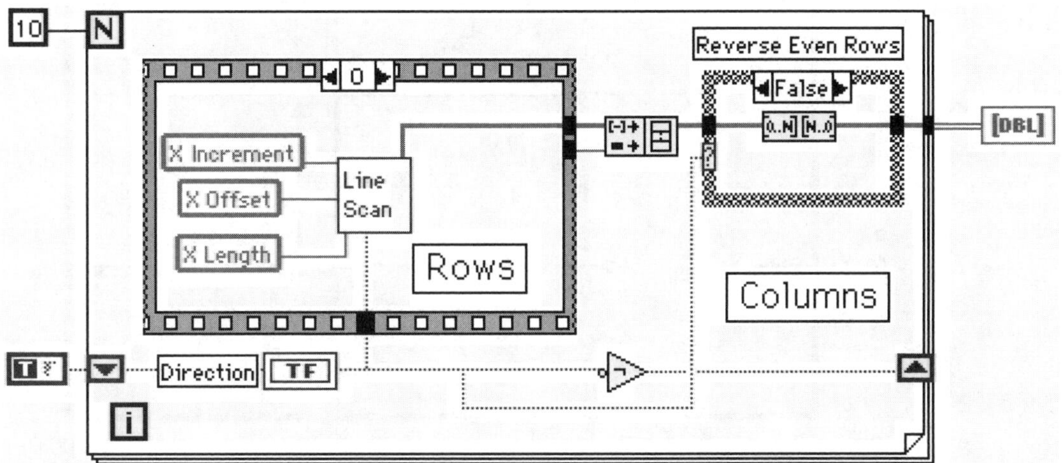

The parameters [X Increment], [X Offset], and [X Length], and similar parameters for the Y direction are passed into the sequence structure by local variable structures.

In analyzing data from a flight, the use of two For . . . Loops, one for the row number and one for the column number, is a useful technique for extracting the data from the array and producing a magnetic field map.

World Coordinates

The world coordinates have been chosen to match the size of the chart recorder platen (10 × 8 inches). For convenience, the ranges are 1000 and 800 for X and Y respectively. In many cases, a smaller area inside the world arena may be chosen for detailed scrutiny. This sub-area is framed by drawing vertical and horizontal lines at the world cordinates X_1, X_2, and Y_1, Y_2, respectively. Inside the sub-area, a new local coordinate system (x,y) is formed where

$0 < x < L_x$ and $0 < y < L_y$. It is inside this area of interest that we will fly over again, taking data at higher resolution.

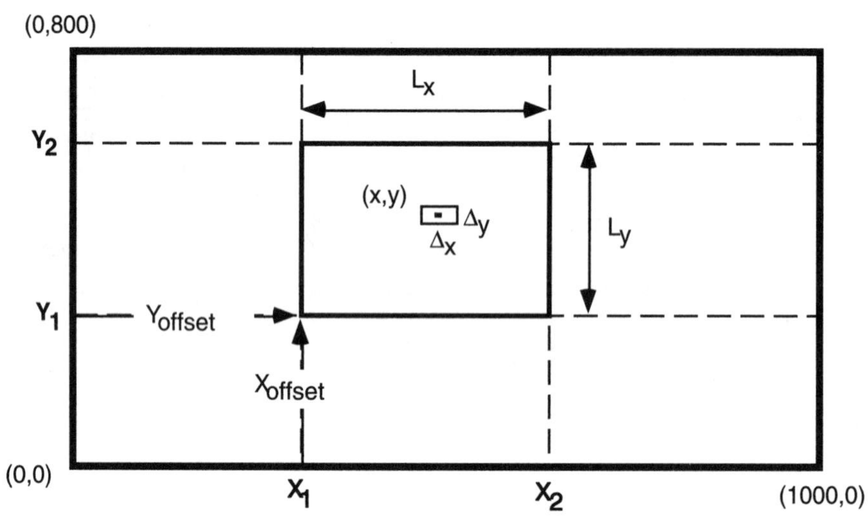

The relationship between the world coordinates (X,Y) and the local coordinates (x,y) are

$$\mathbf{X = Xoffset + x} \qquad \mathbf{Y = Yoffset + y}$$
$$\mathbf{Lx = X2 - X1} \qquad \mathbf{Ly = Y2 - Y1}$$

Inside the sub-area a new search grid is formed with $\Delta_x = L_x / N_x$ and $\Delta_y = L_y / N_y$, where N_x and N_y are the x and y local resolution factors.

The difference between world and local coordinates is important since the DAC requires world (absolute) coordinates, but

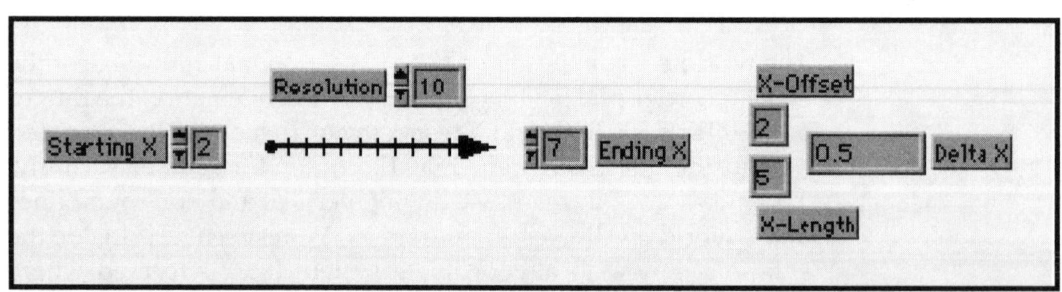

the scan pattern uses local (relative) coordinates. The user can use real-world values and with these equations let LabVIEW do all the coordinate transformations.

One way to define the sub-area is to define the starting (X_1 or Y_1) and ending (X_2 or Y_2) values of the local range in world coordinates. LabVIEW then calculates the offset, length, and increment size.

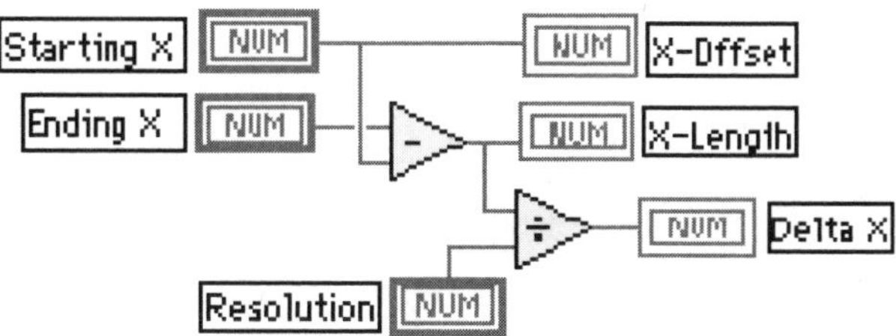

A more natural way to define the scanning sub-area is to use a colored slide control with two limits. From the controls pallete select **Numeric>Horizontal Fill Slide.** Popping up on the slide control, select **Add Slider.** Again, from the pop-up menu an appropriate **Fill** option can be selected. By grabbing hold of the sliders, the range of interest can quickly be chosen. With sliders for both X and Y attached to the graph area, a convenient and user friendly front panel display is formed.

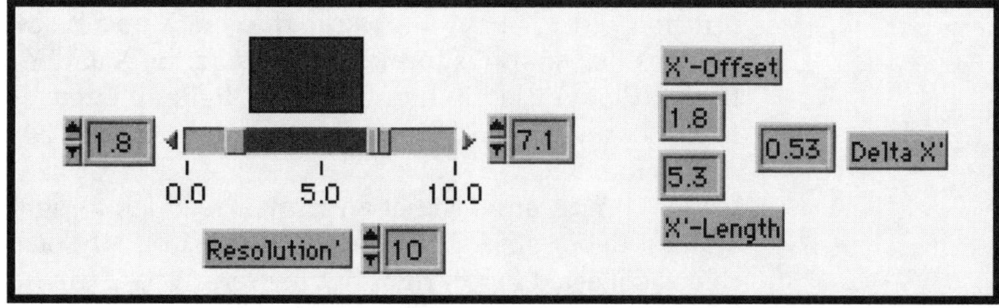

The two outputs of the slider control can be bundled into a numeric cluster, which makes the block diagram less congested.

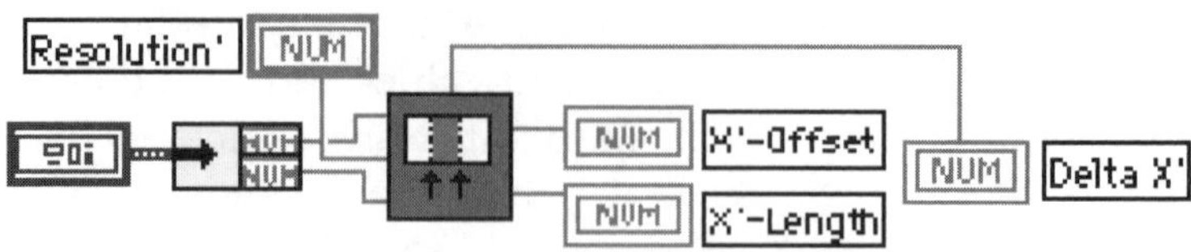

Intensity Graphs

One of the more interesting graphic presentations is a plot of the contents of a spreadsheet file as a false color map. The coordinates of the map are the row and column addresses and the contents of that cell is the intensity. Three dimensional data is presented on a 2D surface. In the flight data sets, spreadsheet data is stored in the same order as the physical grid. The contents of each cell is the magnetic field at a physical point. The resulting intensity graph is a map of the magnetic field over the scanned area.

A file called Magnetic Field.data1 contains the magnetic field data collected on a 50 by 50 grid. The following program Area Scan.vi simulates the scanning routine and displays the magnetic field data as a dynamic intensity graph. The position of the aircraft is plotted dynamically on a position graph so the reader can follow the progress of the aircraft and observe the magnetic field anomaly map as it is uncovered.

A first task is to initialize an array 50 by 50 to zero values. As the data is collected from the (flight) file, each element read is place into the display array using the **Replace Element Array** function. Two For . . . Loops provide the X and Y scan indices. The position graph is formed by passing the X and Y indices to the XY Plot.vi used in the section on digital plotters. It is scaled to a 1000 by 800 grid using the two scale factors, 20 for x-axis and 16 for y-axis.

One complication is that an array is read left to right for rows and top to bottom for columns. In the raster flight plan, the first row is collected left to right but the return pass (row) is right to left. The Reverse function shown in the Case structure on the left reverses the data order in alternate rows.

The following picture shows a snapshot of the action during a scan. The single spot indicates the position of the aircraft within

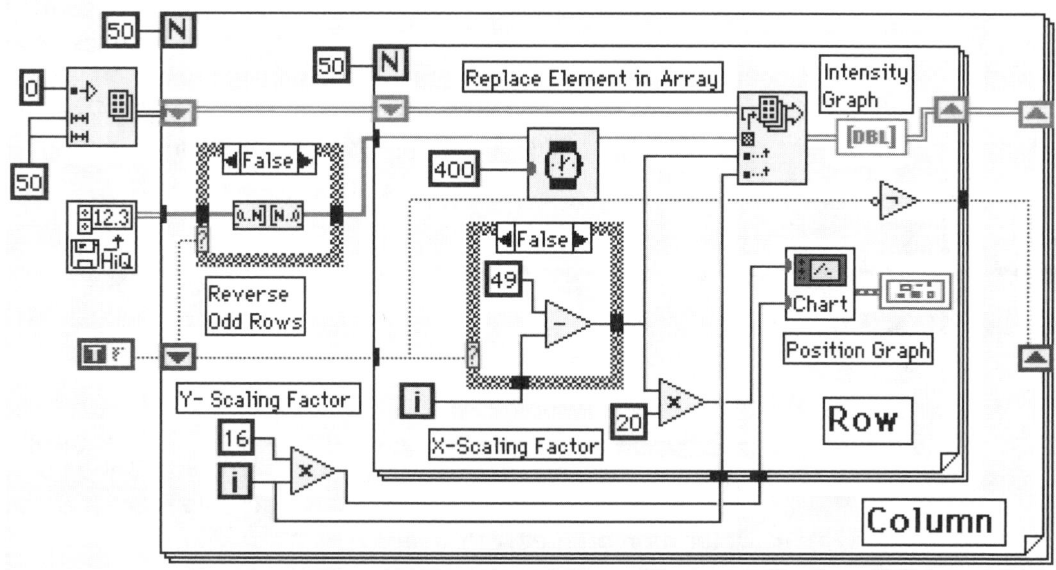

its raster flight plan. The intensity plot is gradually building up showing a large magnetic anomaly. The bright region is an area where the field is dipping strongly into the surface while the dark area is a region where the field comes out of the surface.

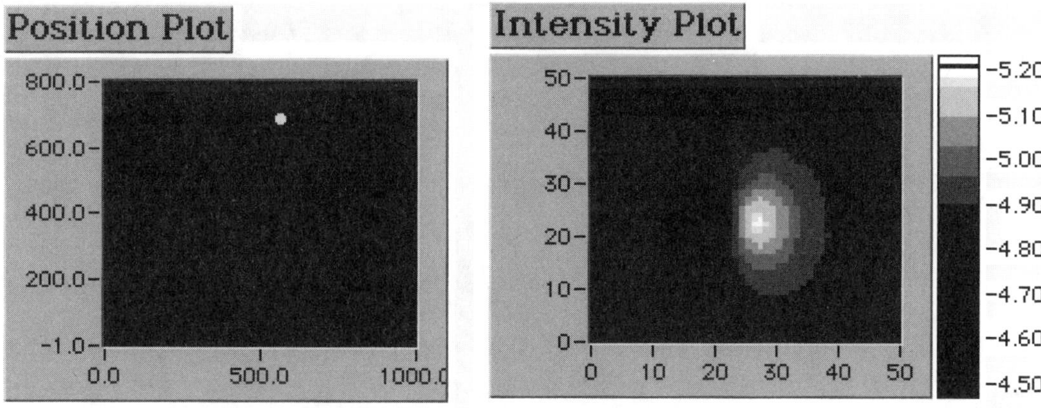

Intensity plots are useful for displaying varying data over a two-dimensional grid. The color can be depth, altitude, temperature, magnetic field, or whatever physical parameter needs to be displayed for a region.

In this demonstration, a small magnet was used as the source. You can even see the fringing field lines and thus visualize the magnet hidden beneath the surface. Use the library VI called Area Scan.vi to view a second data set called Magnetic Field.data2. It reveals a different-shaped magnetic field anomaly.

Dynamic Scanning

In the previous demonstration, the position map and the magnetic anomaly map were correlated in time but not in position. A much more natural and useful front panel would be to combine the two displays on a single display and then place it at the correct location on the world map.

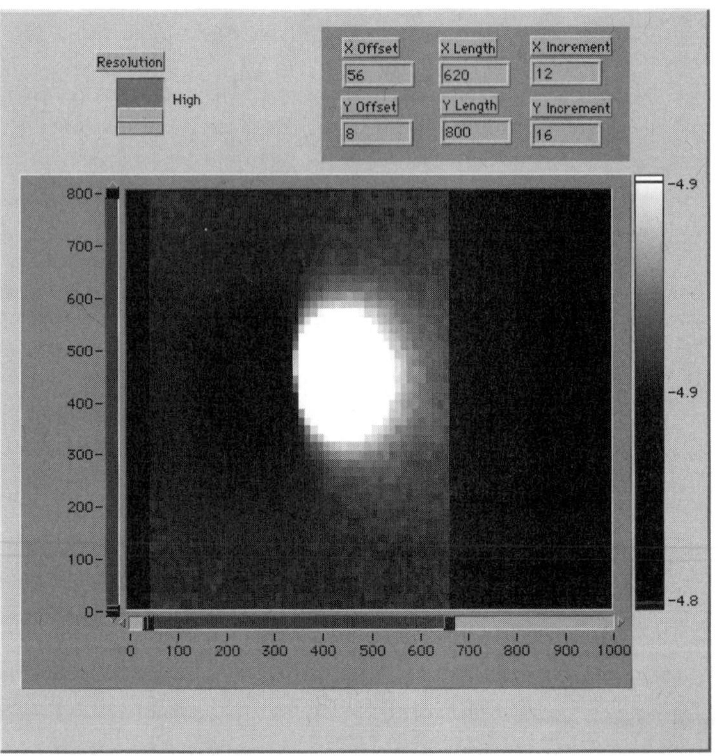

The ability to predefine the scan area allows you to select the flight plan. Two slide controls, one for the X range and the other for the Y range, allow the scan area to be selected in the form of a rectangular box. For each axis, the lower and upper limits can be selected by dragging the slider to a suitable position. A resolution switch defines the number of measurement points in a row and the number of scan lines. Large scan areas with high resolution take a long time to collect, process, and plot. In a usual hunt, a preliminary scan at low resolution is first completed. If an anomaly is found, a smaller scan area is set around its location and a second pass occurs at higher resolution. You can repeat the process until the target is found.

Scan XY Version 2.0

The XY scan version 2.0 called X-YScan.vi can be found in the VI library. Open this program and select the diagram panel shown below. Study the program carefully. Investigate each sub-VI to see how the various program elements, clusters, arrays, numerics, strings, and Booleans are woven together to form the complete tapestry. Note that the program displays the data one row of measurements at a time at the end of each line scan. The Replace element used in the last program has been replaced by the [Replace Array Row] VI. Open it up and see some more array magic.

To run this program in the real world requires a DAQ card, a XY analog plotter, and a Hall effect sensor mounted on the pen holder. If a plotter is not available, the program can still be run by connecting the analog outputs to a dual channel oscilloscope. Set the scope to run on XY mode. The data measured during the scan can be taken from the output of a function generator. Once the scan is completed and the measurements stored safely into a file, the data can be read and plotted with any conventional spreadsheet or 3D plot program. Or you can modify the area plot program AREA.vi to see an intensity plot of your data set. Choosing a sine, triangular, or pulse output can produce same very interesting 3D plots.

If you do not have a DAQ card, then try the challenge below.

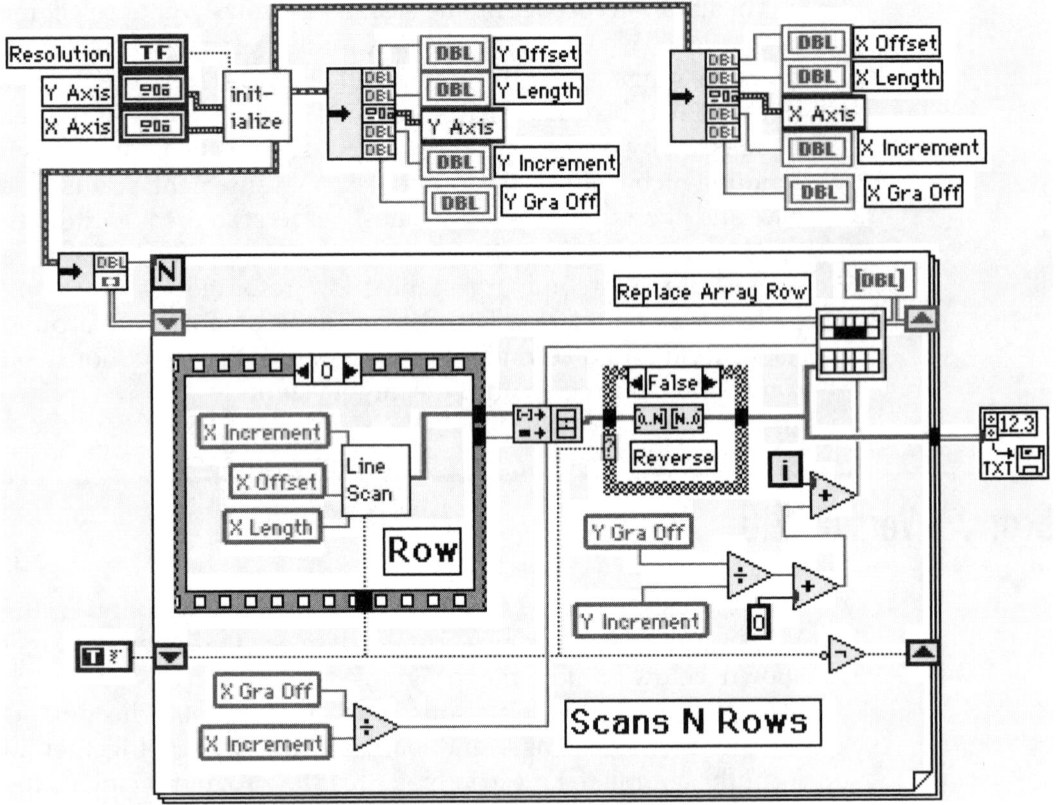

■ LabVIEW Challenge: Find Red October

The magnetic anomaly data for the mysterious mammoth is contained in a file called Red October.data formatted as a 2D data set. *Design a program to read the data and make an intensity plot to find Red October.*

Overview

Some new vehicles come equipped with a built-in electronic compasse. The instruction manual tells you to drive the vehicle in a circle while the compass gets its bearings. In this chapter, a magnetoresistance sensor similar to the sensor used in the vehicle compass is placed on a rotating surface. After the sensor is rotated through 360 degrees, the sensor has learned the direction to the north pole. Angles can now be measured with respect to north or the table can be rotated to a new bearing with an absolute accuracy of a few degrees.

Goals

- Design a stepping motor driver
- Simulate angular position with a radar plot
- Design an algorithm to find north
- Study noise reduction strategies
- Design a real-time compass application

Key Terms

- Magnetoresistance
- Bridge sensor
- MR sensor
- Stepping motor

Electronic Compass

14

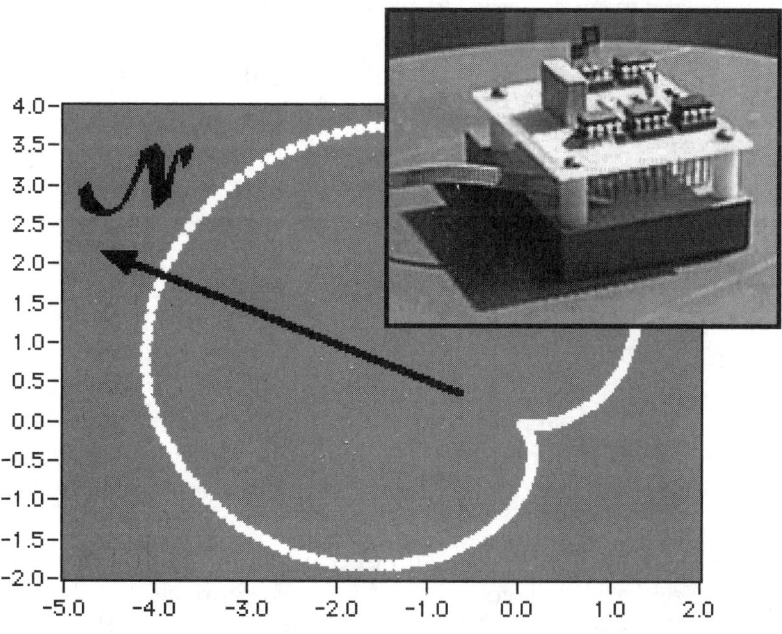

Finding North from the Output of a Magnetoresistance Sensor Rotated through a 360-degree Angle

Deep in the bowels of the earth below the crust and mantle, giant electric currents flow in the outer molten core around the girth of a solid inner core. Moving charges generate currents, which in turn produce the earth's magnet field. Magnetic field lines flow from the southern hemisphere and plunge back into the northern hemisphere in a plane that passes through the magnetic north pole. The component of these magnetic field lines on the surface of the earth points directly towards the North Pole, providing a "fixed" reference point to measure angles. The component of the magnetic field line perpendicular to the earth's surface at the magnetic equator is zero. As one moves northward, the angle increases until, at the magnetic north pole, it is 90 degrees, pointing along the earth's magnetic axis. The earth spins about the imaginary North and South Poles called the geographic poles. The magnetic North Pole, located just north of Hudson Bay in Canada, is about 1300 miles from the geographic North Pole. Over time, the magnetic pole wanders and has even reversed its direction several times over the last million years. However, over the lifetime of this project it can be considered fixed. In Chapter 2, the Hall probe was introduced as a sensor to measure magnetic fields. In this chapter an even more sensitive sensor will be used to detect the earth's magnetic field and become the heart of the electronic compass.

Magnetoresistance Sensor

Magnetoresistance (MR) sensors are extremely sensitive to the application of an external magnetic field, often with a sensitivity 50 times smaller than the earth's magnetic field. With a simple electronic interface, the magnetoresistance signal can be converted into a voltage level in the range of most analog-to-digital converters.

Consider a long film of conducting material placed in a magnetic field. Two voltage leads are connected at the end points of this conductor. The magnetoresitance is just the measured voltage, V between these two points divided by the current, I flowing through the sensor.

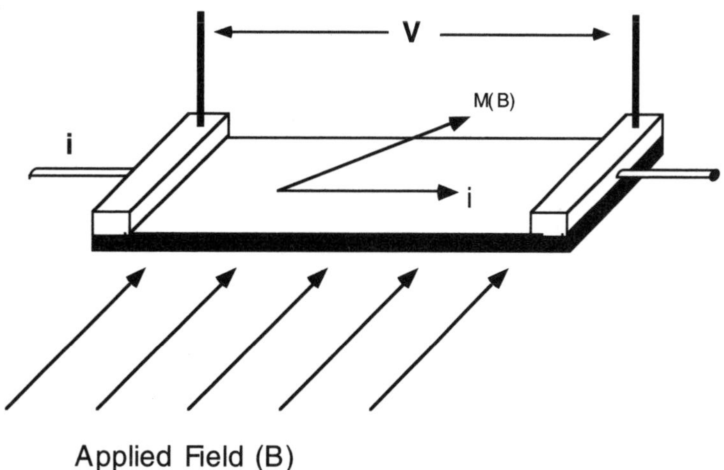

Special alloys of nickel and iron can be prepared so that the magnetic moments inside the film are aligned along the direction of the current flow. When no field is applied, the electrons are deflected or slowed down by the internal magnetization M(0) inside the film. This means that the resistance to current flow is larger than in a normal metal without magnetic moments. When an external field is applied in a direction perpendicular to the current flow, the internal magnetization M(H) rotates towards the applied field. The larger the field, the larger the angle of rotation between the current direction and the magnetization vector. Since the magnetization does not block current flow as strongly in this orientation, the resistance is smaller.

The magnetoresitance is often displayed as a fraction—that is, the change in the sensor resistance when it is placed in a magnetic field divided by the resistance with no field present. In an external field of 100 Gauss, the resistance may change by up to 3 percent.

Permalloy, an alloy of nickel and iron, is the most common material used in MR sensors. It has a relatively high magnetore-

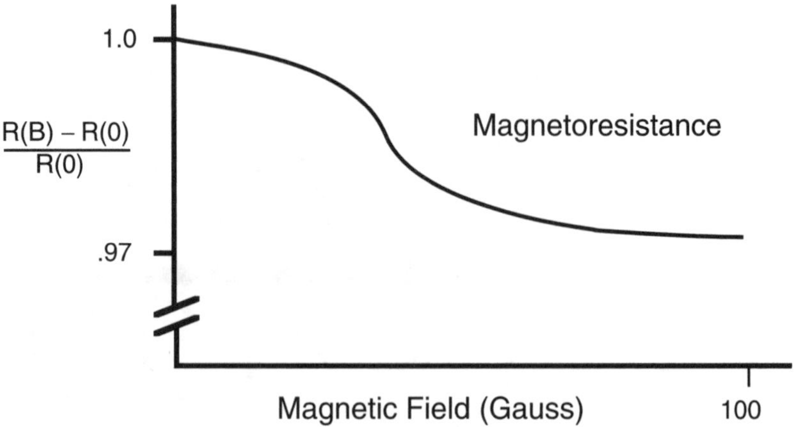

sistance coefficient and can be combined with tradition silicon technology to fabricate miniature magnetic field sensors. An integrated sensor normally consists of four permalloy magnetoresistors arranged in a bridge configuration. A constant exciting voltage, V_b, is placed across the bridge. In the absence of any magnetic field the bridge output is zero. Any magnetic field causes a bridge imbalance and is signaled by a voltage level across the output pins.

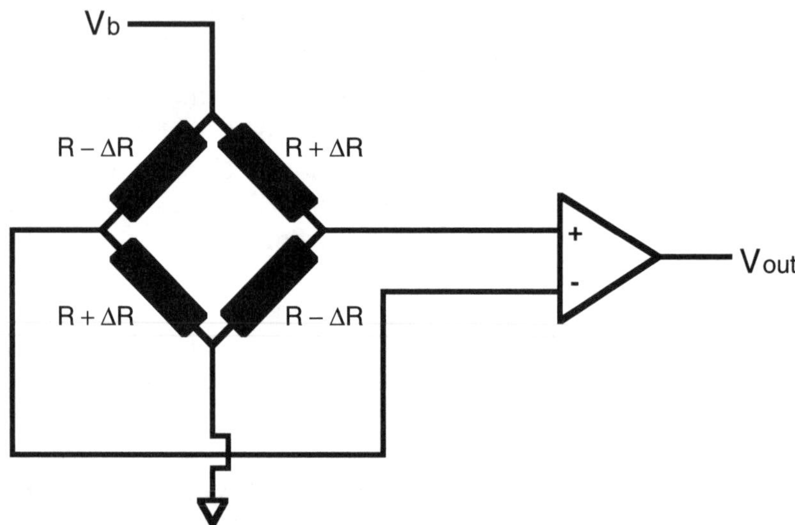

In some devices the MR bridge is integrated with the interface electronics so that the output voltage levels are compatible with most analog-to-digital signal levels. In addition, the bridge sensor is temperature-compensated to first order.

The Philips KMZ10A1 is an extremely sensitive magnetic field sensor employing the magnetoresistance effect of a thin permalloy film. It uses a four-element bridge arrangement and requires an external electronics interface. The sensor package is only a few millimeters on a side.

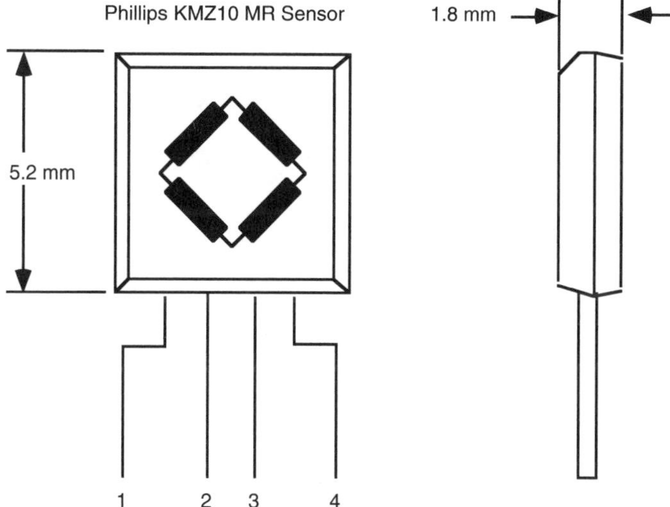

The following circuit uses three operational amplifiers and a constant voltage source to provide the compass sensor package. Aligned in a direction towards the North Pole, the signal has a maximum signal level of about 6 volts and in the opposite direction a minimum level of about 1 volt. This output range falls in the convenient 0 to 10 volt range of most analog-to-digital converters.

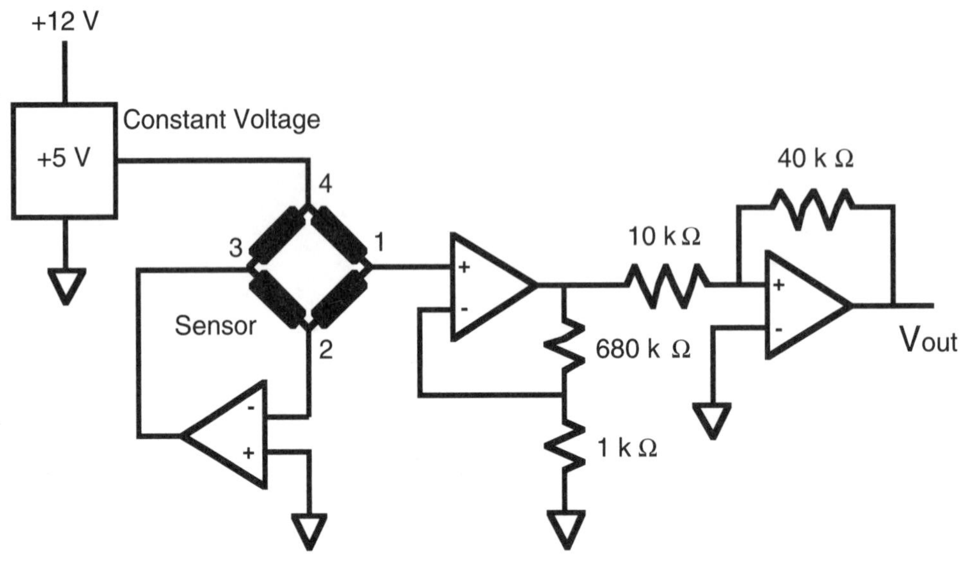

While any digital voltmeter interfaced to LabVIEW over the RS232C or IEEE 488 interface bus could be used to measure the magnetic field, in this chapter LabVIEW's DAQ card is featured.

MR Compass Sensor

In a compass application, the MR compass sensor would be placed on a rotating surface in a plane parallel to the earth's surface. Think of the sensor as attached to a car with its current axis aligned with the direction of motion. As the car turns, the angle between the front of the car and north will change. The MR compass sensor will return a voltage directly proportional to the component of the earth's magnetic field in the direction of the car. The output of the magnetoresistance sensor is very similar to a Hall probe. In the search mode, where the MR sensor is rotated continuously in the plane of the earth's surface, the sensor output follows a sine function. In a simulation VI, each time the data generator MRData.vi is called, it returns a MR measurement 1.8 degrees beyond the last call. Calling this VI 200 times would result in one complete revolution. A maximum signal corresponds to the North Pole and a minimum signal to the South Pole.

Exercise: Design a simple program, GraphMR.vi using the MR data set to display the magnetoresistance output as the sensor is rotated. In this simulation, one is observing a real data set taken using a Philips MR sensor.

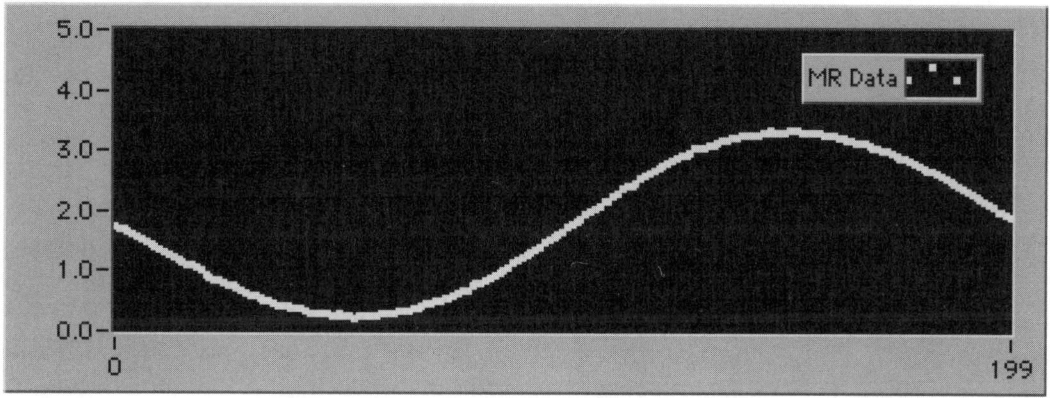

It is clear from the data set that the MR sensor produces a sinusoidal response with a DC offset about 1.7 volts and an amplitude about 1.5 volts. How could you prove that the data set is sinusodial?

Get Your Bearings!

To orient the sensor with the North or South Pole, it is necessary to rotate a real sensor in the plane of the earth's surface through at least one complete revolution. This is accomplished by mounting the sensor on the rotor of a stepping motor and then rotating the sensor while observing the MR signal. A maximum signal corresponds to North and a minimum signal to South. By keeping track of the number of steps it takes to reach the maximum or minimum signal level, the program can learn the North or South Pole directions. Once the program has "gotten its bearings," then other directions can be calculated and the sensor moved to point in a specific direction. The four-phase stepping motor interface discussed in an earlier chapter could be used to rotate the MR sensor, but in this chapter another popular stepping motor interface is featured.

Stepping Motor VI's

Numerous manufacturers make stepping motor chip sets or interface cards that include all the necessary logic to generate the four-phase signals used to drive a stepping motor. Only two input lines are required to control each motor. One line defines the direction of rotation, (clockwise or counterclockwise) and the other line generates a step command. Each time this line is asserted, the stepping motor rotates one step in the direction selected by the state of the other line. The resolution or the step size depends on the stepping motor design and the method of driving the motor phases. In the full-step mode, most motors have 200 steps per revolution while in the half-step mode, it takes 400 steps per revolution (0.9 degrees per step).

A LabVIEW Boolean switch is used to control a single output line on the DAQ card. The output is TTL compatible and can be connected directly to the direction input of the stepping motor controller. The LabVIEW DAQ function **Write to Digital Line.vi** found in the **Functions>Data Acquisition>Digital I/O** subpalette is used to set the direction pin on the parallel port. Connect this pin to the direction input on the stepping motor interface.

The inputs required for this digital output function are the device number (slot number where the DAQ board is installed), the port address (which 8-bit parallel port is to be used on the DAQ card), bit number (which bit is the direction line), and, of course, the direction command (clockwise or counterclockwise).

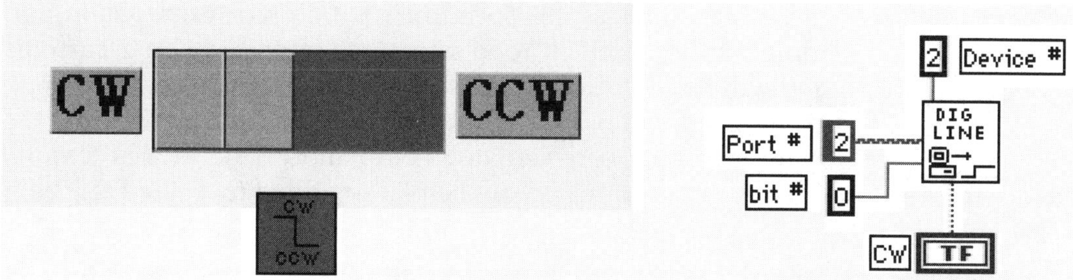

In general, there may be a setup time for the direction signal to be held constant on the interface before a step command can be issued. This is usually small (a few nanoseconds) and LabVIEW's execution time is long enough so that this delay can be ignored.

The step command is often in the form of a positive or negative pulse with the rising edge or falling edge used to trigger the step. If the output line is initialized in a low state, then a positive pulse (step command) is generated with a simple two-frame sequence. Frame 0 sets the output line high and waits for an on-time specified by the manufacturer. Frame 1 resets the output line low. It is this falling edge that causes the motion.

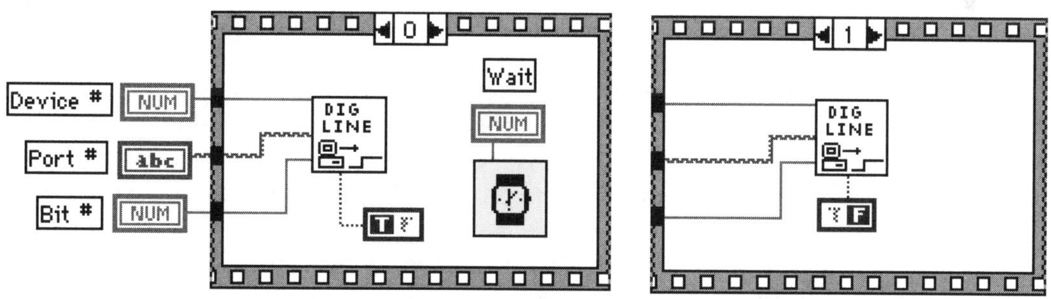

The on-time for the pulse is provided by a Wait function. Note once again that the LabVIEW DAQ function **Write to Digital Line.vi** is used to set a step line on the parallel port. Connect this pin to the step input on the stepping motor interface.

The maximum speed of rotation is determined by the frequency of step commands. Each type of stepping motor has a different maximum speed. Physical and electrical constraints in the design limit of the speed of rotation. For example, a high torque motor used in robots requires large switching currents and will have a much slower rotation rate than a small stepping motor used in a printer. The speed is controlled by defining the minimum time between step commands. To add this feature, another [Wait] function can be placed added to a new Frame 2.

Compass Driver

The direction of the earth's magnetic field is derived from the magnetoresistance output pattern as it is rotated in the plane of the earth's surface. Consider a stepping motor with a resolution of 200 steps per revolution driven slowly in the clockwise direction. At the end of each step, the earth's magnetic field is to be sampled. When a complete rotation of 360 degrees has been realized (200 steps), the motor is rotated quickly back to its home position. The return trip in the counterclockwise direction is required to unwind the sensor cable so that it does not "choke" itself. The return trip can be done at the maximum rotation rate since no measurements are taken on the way back. Four hundred steps form the search mode of the electronic compass (200 steps CW and 200 steps CCW). Events happen in a well defined sequence and a combination of Sequences, For . . . Loops, and While . . . Loops structures are used in the following VIs for the electronic compass. There are two unique operations in the acquisition of the compass data. One is the motion control of the stepper motor and the other is the measurement of the magnetoresistance sensor.

The motion controller **Step Motor.vi** uses an up/down counter to generate step commands. Each time the While . . . Loop cycles a single step is executed. A Boolean control from the front panel [Direction] defines the direction output [CW/CCW] in the lower case structure. The same Boolean defines the direction of the up/down counter and the time between steps (upper case structure). A Wait input has been added to the [Step] VI to set the maximum speed of rotation. The Case structure allows the time between steps for the clockwise and counterclockwise directions to be different. For a resolution of 200 hundred steps per revolution, each step corresponds to 1.8 degrees. The **[Up/Dw Cnt]** counter output is multiplied by this factor to convert the count into an angular position in degrees. This angle is then passed to the Polar Plot.vi, which simulates the rotation of the stepper motor on the front panel.

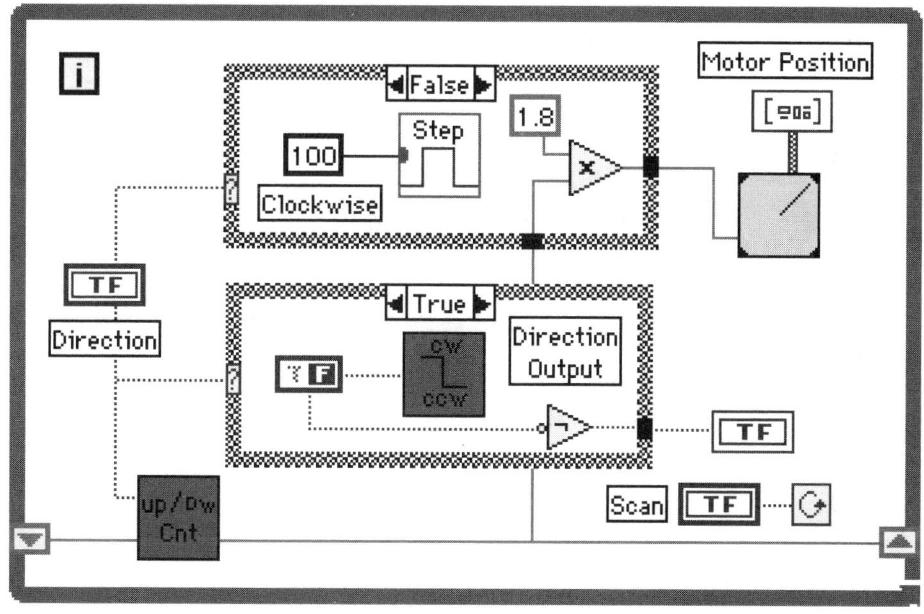

Load and run StepMotor.vi and take the motor for a spin. Observe the difference in speed of rotation of the CW and CCW directions.

The magnetoresistance signal, Read Data.vi, is acquired from a DAQ card analog input channel. Some initialization for the coupling and input configuration, for the input range limits, and for the device number and channel are required. This depends on the DAQ card used and the manufacturer's instructions should be studied.

If a DAQ card is not available, then a digital voltmeter with an RS232 or IEEE 488 interface can be used with one of the device drivers discussed in Chapters 5 and 15. If none of these are available, the data set MR Data.vi can be used. This is a real

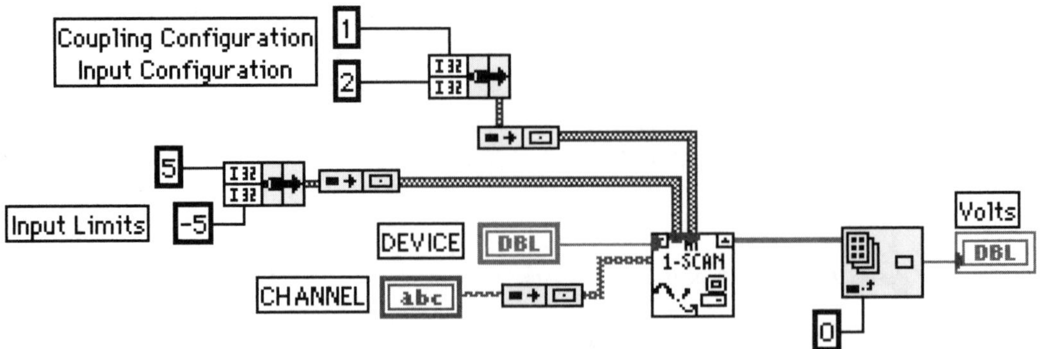

data set taken using the Phillips sensor with the above interface card and measurements taken every 1.8 degrees.

Combining the two operations—motion control and data acquisition—provides the complete compass driver.

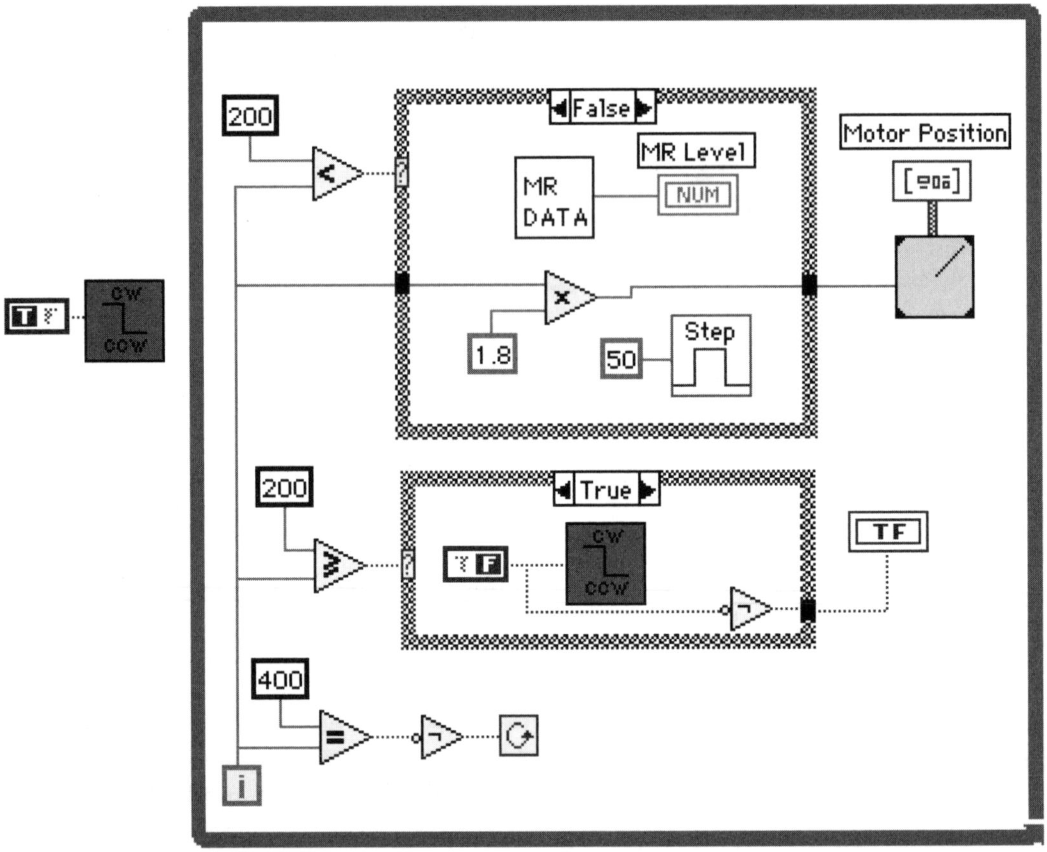

Initially the clockwise direction is selected and set on the direction line of the parallel port [CW/CCW]. Four hundred steps are required to rotate the stepping motor through one complete revolution and back to its initial position. The middle step, number 200, is used to select either the clockwise or counterclockwise action. In the clockwise direction, the magnetoresitance sensor is measured in the VI called [**MR DATA**]. Then the motor is moved one step, [**Step**]. Note that the input to Step.vi has a convenient delay time between measurements. This ensures the sensor is motionless when the magnetoresistance signal is sampled.

For counts 200 or higher, the <|True|> case of the stepping motor circuit is asserted and the direction line is set low so that the stepping motor will rotate (CCW) back to its starting position. Also, the <|True|> state of the upper Case structure is selected and no measurements are taken on the return trip. A smaller delay in [Step] provides a speedy return.

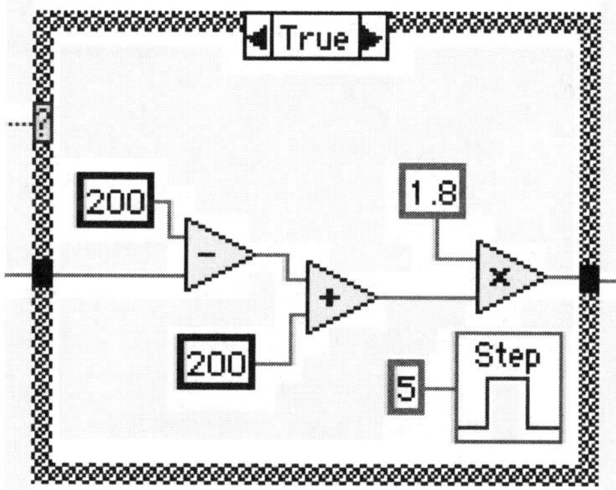

Note the order of the subtract and add functions which decrements the down counter from 360 to 0.

Load and run the simulated scan VI called SS.vi. Observe as the motor rotates in the counterclockwise direction that the magnetoresitance data is sampled on the fly. When the angle reaches 360 degrees, sampling stops and the rotor returns quickly to the initial position.

In this simulation, the magnetoresitance data is taken from an array of data points collected by a real MR sensor. The actual

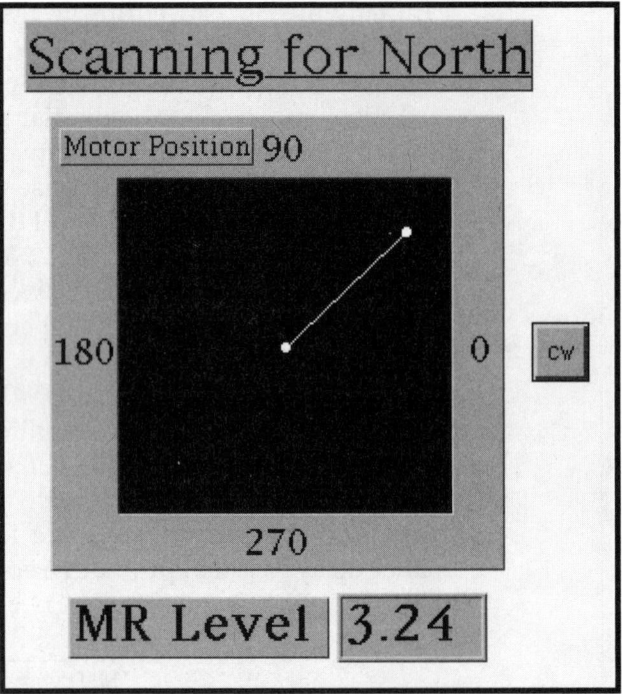

calls to a DAQ card are not present in this VI but can be found in the chapter library, called Read Data.vi.

While useful information is presented with a digital indicator for the MR measurements, the full impact of scanning can only be observed by plotting the data set over at least one full cycle. Modify the simulation so a waveform chart displays the data set as a search is conducted.

This program can also be found in the chapter library as SS(+chart).vi.

Finding North

An interesting way to look at the data set is in the form of a polar plot. Points are represented as a vector of length R and angle θ. A plot of the vector tip as the angle is varied generates a polar plot. In fact, this is the most natural plot for the electronic

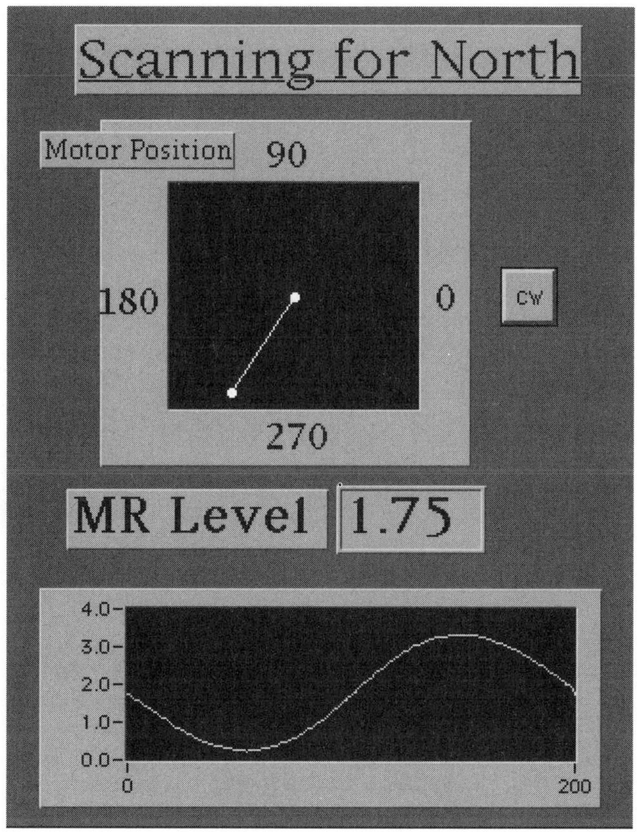

compass since the angle of rotation set by the stepping motor is the polar angle and the magnitude of the earth's magnetic field is the vector's length. The following graph is the polar plot for the MR data set.

The maximum radius from the origin (0,0) defines the north direction. The minimum radius shown here as a dimple in the polar plot is the south direction. Picking the south direction from this diagram is considerably easier than selecting the north direction, where the plot is a broad maximum. The line drawn from the dimple to the maximum radius is the best estimate for the north direction. An interesting question to ponder is "How would one design a computer program to find the North and South Pole directions automatically?"

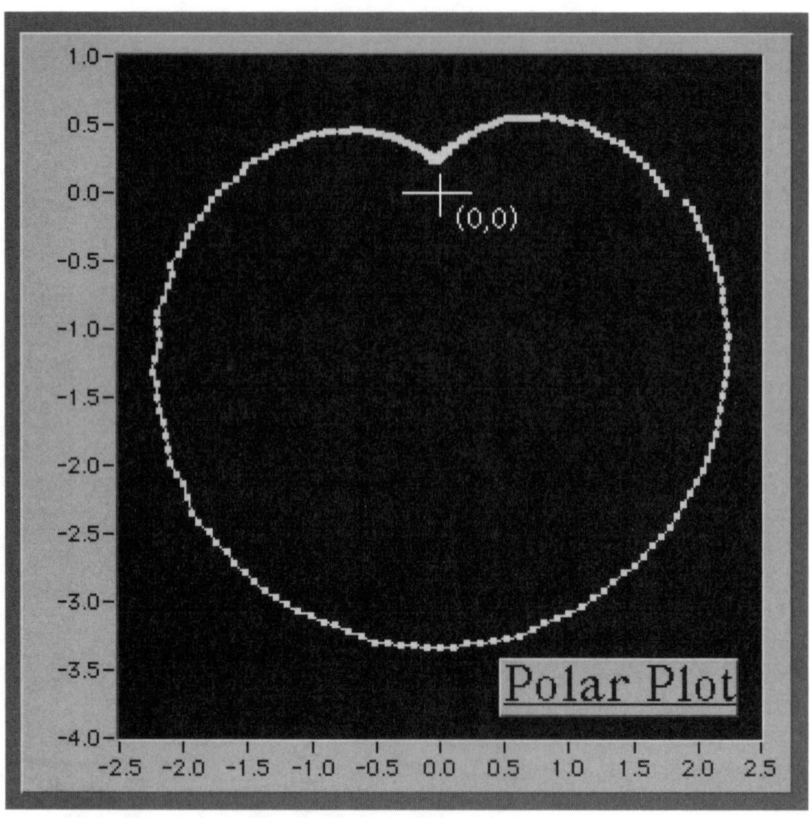

First, let's take a closer look at how the polar plot is formed. Each measured point gives an angle (θ) and a magnitude (B). The simple coordinate transformation $X = B\ \mathrm{Cos}(\theta)$ and $Y = B\ \mathrm{Sin}(\theta)$ produces the rectangular coordinates (X,Y). The sub-VI [Polar to X–Y] accomplishes this task. Note that LabVIEW's **1D Polar to Rectangular** function will also provide this transformation. The next task is to pass these rectangular coordinates on to the LabVIEW **XY Graph** function for plotting. Recall graphs require array inputs, hence an array of MR values must be built. Popping up on the boundary of the While . . . Loop in the previous example and selecting **Enable Indexing** allows the data points to be collected into an array. The first 200 points, measured field values, and angles of the counterclockwise scan are sliced off as a subset. These points are bundled together to satisfy the input requirements of the XY Graph. The following block diagram for SS.(+polarplot).vi adds the polar plot output to the previous numeric and chart outputs.

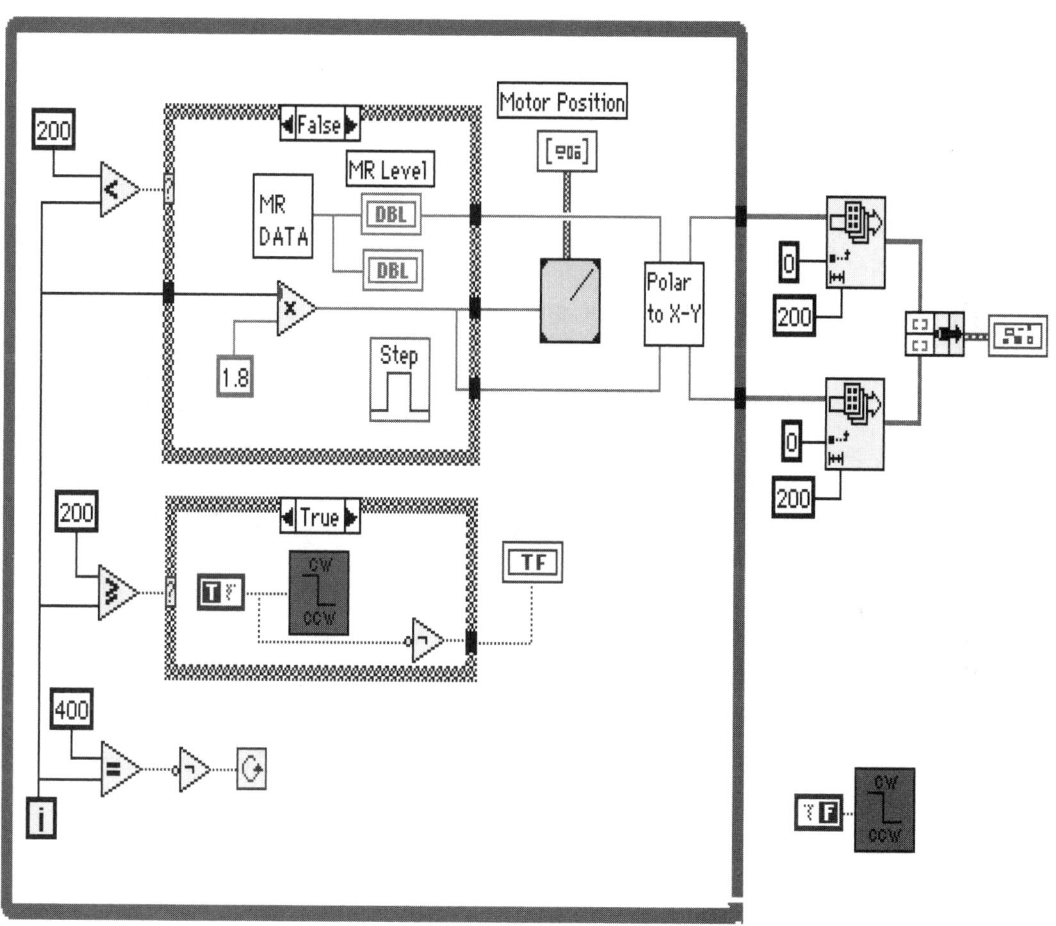

Determination of the maximum and minimum radii can be accomplished using the **Array Max & Min** function found in the **Functions>Arrays** subpalette. It operates on the 1D array of magnetic field values. The index of the maximum and minimum values is automatically calculated. Multiplication of the index number by the scaling factor (1.8 degrees per step) leads to a determination of the angles of the maximum and minimum radii. Any noise in the data set near the these limits will cause an error in determining the direction from the index found in the **Array Max & Min** function. This is especially noticeable for the maximum where the data set displays a broad maximum. Smoothing is in order before placing much credibility on the calculated direction to the North Pole.

Noise Reduction Strategy

The polar and chart plots display some variability in the data set and some form of smoothing algorithm is required. In the Barometer project, a running average over three data points was used with some success in smoothing the data set. The same strategy can be used in the current data set, provided care is taken in calculation of the average index. If five contiguous points are used, then the index of the average value is the center point index. Application of a five-point smoothing algorithm, Smooth5.vi, to the magnetoresitance data set shows a significant reduction in the experimental noise.

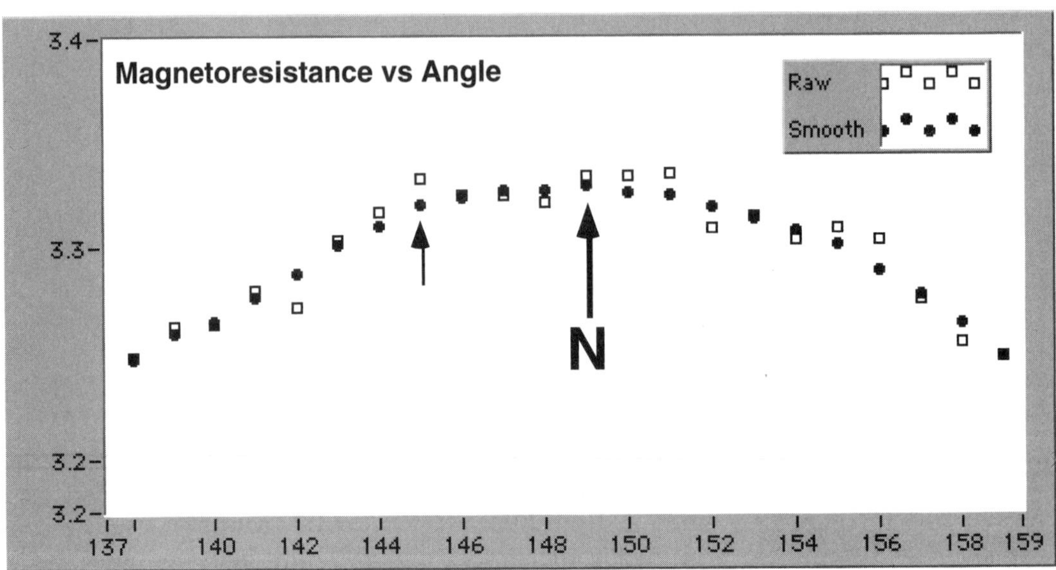

Without smoothing **Array Max & Min** would pick the position (smaller arrow) at index 145 for the best estimate of north. With smoothing, a better estimate is found at index 149, four data points to the right (large arrow), a difference of 7.2 degrees.

LabVIEW Challenge: Real-Time Compass Application

In the previous examples, the data set was collected before any analysis was attempted. In a real-time application, one would wish to see that data plotted on the polar plot as it was collected. *Modify the program SS(+polarplot).vi to collect and display data on the fly. Add a smoothing algorithm to get a better estimate for north.*

Now that the north direction has been determined, it is possible to rotate the sensor into any specified direction. Can you suggest several applications were this feature would be necessary? Can you read from a front panel control a heading and set the direction of the stepping motor to that heading?

Overview

The power of LabVIEW is that many measurements, processes, or techniques can be simulated in a very real way using well designed sub-VIs. Each sub-VI can be tested and debugged independently and then put together into a much larger program. A Sub-VI that simulates port activity can be replaced with the real port driver and the simulation becomes a virtual instrument operating in the real world. This chapter demonstrates the technique using simulated digital multimeters connected over a virtual IEEE 488 instrument network. On completion, the simulated digital voltmeters can be replaced with real digital voltmeters interfaced over a real GPIB instrument bus.

Goals

- Understand the IEEE 488 GPIB instrument bus
- Design an experiment using the GPIB bus
- Build an GPIB driver VI
- Design a VI to simulated a digital voltmeter
- Design an experiment to measure two parameters

Key Terms

- GPIB write function
- GPIB read function
- Command processor
- LED IV characteristic curve

Who Has Seen the Light?

And What Color Is It?

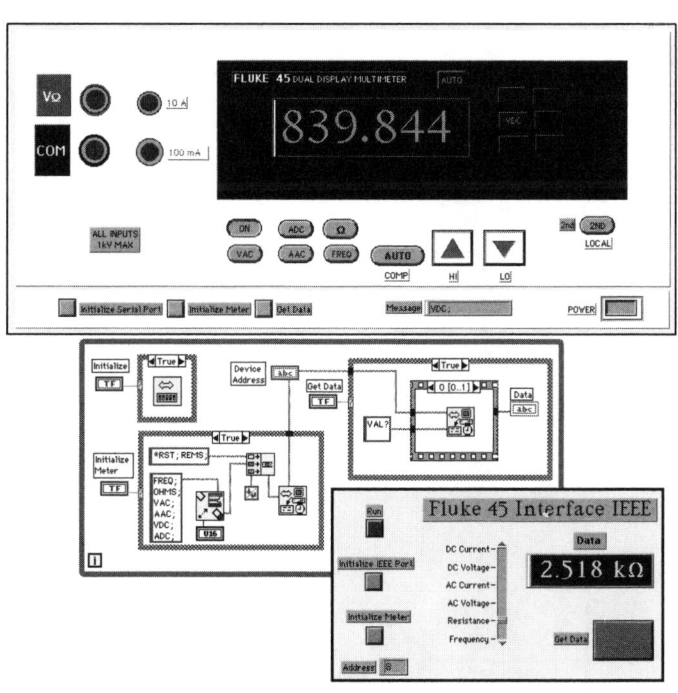

Digital Multimeter with LabVIEW GPIB Driver

Light emitting diodes (LEDs) are common components used in all forms of consumer, industrial, and instrument products. LabVIEW uses LED icons as Boolean indicators. Many shapes, sizes, and styles are available in the Boolean palette. The LED color can be set to any of hundreds or thousands of colors. But what about real light emitting diodes, what colors are available? You have seen red ones, orange ones, yellow ones, green ones, and some you cannot see, the infrared ones. They all have a common electrical characteristic curve but different optical characteristics.

Light emitting diodes are characterized by the current-to-voltage (IV) relationship. The current flowing through the device is plotted on the Y-axis and the voltage across the device is plotted on the X-axis. The resulting curve is called the IV electrical characteristic. In the forward bias region (when V is positive) the current grows exponentially. It is in this region that light visible or invisible is given off. In the reverse bias direction, the current is very small and on a milliamps scale essentially zero. Switching the driving voltage from one region to the other al-

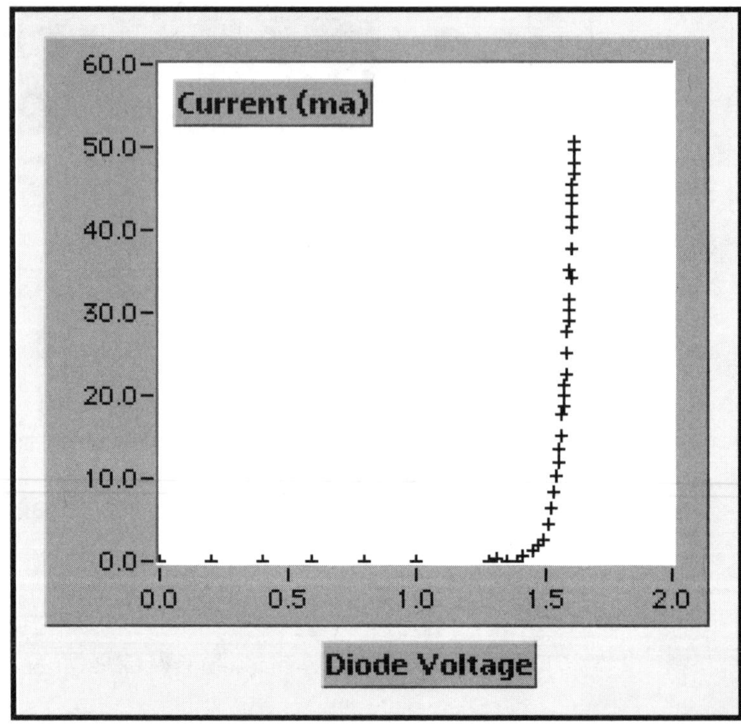

lows light pulses to be generated and hence optical communication (see IR communications, Chapter 9). But is there more information in the IV characteristic curve? Can one tell the color of the light emitting diode from the electrical characteristics?

In the forward bias region, it appears that the current is small until a "threshold" voltage is reached, beyond which the current grows rapidly. It is in this region that light comes out of the light emitting diode. Is this threshold voltage related to the color of the light emitted?

The short answer is yes. When the voltage across a forward biased diode is large enough, electrons and holes can recombine near the junction to give off light. The energy lost by the electron-hole pair recombination is related to the energy gap of the semiconductor material and that energy loss is close to the photon energy of the emitted light. From the conservation of energy:

$$E_g = h\upsilon = hc/\lambda$$

Here E_g is the energy gap of the semiconductor, h is Planck's constant, c is the speed of light, and λ is the photon wavelength. Since the photon energy $h\upsilon$ can be expressed by the wavelength or color of the photon, then the color is related to the apparent "threshold" seen in the IV characteristic curve. A simple experiment will verify this hypothesis.

IV LED Characteristics

A light emitting diode is placed in series with a resistor and a power supply. The resistor is used to limit the current in the circuit so as not to burn out the LED. The voltage across the LED and will be monitored by a digital multimeter configured as a voltmeter. The current in the circuit and hence the LED current will be monitored by a second digital multimeter configured as a current meter. The voltage is changed in steps. After each step, the reading on the digital meters is measured. To automate these measurements, the digital multimeters are connected over an IEEE 488 network to a computer with a GPIB controller. We will use LabVIEW to view the data points as they are collected and then take a snapshot of the data and plot the IV characteristic.

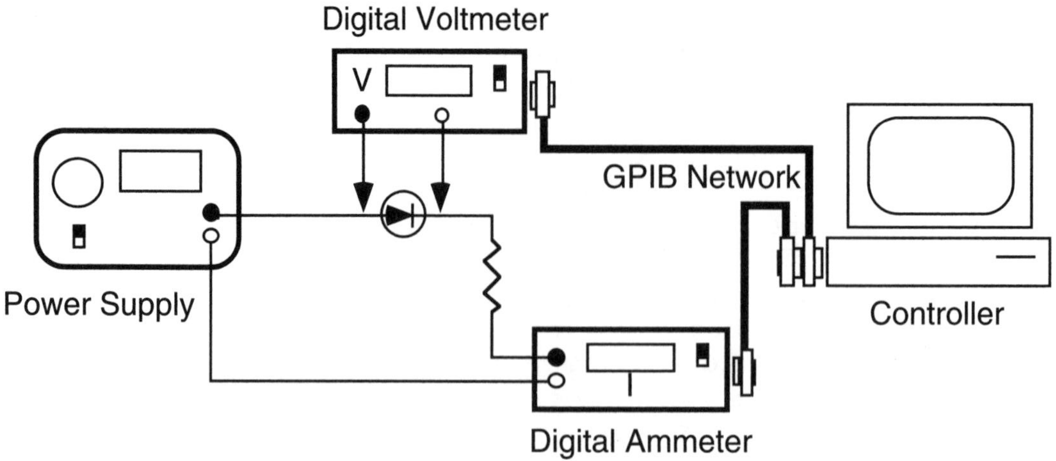

■ LabVIEW Simulation: IV Characteristic Curve

The program DiodeExp.vi is a LabVIEW simulation for the measurement of the IV characteristic curve of a light emitting diode. On the front panel are the digital meters, oscilloscope, and chart recorder outputs. On the block diagram following, note how the components mirror the above schematic circuit.

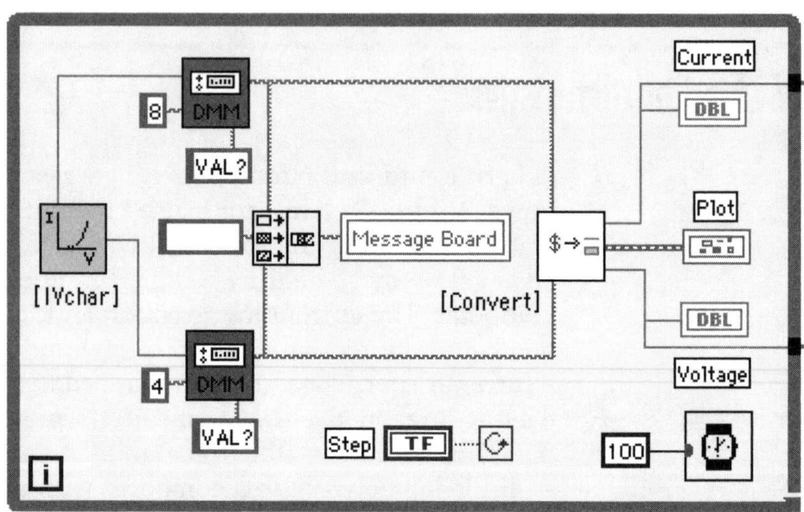

The icon called [IVchar] contains all the experimental apparatus: the power supply, the LED, and the resistor. Each time this

VI is called another voltage step is taken and the current I and voltage V are available as a numeric outputs. These outputs are in turn connected to the input of a digital multimeter [DMM]. One at address 8 is configured as a digital ammeter and the other at address 4 is configured as a voltmeter. Each time a digital meter is sent the command "VAL?" over the GPIB bus, it responds with a front panel reading (either the current or the voltage) in the form of a string message. For convenience, these messages are forwarded to a front panel message board via a local variable [Message Board]. The messages are also passed on to a VI [$→ –], which converts the ASCII strings into numbers and an array suitable for plotting.

Before continuing with the project, a brief review of the IEEE 488 or GPIB instrument network is presented. If you are familiar with the GPIB, then skip ahead to the section on the LabVIEW simulation of a GPIB controlled digital multimeter.

Interfacing IEEE 488 Instruments

In the early 1960s Hewlett Packard developed a high-speed instrument bus (originally called the (HP-IB) which soon become an ad hoc industry standard for intercomputer communication and peripheral control. By 1975, the HP-IB became the IEEE Standard 488 and was renamed by its users as the General Purpose Interface Bus (GPIB). In 1987 the standard was revised to the ANSI/IEEE Standard 488.1 and was widely used by all instrument manufacturers. At the same time, the IEEE 488.2 (latest version) established stronger rules for GPIB communication. A new set of functions, the GPIB 488.2 command set made programming easier and more uniform. Our ability to communicate over the IEEE 488 instrument bus opens up a whole new world of research, industrial and manufacturing instruments that can be controlled, manipulated, and made to "sing and dance."

■ What Is the GPIB?

The GPIB is a combination of hardware and software that works seamlessly to provide a high-speed communication network between computers and peripheral devices. It uses two 8-bit parallel ports to transfer commands and data, at data at rates up to 1 megabyte per second. A low-level handshaking protocol provides built-in acknowledge signals to ensure high reliability of data transfers. Up to fifteen external devices can be connected into the network, each device possessing its own unique address. All commands to devices and responses from devices are in the form of ASCII strings. The GPIB carries device specific messages and interface messages. A detailed knowledge of the device command structure is essential for successful communications. Once determined, LabVIEW's GPIB drivers make program design straightforward and similar to the RS232 interface in Chapter 5.

GPIB Architecture

The General Purpose Interface Bus uses a master-slave configuration for controller to "peripheral" communications. At any one time only one device can be the system controller, the so-called Controller In Charge (CIC). The system controller manages the flow of information by sending commands to all devices on the network saying who can talk and who can listen. The Controller is the master, the other devices the slaves. Slaves can be one of two types, a Talker or a Listener. A Talker on command from the Controller sends messages to one or more Listeners. A Listener receives data until disconnected by the Controller. Some devices such as a digital multimeter may be both a Talker and a Listener.

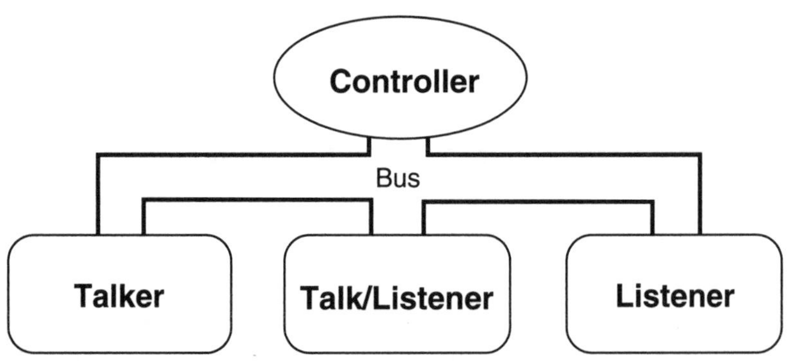

The role of the Controller is to supervise, monitor, and ensure that no data packets collide on the bus. When the Controller, sees that a device (slave) wishes to send a message to some other device, the Controller checks to see if the bus is free. If so, the Talker is connected to the Listener and all other devices are told not to use the bus. Once the message is transmitted, the Controller can unaddress the Talker and Listener so that other devices can use the bus. In essence, low-level signals on the bus allow the Controller to be in intimate contact with Talkers and Listeners to see when a link is requested, when the data is sent, when an acknowledgment is returned, and when the conversation is finished. In case of emergency, such as a data collision on the bus, the Controller can reassert its authority by unaddressing all devices and bringing the bus back to a known state.

GPIB Hardware

Controllers reside inside a computer usually in the form of a plug-in card. Only one controller can be the CIC and it will have the GPIB default address 0. All other GPIB devices, Talkers, Listeners, or Talker/Listeners will have a non-zero address from 1 to 15. Most peripheral devices have the GPIB interface built into the instrument. The only external GPIB hardware is a unique GPIB connector on the back of the instrument and a switch to set the device address.

The standard connector is an Amphenol or Cinch Series 57 with a plug on one side and a receptacle on the back side.

The pin assignment for the GPIB connector shown on next page displays the dual nature of the 16-bit parallel bus. Eight lines labeled DIO1 to DIO8 are the bidirectional data input/output lines. The other eight lines are divided into two types: Interface Management lines (EOI, IFC, SRQ, ATN, and REN) and the Transfer Control lines (DAV, NRFD, and NDAC). The meaning of each of these lines is given in the appendix.

Each line uses negative true logic with the standard transistor-transistor logic (TTL) drivers. For example, when the Controller drives attention true, a low TTL signal ($V < 0.8$ volts) appears on the ATN line and the Controller uses the data lines (DIO1–DIO8) to send commands. When the Controller drives

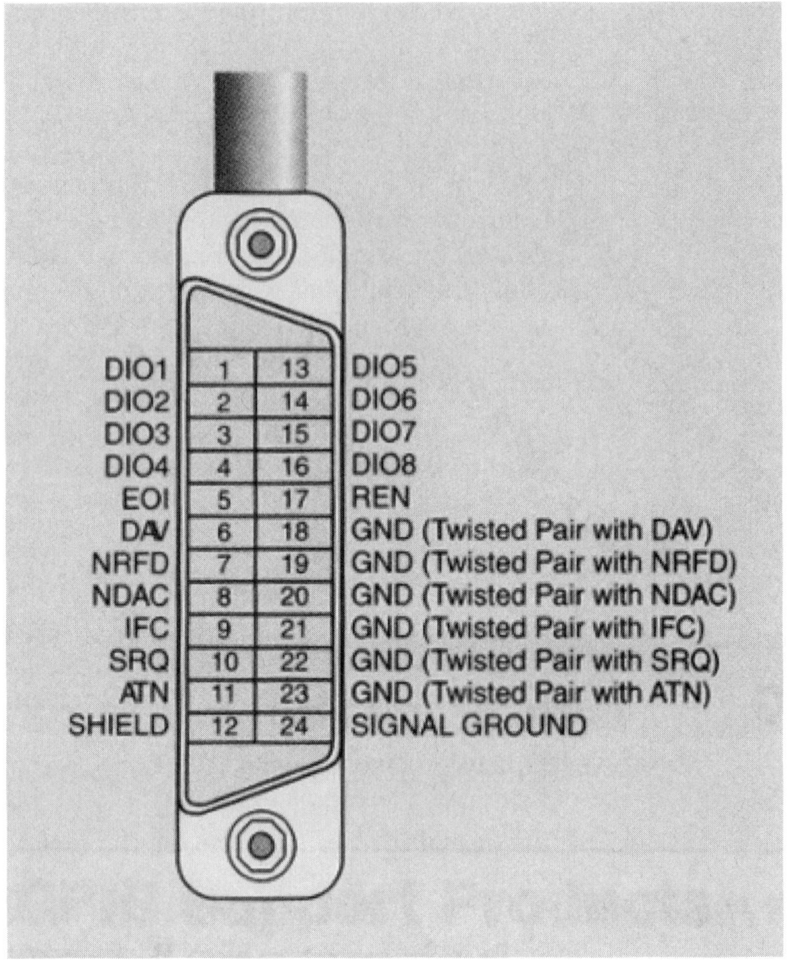

the ATN line false, a high TTL signal ($V > 2.0$ volts) is set and a Talker can send messages.

One of the reasons for the popularity and success of the GPIB is the superior noise rejection capability of a GPIB network. It is due in part to the GPIB cable design. The eight bidirectional data lines and the handshaking lines (EOI and REN) are encased in a woven ground sleeve at the center of the cable. The other six control lines, each one as a twisted pair with a ground-wire, are situated between the outer and inner woven ground sleeves. The entire structure is encased in a rubber-like plastic insulation cover. The ends are then rigidly and electrically connected to the solid metal GPIB connector.

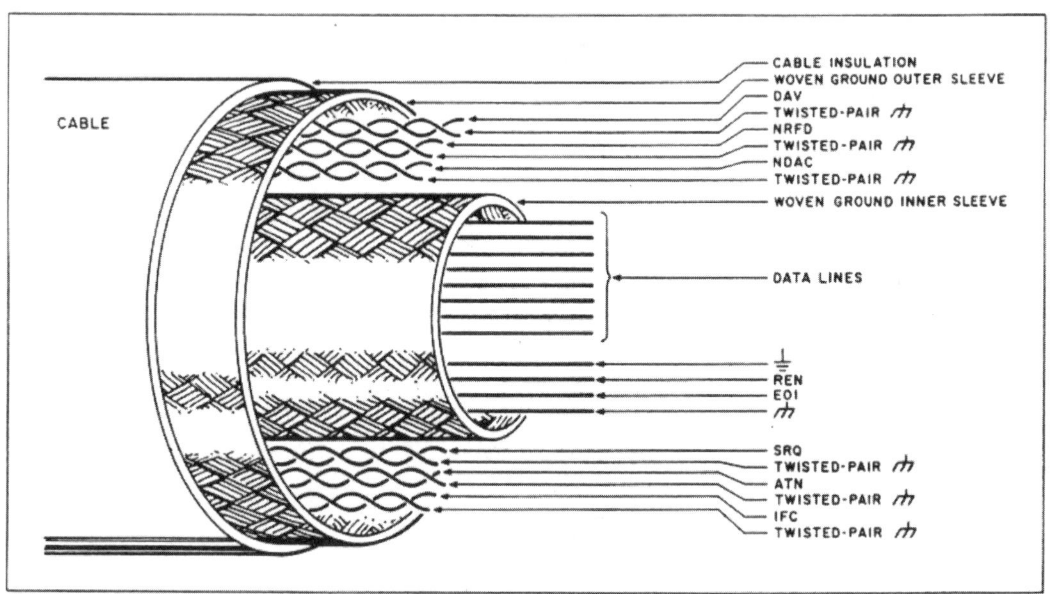

The GPIB network is always run with a controller that resides inside a general purpose or dedicated microcomputer. The devices can be connected to the controller with the GPIB cables in one of two configurations, linear or star.

The master/slave concept is best demonstrated by the star configuration where all the devices are connected to one point that is at the Controller site. The controller monitors the data traffic, routes all messages between devices, and ensures that no device speaks out of turn. A linear configuration is also possible as a polling network where messages are passed down the network from one device to the next. Requests from one device are passed up the network from one device to the next to the Controller. For example, suppose Device C wishes to send a data file to Device A. Device C sends a message request to the Controller by first sending it to Device B. It passes the request on to Device A and in turn it passes the request onto the Controller. The Controller sends a message to Device A to become a Listener and

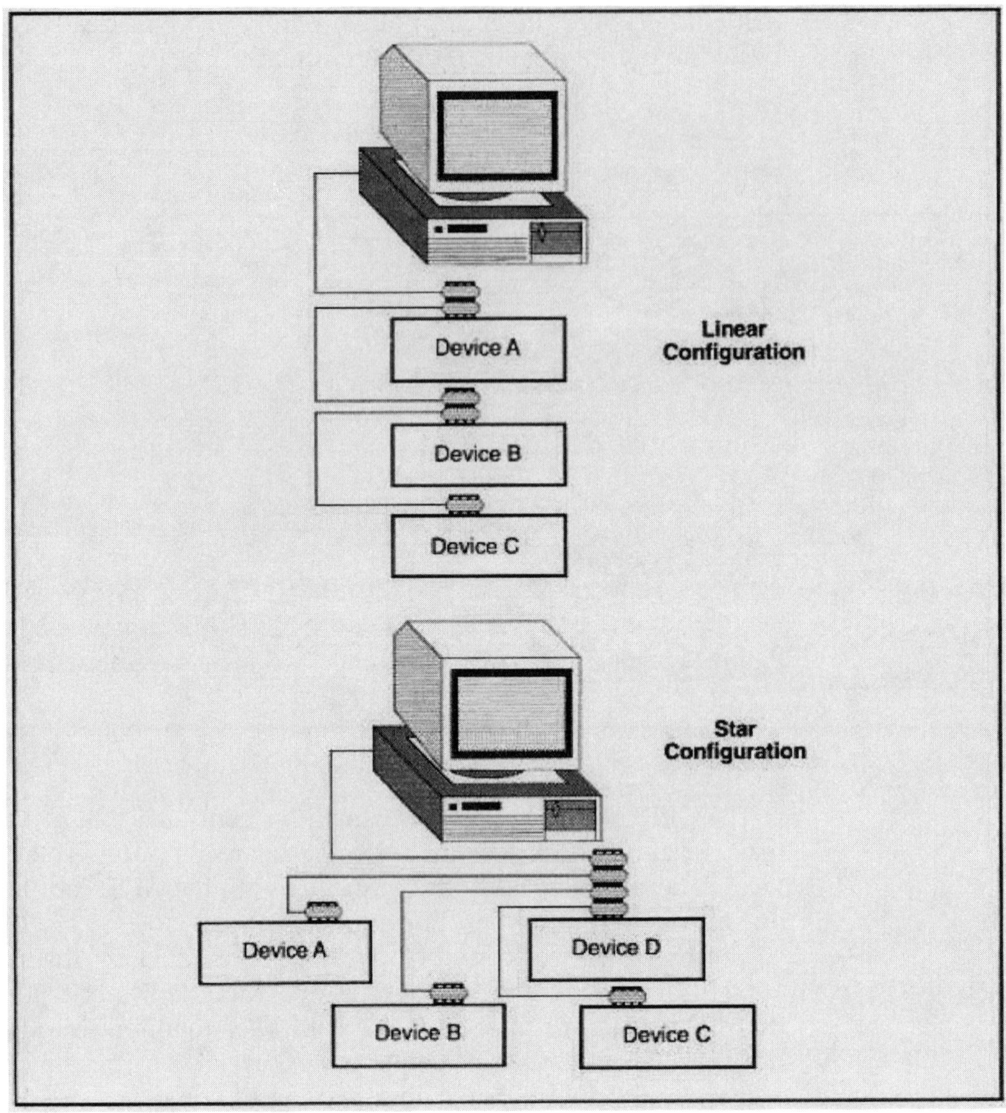

sends a message to Device C via Device A and B to become a Talker. The GPIB control language is rich and varied enough to support protocols for both network configurations. Some limitations on the physical layout of the network exist. These include:

a maximum separation between devices of 4 meters

an average separation between devices of 2 meters

a maximum network cable length of 20 meters

no more than 15 devices connected to the bus

Compliance will ensure that data transfer rates upwards of 1 million bytes per second are possible.

Message Protocol

All GPIB messages take the form of ASCII encoded strings. There are two types of messages, device specific and interface control. Device specific messages are data transfers that contain commands for programming instructions, measurement results, status, and data. Each IEEE 488 device will have its own unique instruction set. A detailed knowledge of the structure and meaning of the command set is essential for successful GPIB communications. Interface messages are more standard and perform such tasks as initializing the bus, addressing and unaddressing devices, and setting devices to remote or local control.

The usual message protocol will be in the form of a task specific message. The controller tells the device what operation to execute and the device responds with the status of the operation or with a data transfer.

Elementary GPIB Driver

The most elementary GPIB driver consists of a write operation or command to an addressed device followed by a response from that device. LabVIEW provides the standard set of GPIB functions found in the **Functions>Instrument IO>GPIB** subpalette. The **GPIB Write.vi** sends a string message (data) out the GPIB port to a remote device at some string address. The VI also has a built-in timeout feature that returns control back to the LabVIEW program even if the device refuses to answer. Several other features such as error reporting are available but will not be used in this interface.

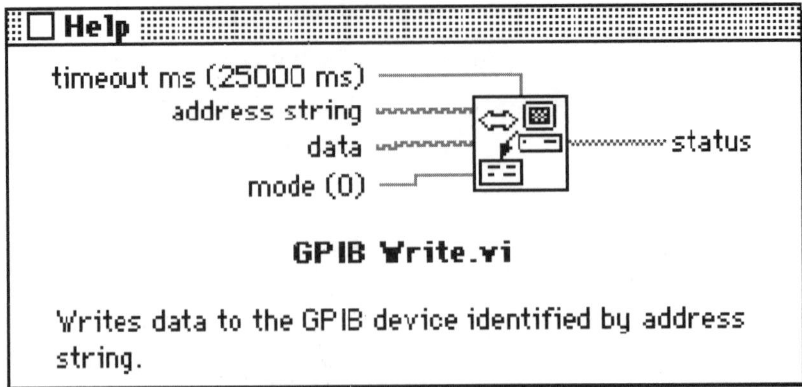

Interaction with a remote device is by its nature, a sequential process, and as such our interface uses a two-frame sequence structure.

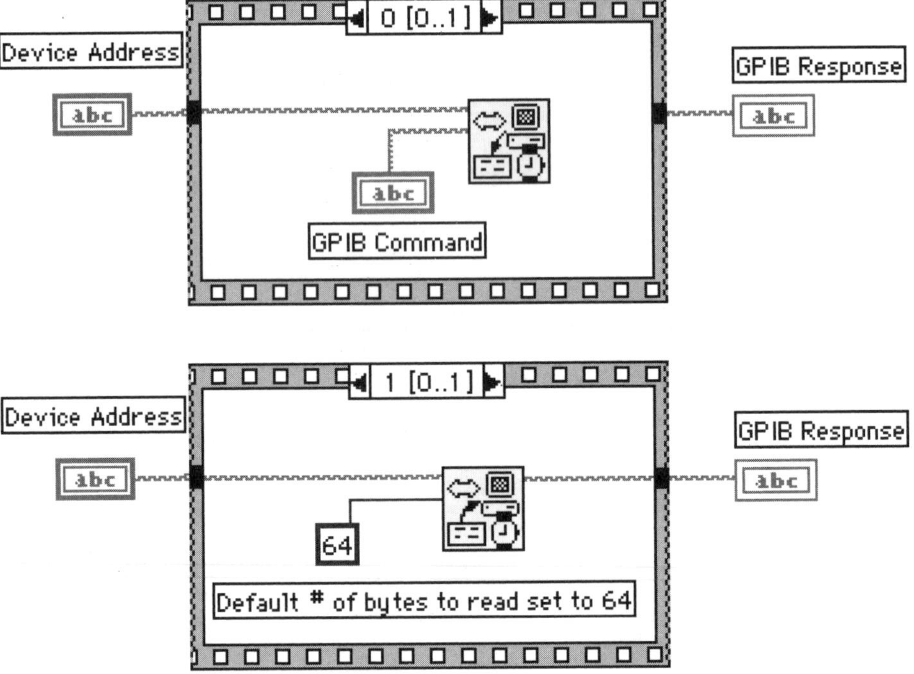

Frame 0 sends the GPIB command to a remote user at some GPIB device address. Frame 1 waits for a response, then reports it to a front panel string display. GPIB does not need to read the input buffer as in the RS232 interface. The number of characters is embedded in the response message. The default size, however, must be greater than the maximum message length.

The front panel requires only a valid address and command. In the following example, the command message "*RST; REM; VAC" is sent over the GPIB bus to a digital multimeter at address 13. Loosely translated this would read,

"**R**e**S**e**T** the digital multimeter at address 13";

"enable the device for **REM**ote operation";

and "set the input to read **V**olts **AC**."

The device responds with a status message "DVM ready at address 13."

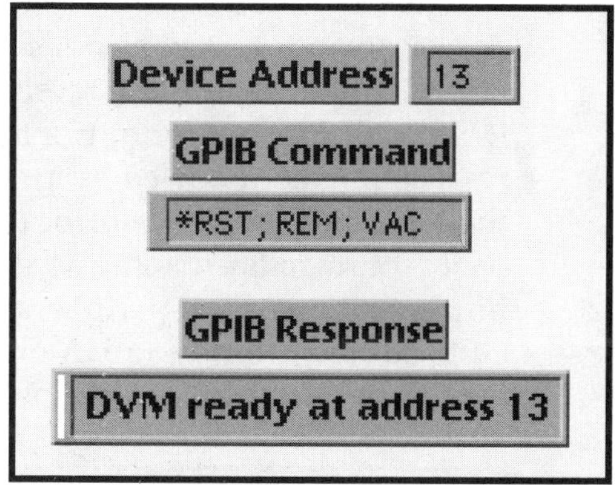

The digital voltmeter is now ready to receive measure commands. Note there is no initialization in the GPIB elementary driver. The act of turning on a remote instrument is often sufficient to perform the GPIB initialization. If not, LabVIEW provides a GPIB initialization function to accomplish this operation.

LabVIEW also provides a slightly more complex GPIB driver routine found in the LabVIEW/examples/inst/smplgpib folder called **LabVIEW<->GPIB.vi.** It contains both a timeout feature and some error reporting capability. The next step is to convert either the elementary or the LabVIEW GPIB driver into a sub-VI.

Each GPIB device has its own device specific command structure. The elementary or LabVIEW driver can be used as a diagnostic tool to test an external device and observe the operation of the command set. Once the command set is understood, a GPIB driver specific to this instrument can be designed.

GPIB-Controlled Digital Multimeter

A digital multimeter with a GPIB interface card can be controlled over the IEEE 488 instrument bus by receiving commands from a central controller (CIC). Intuitive and simple device specific commands allow peripheral devices to be configured into a particular mode, make measurements, cause action, and report the status of activities in progress. When the digital multimeter is first turned on, its operating system polls all inputs to see which are connected. If no messages are received on the external ports, then the DMM will function from front panel commands. However, if a reset command followed by a remote command is received, then the DMM will relinquish local control, shutting out any front panel commands and only accepting remote commands over the instrument bus. Each peripheral device will have its own unique GPIB address from 1 to 15. A proper GPIB command will include both the device address and the ASCII command.

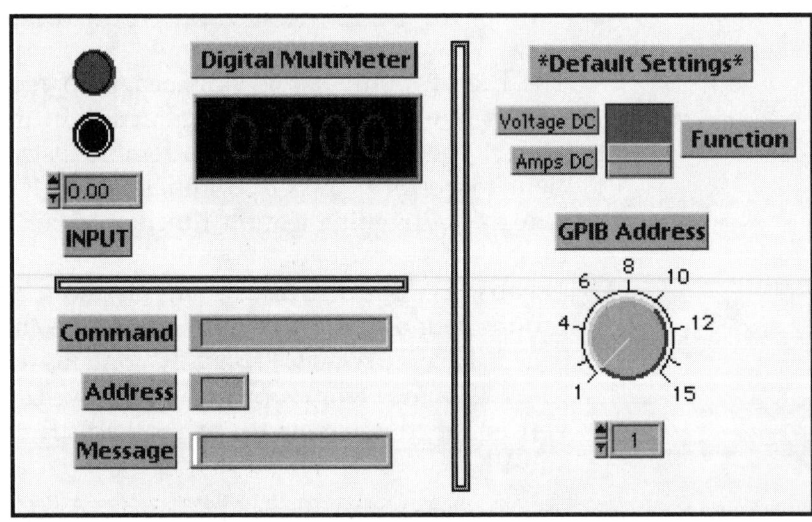

In operation, our simplified digital multimeter, called DMM.vi, has only two modes of operation, Voltage DC and Current DC. The voltage setting reads volts while the current setting reads amps scaled for milliamp inputs. If no GPIB commands are sent, the Message box will display either "Bad Address" or "Error," and the input will appear on the front panel meter. For GPIB operation, device-specific commands must be sent to the DMM via the Command input together with a valid address. The GPIB address for this DMM can be set on the LabVIEW front panel by rotating the GPIB Address dial with the operating tool to the assigned device address. The block diagram for DMM.vi displays how the GPIB operations proceed.

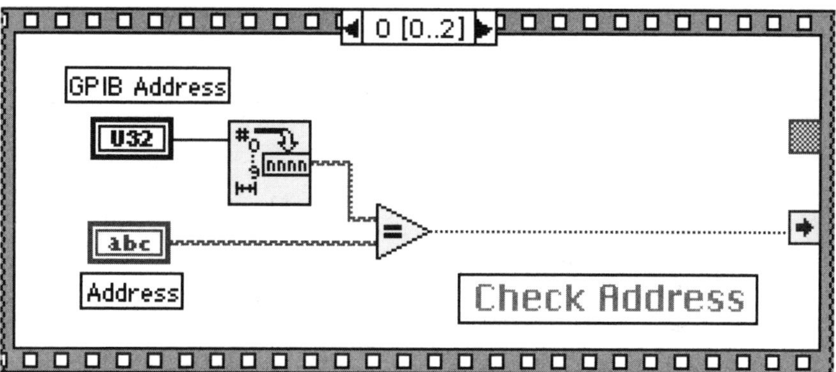

Frame 0 compares the command address received over the GPIB bus with the DMM's address. The result is passed on via a local variable to frame 2.

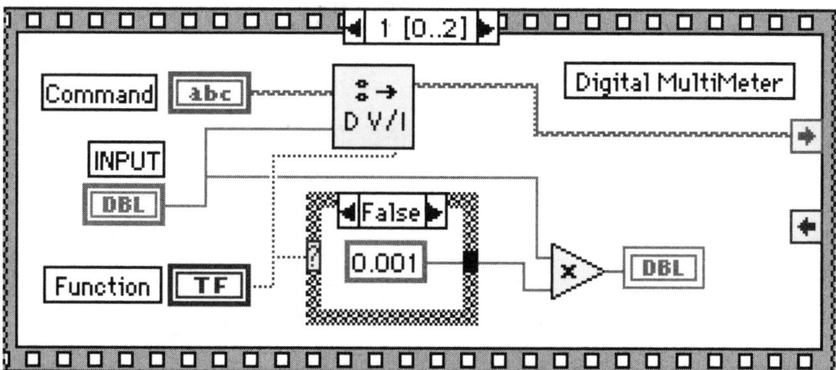

Frame 1 does most of the work. The GPIB command is passed on to the command processor [D V/I] together with the front panel input signal level, a numeric, and the selected DMM function, a Boolean. True selects the voltage mode with a scaling factor of 1 and a false selects the current mode with a scaling factor of 0.001. This latter factor reflects the fact that LED currents are in the milliamp range. The command processor checks to see if the input GPIB command is in the command set and if it agrees with the default setting. If so, then the front panel reading is converted into an ASCII string and passed onto frame 2.

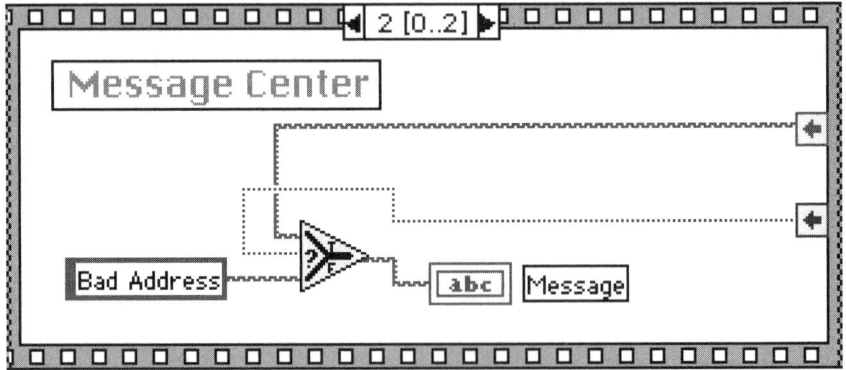

Frame 2 reports the status and/or data to the GPIB controller. First, if the address is incorrect, then the message reads "Bad Address." If the address is correct, then the command processor message is forwarded. If the GPIB command was incorrect or the wrong mode is requested, then the message will read "Error." If the GPIB command is correct, then the front panel message will display either an acknowledgment that the DMM is ready for the front panel reading or the DMM value.

The following device commands have been implemented:

*RST	reset the device to an initial state
REM	place the meter in remote operation
VDC	input set for volts DC
ADC	input set for current DC
VAL?	send the front panel reading.

To simplify the command structure, the above commands have been grouped together with semicolons to give only three valid commands:

*RST; REM; VDC

*RST; REM; ADC

VAL?

Load the program DMM.vi. Try various combinations of the GPIB commands to get a feel for how device specific commands operate. Manually set the input, the function, and the GPIB address. Remote commands are entered into the [Command] and [Address] string control. Responses from the digital multimeter show up in the [Message] display.

On the block diagram, the command processor [D VI] can be opened up to reveal how the commands are processed in this simulation. Its block diagram shown below displays some interesting constructions.

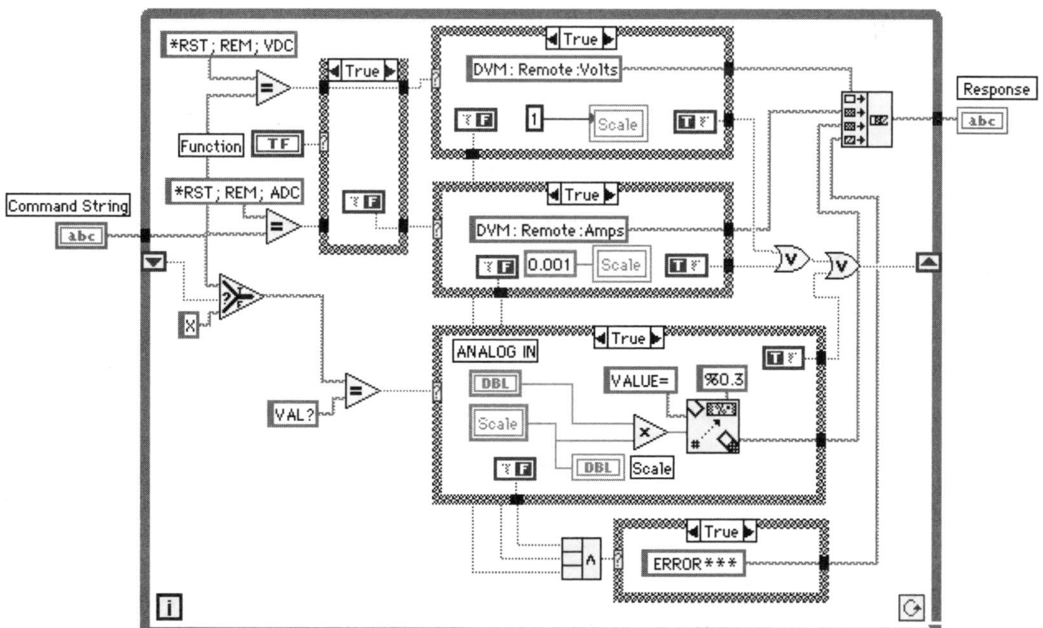

The front panel [Function] selects which case (voltage or current) will be active. If voltage is chosen, the scale is set to 1, while for current the scale is 0.001. The command VAL? activates a middle case that reads from the front panel the input level [Analog In] and applies the appropriate scaling factor before converting the result into a string response. Various Boolean operations are used to ensure that only valid commands in the correct sequence are executed. Trace the data flow and observe how the logic operations work.

Measuring the IV Characteristic Curve

Two digital multimeters are used to measure the IV characteristic curve of a light emitting diode. One is configured for voltage and has the GPIB address set to 4 while the other DMM configured for current has the GPIB address 8. The heart of the program shown earlier and reproduced below shows how the data is collected in frame 2.

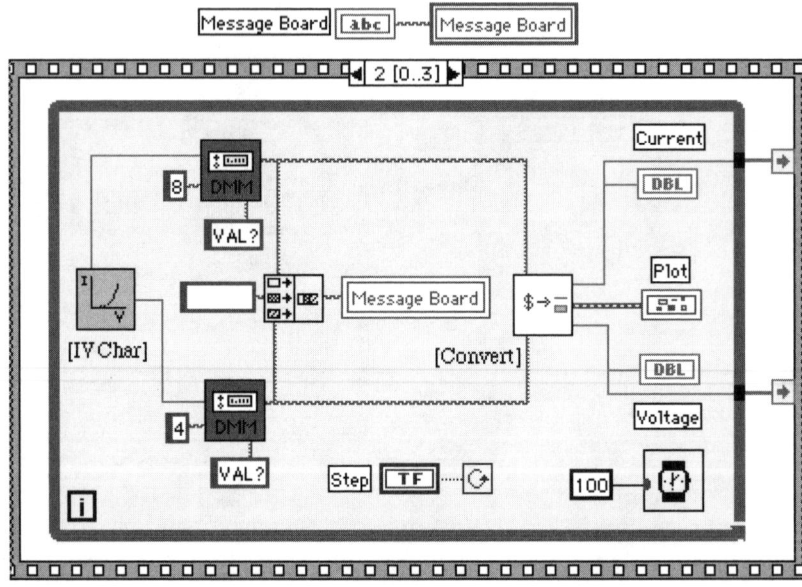

Each time the command VAL? is sent to the digital voltmeters, a string value of the current and voltage is measured. The string response is passed to the front panel using a local variable [Message Board]. The string is also passed to Convert.vi where it is converted into a single numeric value and a point array value. The numeric value is passed on to a front panel numeric display and also collected as an array of values at the While . . . Loop boundary. The point array goes to an XY graph, which simulates an oscilloscope display. When the program DiodeExp.vi is run, the individual (x,y) points appear on the display one point at a time. To see the entire data set, a capture feature is required. When the front panel switch is set to Plot, measurement stops and the processing continues in the next frame.

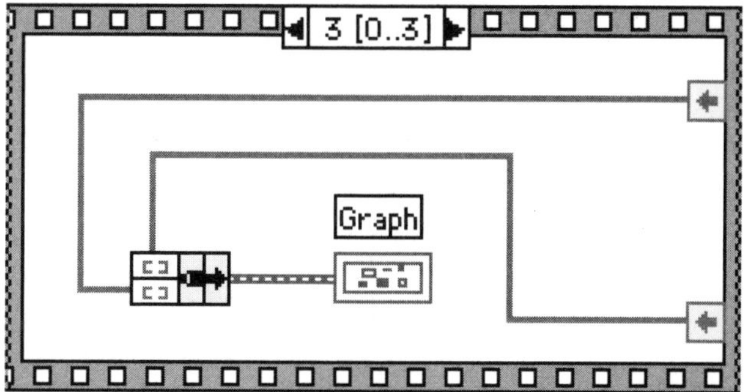

Frame 3 collects the array of data points and plots it on the front panel as a XY graph.

The sequence structure was chosen to ensure that the GPIB commands occur in the correct order. For example, one cannot read remotely a real digital voltmeter until it has been reset, set to remote, and a function selected. Frame 0 sends the ASCII string command message to the digital multimeters. Clearly from the earlier discussion of the command processor DVI.vi, an incorrect command would yield an error message. Frame 1 not shown as it just adds a delay, so the reader can observe the action.

Load the measuring program DiodeExp.vi found in the chapter library. On running the program, the individual measure-

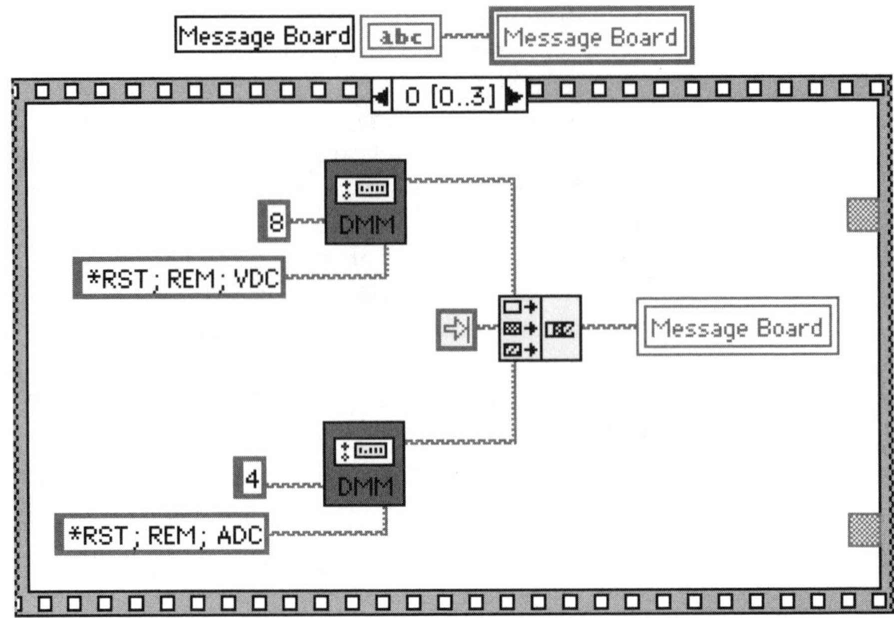

ments of current and voltage can be seen. After the output begins to repeat, throw the switch to plot and the IV characteristic curve for a light emitting diode appears in the graph display. We are now ready to test the hypothesis that the threshold voltage gives an estimate of the color of the light emitted.

Analysis of the IV Characteristic Curve

The current in the IV characteristic curve rises rapidly beyond a threshold voltage. Note that for current values larger than about 50 percent of the maximum current, the characteristic is approximately linear. By fitting a straight line to these points, the intercept on the voltage axis gives an estimate of the threshold voltage. In this case the intercept is about 1.56 volts.

Now using the hypothesis that the threshold voltage converted into energy (eV_{Th}) is equal to the photon energy ($h\upsilon$) and with the relationship between the photon frequency, speed of

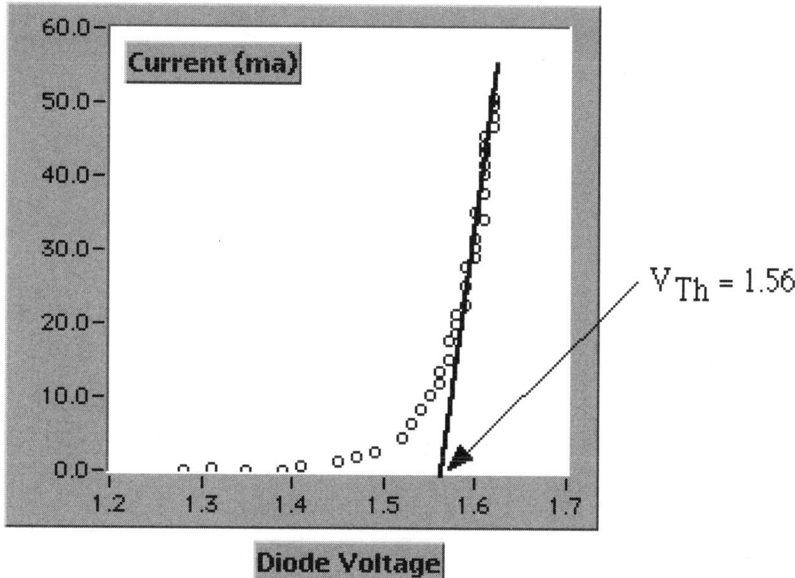

light, and wavelength ($\upsilon = c/\lambda$), then the wavelength of the emitted light can be found.

$$eV_{Th} = h\upsilon = hc/\lambda$$

The elementary charge [e], Plank's constant [h], and the speed of light [c] can be found as icons in the **Numerics>Additional Numeric Constants** subpalette. In the above case, the threshold voltage corresponds to a wavelength of 795 nanometers or 7949 Angstroms and the color is deep red.

■ LabVIEW Challenge: Linear Fit to IV Data Set

Design an add-on function to DiodeExp.vi that fits a straight line to the large current values. You might find a review of the Linear Fit.vi used in Chapter 7 useful. The intercept of the linear fit with the voltage axis yields the threshold voltage. Further, with the LabVIEW icons for the fundamental constants h, c, and e, the wavelength can be calculated directly and reported on the front panel.

An unknown LED is placed in the measuring circuit, the voltage is stepped through the forward biased region, and the data are recorded in an experiment called DiodeExp2.vi. What a soft and unusual color! I have seen the light and I know what color it is.

■ LabVIEW Challenge: What Color Is That LED?

What color is the light emitting diode in DiodeExp2.vi?

How to Use This Project

The original data set was taken point by point and the values of current and voltage entered manually into a spreadsheet. The spreadsheet data was then converted into a LabVIEW 2D array. This array is the source of data used in the experiment simulation, IVchar.vi The data sets in both DiodeExp.vi and DiodeExp2.vi are real and the noise is as observed in the experiment. The second time the experiment was conducted, two digital multimeters recorded the current and the voltage values under LabVIEW control over a GPIB instrument bus. A third time the experiment was conducted, a DAQ card was used. Two analog inputs measured the current and voltage signals and an analog output was used to control the power supply.

This project is a fine example of the flexibility in the LabVIEW approach. Once the analysis VI is designed, tested, and debugged, it just becomes another function (sub-VI). The data set can come from a wide variety of sources, spreadsheets, arrays, and external digital meters connected to a LabVIEW controller over the RS232, the IEEE 488 instrument bus, or the Web using TCP/IP protocol and, of course, an internal DAQ card.

Appendix C: GPIB Signals and Lines

The interface system consists of sixteen signal lines and eight ground-return or shield-drain lines. The sixteen signal lines are divided into three groups:

Eight data lines
Three handshake lines
Five interface management lines

Data Lines

The eight data lines, DIO1 through DIO8, carry both data and command messages. All commands and most data use the 7-bit ASCII or International Standards Organization (ISO) code set, in which case the eighth bit, DIO8, is unused or is used for parity.

Handshake Lines

Three lines asynchronously control the transfer of message bytes among devices This process is called a three-wire interlocked handshake, and it guarantees that message bytes on the data lines are sent and received without transmission error.

NRFD (Not Ready for Data). NRFD indicates whether a device is ready to receive a message byte. All devices drive NDAC when they receive commands, and Listeners drive it when they receive data messages.

NDAC (Not Data Accepted). NDAC indicates whether a device has accepted a message byte. All devices drive NDAC when they receive commands, and Listeners drive it when they receive data messages.

DAV (Data Valid). DAV tells whether the signals on the data lines are stable (valid) and whether devices can accept them safely. The controller drives DAV when sending commands, and the Talker drives it when sending data messages.

Interface Management Lines

Five lines manage the flow of information across the interface.

ATN (Attention). The Controller drives ATN true when it uses the data lines to send comands and drives ATN false when a Talker can send data messages.

IFC (Interface Clear). The System Controller drives the IFC lines to initialize the bus and become CIC.

REN (Remote Enable). The System Controller drives the REN line, which places devices in remote or local program mode.

SRQ (Service Request). Any device can drive the SRQ line to asynchronously request service from the Controller.

EOI (End or Identify). The EOI line has two purposes. The Talker uses the EOI line to mark the end of a message string. The Controller uses the EOI line to tell devices to respond in a parallel poll.

Glossary

Absolute Path File or directory path that describes the location relative to the top of level of the file system.

Active Window Window that is currently set to accept user input, usually the frontmost window. The title bar of an active window is highlighted. You make a window active by clicking on it, or by selecting it from the Windows menu.

A/D Analog-to-digital conversion. Refers to the operation electronic circuitry does to take a real-world analog signal and convert it to a digital form (as a series of bits) that the computer can understand.

ADC See A/D.

ANSI American National Standards Institute.

Array Ordered, indexed set of data elements of the same type.

Array Shell Front panel object that houses an array. It consists of an index display, a data object window, and an optional label. It can accept various data types.

Artificial Data Dependency Condition in a dataflow programming language in which the arrival of data, rather than its value, triggers execution of a node.

ASCII American Standard Code for Information Interchange.

Asynchronous Execution Mode in which multiple processes share processor time. For example, one process executes while others wait for interrupts during device I/O or while waiting for a clock tick.

Auto-Indexing Capability of loop structures to disassemble and assemble arrays at their borders. As an array enters a loop with auto-indexing enabled, the loop automatically disassembles it with scalars extracted from one-dimensional arrays, one-dimensional arrays extracted from two-dimensional arrays, and so on. Loops assemble data into arrays as they exit the loop according to the reverse of the same procedure.

Autoscaling Ability of scales to adjust to the range of plotted values. On graph scales, this feature determines maximum and minimum scale values, as well.

Autosizing Automatic resizing of labels to accommodate text that you enter.

Block Diagram Pictorial description or representation of a program or algorithm. In LabVIEW, the block diagram, which consists of executable icons called nodes and wires that carry data between the nodes, is the source code for the VI. The block diagram resides in the block diagram window of the VI.

Boolean Controls Front panel objects used to manipulate and display or input and

and Indicators output Boolean (TRUE or FALSE) data. Several styles are available, such as switches, buttons and LEDs.

Breakpoint A pause in execution. You set a breakpoint by clicking on a VI, node, or wire with the Breakpoint tool from the Tools palette.

Breakpoint Tool Tool used to set a breakpoint on a VI, node, or wire.

Broken VI VI that cannot be compiled or run; signified by a broken arrow in the run button.

Bundle Node Function that creates clusters from various types of elements.

Byte Stream File File that stores data as a sequence of ASCII characters or bytes.

Case One subdiagram of a Case Structure.

Case Structure Conditional branching control structure, which executes one and only one of its subdiagrams based on its input. It is the combination of the IF, THEN, ELSE, and CASE statements in control flow languages.

Channel Pin or wire lead to which an analog signal is read from or applied.

Chart See scope chart, strip chart, and sweep chart.

CIN See Code Interface Node.

Cloning To make a copy of a control or some other LabVIEW object by clicking the mouse button while pressing the <ctrl> (Windows); <option> (Macintosh); <meta> (Sun); or <alt> (HP-UX) key and dragging the copy to its new location.

(Sun and HP-UX) You can also clone an object by clicking on the object with the middle mouse button and then dragging the copy to its new location.

Cluster A set of ordered, unindexed data elements of any data type including numeric, Boolean, string, array, or cluster. The elements must be all controls or all indicators.

Cluster Shell Front panel object that contains the elements of a cluster.

Code Interface Node (CIN) Special block diagram node through which you can link conventional, text-based code to a VI.

Coercion The automatic conversion LabVIEW performs to change the numeric representation of a data element.

Coercion Dot Glyph on a node or terminal indicating that the numeric representation of the data element changes at that point.

Color Tool Tool you use to set foreground and background colors.

Color Copy Tool Copies colors for pasting with the Color tool.

Compile Process that converts high-level code to machine-executable code. LabVIEW automatically compiles
VIs before they run for the first time after creation or alteration.

Conditional Terminal The terminal of a While Loop that contains a Boolean value that determines whether the VI performs another iteration.

Connector Part of the VI or function node that contains its input and output terminals, through which data passes to and from the node.

Connector Pane Region in the upper right corner of a front panel window that displays the VI terminal pattern. It underlies the icon pane.

Constant See universal constant and user-defined constant.

Continuous Run Execution mode in which a VI is run repeatedly until the operator stops it. You enable it by clicking on the continuous run button.

Control Front panel object for entering data to a VI interactively or to a subVI programmatically.

Control Flow Programming system in which the sequential order of instructions determines execution order. Most conventional text-based programming languages, such as C, Pascal, and BASIC, are control flow languages.

Controls Palette Palette containing front panel controls and indicators.

Conversion Changing the type of a data element.

Count Terminal The terminal of a For Loop whose value determines the number of times the For Loop executes its subdiagram.

CPU Central Processing Unit.

Current VI VI whose front panel, block diagram, or Icon Editor is the active window.

Custom PICT Controls Controls and indicators whose parts can be replaced by graphics and indicators you supply.

D/A Digital-to-analog. The opposite operation of an A/D.

Data Acquisition (DAQ) Process of acquiring data, usually by performing an analog-to-digital (A/D) conversion. Its meaning is sometimes expanded to include data generation (D/A).

Data Dependency Condition in a dataflow programming language in which a node cannot execute until it receives data from another node. See also artificial data dependency.

Data Flow Programming system consisting of executable nodes in which nodes execute only when they have received all required input data and produce output automatically when they have executed. LabVIEW is a dataflow system.

Data Logging Generally, to acquire data and simultaneously store it in a disk file. LabVIEW file I/O functions can log data.

Data Storage Formats The arrangement and representation of data stored in memory.

Data Type Descriptor Code that identifies data types, used in data storage and representation.

Datalog File File that stores data as a sequence of records of a single, arbitrary data type that you specify when you create the file. While all the records in a datalog file must be of a single type, that type can be complex; for instance, you can specify that each record is a cluster containing a string, a number, and an array.

DC Direct Current. The opposite of AC (Alternate Current). Refers to a very low-frequency signal, such as one that varies less than once a second.

Device A plug-in DAQ board

Device Number Number assigned to a device (DAQ board) in the NI-DAQ configuration utility.

Description Box Online documentation for a LabVIEW object.

Destination Terminal See sink terminal.

Dialog Box An interactive screen with prompts in which you specify additional information needed to complete a command.

Differential Measurement Way to configure a device to read signals in which the inputs need not be connected to a reference ground. The measurement is made between two input channels.

Dimension Size and structure attribute of an array.

DMA Direct Memory Access. A method by which you can transfer data to computer memory from a device or memory on the bus (or from computer memory to a device) while the processor does something else. DMA is the fastest method of transferring data to or from computer memory.

Drag To drag the mouse cursor on the screen to select, move, copy, or delete objects.

Empty Array Array that has zero elements, but has a defined data type. For example, an array that has a numeric control in its data display window but has no defined values for any element is an empty numeric array.

EOF End-of-File. Character offset of the end of file relative to the beginning of the file (that is, the EOF is the size of the file).

Execution Highlighting Feature that animates VI execution to illustrate the data flow in the VI.

FFT Fast Fourier transform.

File Refnum An identifier that LabVIEW associates with a file when you open it. You use the file refnum to specify that you want a function or VI to perform an operation on the open file.

Flattened Data Data of any type that has been converted to a string, usually, for writing it to a file.

For Loop Iterative loop structure that executes its subdiagram a set number of times. Equivalent to conventional code: For i=0 to n-1, do....

Formula Node Node that executes formulas that you enter as text. Especially useful for lengthy formulas that would be cumbersome to build in block diagram form.

Frame Subdiagram of a Sequence Structure.

Free Label Label on the front panel or block diagram that does not belong to any other object.

Front Panel The interactive user interface of a VI. Modeled from the front panel of physical instruments, it is composed of switches, slides, meters, graphs, charts, gauges, LEDs, and other controls and indicators.

Function Built-in execution element, comparable to an operator, function, or statement in a conventional language.

Functions Palette Palette containing block diagram structures, constants, communication features, and VIs.

G The LabVIEW graphical programming language.

Global Variable Non-reentrant subVI with local memory that uses an uninitialized shift register to store data from one execution to the next. The memory of copies of these subVIs is shared and thus can be used to pass global data between them.

Glyph A small picture or icon.

GPIB General Purpose Interface Bus. Also known as HP-IB (Hewlett-Packard Interface Bus) and IEEE 488.2 bus (Institute of Electrical and Electronic Engineers standard 488.2), it has become the world standard for almost any instrument to communicate with a computer. Originally developed by Hewlett-Packard in the 1960s to allow their instruments to be programmed in BASIC with a PC, now IEEE has helped define this bus with strict hardware protocols that ensure uniformity across instrument.

Graph Control Front panel object that displays data in a Cartesian plane.

Ground the common reference point in a system; i.e., ground is at 0 volts.

Help Window Special window that displays the names and locations of the terminals for a function or subVI, the description of controls and indicators, the values of universal constants, and descriptions and data types of control attributes. The window also accesses LabVIEW's Online Reference.

Hex Hexadecimal. A base-16 number system.

Hierarchical Palette Menu that contains palettes and subpalettes.

Hierarchy Window Window that graphically displays the hierarchy of VIs and subVIs.

Housing Nonmoving part of front panel controls and indicators that contains sliders and scales.

Hz Hertz. Cycles per second.

Icon Graphical representation of a node on a block diagram.

Icon Editor Interface similar to that of a paint program for creating VI icons.

Icon Pane Region in the upper right corner of the front panel and block diagram that displays the VI icon.

IEEE Institute for Electrical and Electronic Engineers.

Indicator Front panel object that displays output.

Inf Digital display value for a floating-point representation of infinity.

Instrument Driver VI that controls a programmable instrument.

I/O Input/Output. The transfer of data to or from a computer system involving communications channels, operator input devices, and/or data acquisition and control interfaces.

Iteration Terminal The terminal of a For Loop or While Loop that contains the current number of completed iterations.

Label Text object used to name or describe other objects or regions on the front panel or block diagram.

Labeling Tool Tool used to create labels and enter text into text windows.

LabVIEW Laboratory Virtual Instrument Engineering Workbench.

LED Light-emitting diode.

Legend Object owned by a chart or graph that display the names and plot styles of plots on that chart or graph.

Line The equivalent of an analog channel—a path where a single digital signal is set or retrieved.

Marquee A moving, dashed border that surrounds selected objects.

Matrix Two-dimensional array.

MB Megabytes of memory.

Menu Bar Horizontal bar that contains names of main menus.

Modular Programming Programming that uses interchangeable computer routines.

NaN Digital display value for a floating-point representation of not a number, typically the result of an undefined operation, such as log(–1).

NI-DAQ Driver software for National Instruments DAQ boards and SCXI modules. This software includes a configuration utility to configure the hardware, and also acts as an interface between LabVIEW and the devices.

Nodes Execution elements of a block diagram consisting of functions, structures, and subVIs.

Nondisplayable Characters ASCII characters that cannot be displayed, such as newline, tab, and so on.

Not-a-Path A predefined value for the path control that means the path is invalid.

Not-a-Refnum A predefined value that means the refnum is invalid.

Numeric Controls and Indicators Front panel objects used to manipulate and display or input and output numeric data.

NRSE Nonreferenced single-ended.

NRSE Measurement All measurements are made with respect to a common reference. This reference voltage can vary with respect to ground.

Nyquist Frequency One-half the sampling frequency. If the signal contains any frequencies above the Nyquist frequency, the resulting sampled signal will be aliased, or distorted.

Object Generic term for any item on the front panel or block diagram, including controls, nodes, wires, and imported pictures.

Object Pop-Up Menu Tool Tool used to access an object's pop-up menu.

Octal Numbering system, base-eight.

Operating Tool Tool used to enter data into controls as well as operate them. Resembles a pointing finger.

Palette Menu of pictures that represent possible options.

Platform Computer and operating system.

Plot A graphical representation of an array of data shown either on a graph or a chart.

Polymorphism Ability of a node to automatically adjust to data of different representation, type, or structure.

Pop Up To call up a special menu by clicking (usually on an object) with the right mouse button (on Window, Sun, and HP-UX) or while holding down the command key (on the Macintosh).

Pop-Up Menus Menus accessed by popping up, usually on an object. Menu options pertain to that object specifically.

Port A collection of digital lines that are configured in the same direction and can be used at the same time.

Positioning Tool Tool used to move, select, and resize objects.

Probe Debugging feature for checking intermediate values in a VI.

Probe Tool Tool used to create probes on wires.

Programmatic Printing Automatic printing of a VI front panel after execution.

Pseudocode Simplified language-independent representation of programming code.

Pull-Down Menus Menus accessed from a menu bar. Pull-down menu options are usually general in nature.

Reentrant Execution Mode in which calls to multiple instances of a subVI can execute in parallel with distinct and separate data storage.

Representation Subtype of the numeric data type, of which there are signed and unsigned byte, word, and long integers, as well as single-, double-, and extended-precision floating-point numbers, both real and complex.

Resizing Handles Angled handles on the corner of objects that indicate resizing points.

Ring Control Special numeric control that associates 32-bit integers, starting at 0 and increasing sequentially, with a series of text labels or graphics.

RS-232 Recommended Standard #232. A standard proposed by the Instrument Society of America for serial communications. It's used interchangeably with the term "serial communication," although serial communications more generally refers to communicating one bit at a time. A few other standards you might see are RS-485, RS-422, and RS-423.

RSE Referenced single-ended.

RSE Measurement All measurements are made with respect to a common ground; also known as a grounded measurement.

Sample A single analog input or output data point.

Scalar Number capable of being represented by a point on a scale. A single value as opposed to an array. Scalar Booleans and clusters are explicitly singular instances of their respective data types.

Scale Part of mechanical-action, chart, and graph controls and indicators that contains a series of marks or points at known intervals to denote units of measure.

Scope Mode Mode of a waveform chart modeled on the operation of an oscilloscope.

Scroll Tool Tool used to scroll windows.

SCXI Signal Conditioning eXtensions for Instrumentation. A high-performance signal conditioning system devised by National Instruments, using an external chassis that contains I/O modules for signal conditioning, multiplexing, etc. The chassis is wired into a DAQ board in the PC.

Sequence Local Terminal that passes data between the frames of a Sequence Structure.

Sequence Structure Program control structure that executes its subdiagrams in numeric order. Commonly used to force nodes that are not data-dependent to execute in a desired order.

Shift Register Optional mechanism in loop structures used to pass the value of a variable from one iteration of a loop to a subsequent iteration.

Sink Terminal Terminal that absorbs data. Also called a destination terminal.

Slider Moveable part of slide controls and indicators.

Source Terminal Terminal that emits data.

State Machine A method of execution in which individual tasks are separate cases in a Case Structure that is embedded in a While Loop. Sequences are specified as arrays of case numbers.

String Controls and Indicators Front panel objects used to manipulate and display or input and output text.

Strip Mode Mode of a waveform chart modeled after a paper strip chart recorder, which scrolls as it plots data.

Structure Program control element, such as a Sequence, Case, For Loop, or While Loop.

Subdiagram Block diagram within the border of a structure.

subVI VI used in the block diagram of another VI; comparable to a subroutine.

Sweep Mode Similar to scope mode—except a line sweeps across the display to separate old data from new data.

Terminal Object or region on a node through which data passes.

Tool Special LabVIEW cursor you can use to perform specific operations.

Toolbar Bar containing command buttons that you can use to run and debug VIs.

Tools Palette Palette containing tools you can use to edit and debug front panel and block diagram objects.

Top-Level VI VI at the top of the VI hierarchy. This term distinguishes the VI from its subVIs.

Trigger a condition for starting or stopping a DAQ operation.

Tunnel Data entry or exit terminal on a structure.

Typecast To change the type descriptor of a data element without altering the memory image of the data.

Type Descriptor See data type descriptor.

Universal Constant Uneditable block diagram object that emits a particular ASCII character or standard numeric constant, for example, pi.

User-Defined Constant Block diagram object that emits a value you set.

VI See virtual instrument.

VI Library Special file that contains a collection of related VIs for a specific use.

Virtual Instrument LabVIEW program; so called because it models the appearance and function of a physical instrument.

While Loop Loop structure that repeats a section of code until a condition is met. Comparable to a Do loop or a Repeat-Until loop in conventional programming languages.

Wire Data path between nodes.

Wiring Tool Tool used to define data paths between source and sink terminals.

Program Libraries: Sensors, Transducers & LabVIEW

Preface

Data Flow.vi	Demonstration of the data flow architecture
Readme	Library Formats (PC and MAC)

Chapter 1 LabVIEW Environment

Amplifier.vi	Simulation: Analog amplifier
DAC.vi	Eight bit digital- to-analog converter
DAC1.vi	Single bit digital-to- analog converter
DAC2.vi	Two bit digital- to-analog converter
Vector Add.vi	Vector addition of two vectors A+B=C

Chapter 2 Invisible Fields

Gauss2.vi	Hall Effect Sensor with Noise and Offset (subVI)
Gaussmeter.vi	Simulation: Gaussmeter
Gaussmeter2.vi	Gaussmeter as a subVI

Hallchart.vi	Hall signal as sensor is rotated in a magnetic field
Hallchart2.vi	Hall signal as 'real' sensor is rotated in a magnetic field
Motor.vi	Magnetic Field of a muffin fan
Hall.dat	*Spreadsheet data file: Hall probe measurements*

Chapter 3 Random Numbers

DAC0.vi	8-bit DAC (subVI used in Shift, Rotate and PRNG)
Encoder.vi	LabVIEW simulation: A Die
Heads.vi	LabVIEW simulation: Flipping a coin
Heads2.VI	Counting the number of heads in N coin tosses
PRNG.vi	Pseudo-Random Number Generator
PRND2.vi	Calling PRNG N times
Random.vi	Chart output of LabVIEW's Random function
Random6.vi	Modulo 6 counter with start/stop switch
Roll'em.vi	LabVIEW simulation: A Virtual Die
Rotate.vi	Visualization: The Rotate operation
Shift.vi	Visualization: The Shift operation
Star.vi	Catch a Falling Star
Time.vi	Program to determine execution time of subVIs

Chapter 4 Calling all Ports

7-segment.vi	LabVIEW simulation: 7 Segment Display
Encoder.vi	Numeric to 7 segment display encoder
Hex.vi	Hexidecimal to 7 segment display encoder
Phase.vi	LabVIEW encoder for stepping motor phases
Polarplot.vi	'Radar' type plot (subVI)
SteppingMotor.vi	LabVIEW driver for a four phase stepping motor
Stop.vi	LabVIEW simulation: Two way Stop Lights Intersection
Up/Down.vi	LabVIEW simulation: Up/down counter (subVI)

Chapter 5 Serial Communication

Pport.vi	Sub-VI used in Pport8.vi
Pport1.vi	LabVIEW driver for microcontroller (single bit)

Pport8.vi	LabVIEW driver for microcontroller (eight bits)
RS232.vi	LabVIEW driver for RS232 serial port
Selector.vi	Selects V,S,T or R using nested case structures(subVI)
Temp Controller.vi	LabVIEW simulation: Temperature controller
Temp_data.vi	LabVIEW simulation: Temperature measurement

Chapter 6 String Along with Us

Ascii Read.vi	ASCII to numeric command processor (subVI)
Ascii Read2.vi	ASCII to numeric (subVI used in XYPlotter2.vi))
Build String.vi	Demonstration: Building a String Message
Joystick.vi	Virtual Joystick (subVI used in XYPlotter.vi)
Lumin.vi	Chemical Luminescence Sensor (subVI)
Parse String.vi	Demonstration: Extracting character(s) from a string
PlotLum.vi	LabVIEW simulation: Chemical Luminescence
XYPlot.vi	LabVIEW simulation: Digital plotter
XYPlotter1.vi	Digital plotter with Home and Move commands
XYPlotter2.vi	Digital plotter with Home, Move and Draw commands
XYPlotter3.vi	Digital plotter with Joystick controller

Chapter 7 Arrays of Light

Arc.vi	Plots the locus of the points that define an arc
Circle1.vi	Plots a circle or part of a circle
Counts6.vi	Visualization: The number of 1's – 6's in N die rolls
Exp2.vi	Heating Curve measured with a Pt Resistance thermometer
Polarplot.vi	'Radar' plot (subVI)
Polyfit.vi	Second order polynomial fit to Pt calibration data
Pt Term.vi	LabVIEW simulation: Heating curve (subVI)
Random6.vi	Modulo 6 counter (0-5) .
Pt Calibration.data	*Spreadsheet file: Platinum resistance data*

Chapter 8 Some Like It Hot

Coffee+.vi	Data set for coffee cup with lid
Coffee.vi	Data set for coffee cup without lid

Cooling.vi	Cooling curve for coffee.vi data set
Polyfit.vi	Sixth order fit to Chromel/constantan thermocouple data set
Dark.data	*Spreadsheet data: Dark clothing heating curve*
Light.data	*Spreadsheet data: Light clothing heating curve*
Chrom/Const.data	*Calibration data set for chromel/constantan TC*

Chapter 9 IR Communications

IR Sensing.vi	Proximity sensor using ultrasonic ranging sensor
Clear Buffer.vi	Clears the serial input buffer (subVI)

IR Chat.llb

Chat.vi	LabVIEW Chat line using IR transmitter/receiver modules
Clear buffer.vi	Clears the serial input buffer (subVI)
Get string.vi	Gets a string on the serial input buffer (subVI)
Send string.vi	Sends a string to the serial output buffer (subVI)

Friend/Foe.llb

Clear Buffer.vi	Clears the serial input buffer (subVI)
DAC.vi	8-bit Digital-to-Analog converter (subVI)
FF Answer.vi	Friend/For response program
FF ASK.vi	Friend/Foe send program
Friend/Foe Sim.vi	Runs friend/foe program on one computer*
Num to Boolean.vi	8-bit Analog-to-Digital converter (subVI)
PRNG.vi	Pseudo-random number generator (subVI)
Receive string.vi	Gets a string on the serial input buffer (subVI)
Send string.vi	Sends a string to the serial output buffer (subVI)

Chapter 10 The Barometer

Av3.vi	Averages three inputs (subVI)
Barometer1.0.vi	Electronic Barometer with logging
Barometer2.0.vi	Weather Predicting Barometer
Barometer2.1.vi	Weather Predicting Barometer (new front panel)
Barometer2.2.vi	Weather Predicting Barometer (another front panel)

* - Assumes the system clock has the format d/m/y

Better Weather.vi	Predicts the arrival of better weather (subVI)
HiLo.vi	Window Comparator
Pressure Range.vi	Displays the conventional pressure ranges
Pressure Record.vi	Chart display of simulated weather patterns
Pressure.vi	LabVIEW simulation: Pressure sensor
Slope.vi	Calculates the rate of change of pressure (subVI)
Smoothing.vi	Smooths a data set using a 3 point average (subVI)
Storm Warning.vi	Predicts the arrival of stormy weather (subVI)

Chapter 11 Video Surveillance

Client.vi	TCP/IP Client to control Joy Stick
Controller Simulator.vi	Latched Joy Stick
Joy Stick.vi	Virtual Joy Stick (subVI)
Latch.vi	Data latch using Case structure (subVI)
Pan/Tilt Sim.vi	LabVIEW simulation: Pan & tilt operations
Relay0.vi	Pan/tilt relays (subVI used in Client.vi)
Relays.vi	LabVIEW simulation: Pan/tilt relays
Server.vi	TCP/IP Server for pan/tilt controller
Sticky Limits.vi	Detects the max/min limits of pan/tilt position
String Convert.vi	Converts ASCII strings to numerics

Chapter 12 Beer's Law

Beer's Law.vi	Beer's Law Analysis
Help2.vi	Help VI uses dialog boxes (subVI)
Make Graph.vi	Linear Fit to Calibrated Values
Math.vi	Calculates parameter (I-Ib)/Io
Read Calibrate.vi	Gets calibration data from an array
Read Initialize.vi	Gets initialization data from an array
Read Unknown.vi	Gets unknown measurement from an array

Chapter 13 Hunt for Red October

Scan.llb

Area Scan.vi	LabVIEW Simulataion: MAD Detection
Convert.vi	Adds the Scaling Factor into X & Y Coordinates

Get(V).vi	SubVI: Gets voltage from DAQ card
Inc Translate.vi	SubVI: Generates voltage on DAQ card
Sample.VI	SubVI: Measures voltage and Increments XorY
Sliders.vi	Method to enter Scan Profile
XYPlotter.vi	SubVI: Simulation of XY recorder
Y->Y'.vi	SubVI: Incremental Move in Y
Magnetic Field.data1	*Spreadsheet Data File: Magnet Field Profile1*
Magnetic Field.data2	*Spreadsheet Data File: Magnet Field Profile2*
Red October.data	*Spreadsheet Data File: Magnet Field Profile3*

Magnet.llb

AI from DAQ 1chan	LabVIEW DAQ VI
Convert.vi	Adds the Scaling Factor into X & Y Coordinates
Get(V).vi	SubVI: Gets voltage from DAQ card
Initialize2.vi	Sets up scan coordinates to match intensity plot coordinates
Measure voltage.vi	Gets the Hall Voltage from DAQ card
Replace array.vi	SubVI: replays array row with new row
Reset.vi	Set new scan location
Round down.vi	Round off subVI
Sample.vi	SubVI: Measures voltage and Increments XorY
X-line scan.vi	Measure and step in X N times
X-Y Scan.vi	Main Scanning Program (Requires DAC card and Interfaces)
X measure+increment.vi	Measure and step in X
X-increment.vi	Sends X increment to DAC
Y->Y'.vi	SubVI: Incremental Move in Y
Y measure+increment.vi	Measure and step in Y
Y-increment.vi	Sends Y increment to DAC

Chapter 14 Electronic Compass

Counter.vi	Simlates the action of an up/down counter
Direction.vi	Simulates CW or CCW direction (sub VI)
Direction2.vi	Selects CW or CCW direction on DAQ card
GraphMR.vi	Magnetoresistance data as sensor is rotated
MRdata.vi	Generates Magnetoresistance Sensor data
Polar to XY.vi	Polar to rectangular coordinate transformation
Polarplot.vi	Polar plot (sub VI)

Read Data.vi	Reads voltage on DAQ card
Smooth.vi	Smoothing algorithm (sub VI)
Smooth5.vi	Smoothing algorithm using 5 points
SS(+chart)1.vi	Simulated Scan with chart output of MR signal
SS(+polar).vi	Simulated Scan with polar plot output
SS(+real t)2.vi	Simulate Scan using DAQ card
SS.vi	Simulate Scan of MR sensor
Step Motor.vi	Simulates rotating sensor in magnetic field
Step.vi	Simulates a rotation step (sub VI)
Step2.vi	Sets a rotation step on DAQ card

Chapter 15 Who has Seen the Light?

DiodeExp.vi	Measures the IV Characteristics of a red LED
DiodeExp2.vi	IV Characteristics of unknown LED
DMM.vi	Simulation: Digital MultiMeter
DVI.vi	Simulation: Digital Voltage/Current meter
Fluke1.vi	Simulation: Fluke 45 DMM
Fluke2.vi	Simulation: Fluke 45 DMM
GPIB_Driver.vi	Driver for IEEE 488 instruments
Ivchar.vi	Generates the IV red LED data set (subVI)
Ivchar2.vi	Generates the unknown data set (subVI)
IVTable2.vi	Unknown data array
Osc.vi	Simulates an oscilloscope display
RedLED.vi	Red LED data array
String->Num.vi	Convert a string variable to numeric variable

Index

A

B

C

D

E

F

G

H

I

J

K

L

M

N

O

P

R

S

T

U

V

W

X

Y

LICENSE AGREEMENT AND LIMITED WARRANTY

READ THE FOLLOWING TERMS AND CONDITIONS CAREFULLY BEFORE OPENING THIS DISK PACKAGE. THIS LEGAL DOCUMENT IS AN AGREEMENT BETWEEN YOU AND PRENTICE-HALL, INC. (THE "COMPANY"). BY OPENING THIS SEALED DISK PACKAGE, YOU ARE AGREEING TO BE BOUND BY THESE TERMS AND CONDITIONS. IF YOU DO NOT AGREE WITH THESE TERMS AND CONDITIONS, DO NOT OPEN THE DISK PACKAGE. PROMPTLY RETURN THE UNOPENED DISK PACKAGE AND ALL ACCOMPANYING ITEMS TO THE PLACE YOU OBTAINED THEM FOR A FULL REFUND OF ANY SUMS YOU HAVE PAID.

1. **GRANT OF LICENSE:** In consideration of your payment of the license fee, which is part of the price you paid for this product, and your agreement to abide by the terms and conditions of this Agreement, the Company grants to you a nonexclusive right to use and display the copy of the enclosed software program (hereinafter the "SOFTWARE") on a single computer (i.e., with a single CPU) at a single location so long as you comply with the terms of this Agreement. The Company reserves all rights not expressly granted to you under this Agreement.

2. **OWNERSHIP OF SOFTWARE:** You own only the magnetic or physical media (the enclosed disks) on which the SOFTWARE is recorded or fixed, but the Company retains all the rights, title, and ownership to the SOFTWARE recorded on the original disk copy(ies) and all subsequent copies of the SOFTWARE, regardless of the form or media on which the original or other copies may exist. This license is not a sale of the original SOFTWARE or any copy to you.

3. **COPY RESTRICTIONS:** This SOFTWARE and the accompanying printed materials and user manual (the "Documentation") are the subject of copyright. You may not copy the Documentation or the SOFTWARE, except that you may make a single copy of the SOFTWARE for backup or archival purposes only. You may be held legally responsible for any copying or copyright infringement which is caused or encouraged by your failure to abide by the terms of this restriction.

4. **USE RESTRICTIONS:** You may not network the SOFTWARE or otherwise use it on more than one computer or computer terminal at the same time. You may physically transfer the SOFTWARE from one computer to another provided that the SOFTWARE is used on only one computer at a time. You may not distribute copies of the SOFTWARE or Documentation to others. You may not reverse engineer, disassemble, decompile, modify, adapt, translate, or create derivative works based on the SOFTWARE or the Documentation without the prior written consent of the Company.

5. **TRANSFER RESTRICTIONS:** The enclosed SOFTWARE is licensed only to you and may not be transferred to any one else without the prior written consent of the Company. Any unauthorized transfer of the SOFTWARE shall result in the immediate termination of this Agreement.

6. **TERMINATION:** This license is effective until terminated. This license will terminate automatically without notice from the Company and become null and void if you fail to comply with any provisions or limitations of this license. Upon termination, you shall destroy the Documentation and all copies of the SOFTWARE. All provisions of this Agreement as to warranties, limitation of liability, remedies or damages, and our ownership rights shall survive termination.

7. **MISCELLANEOUS:** This Agreement shall be construed in accordance with the laws of the United States of America and the State of New York and shall benefit the Company, its affiliates, and assignees.

8. **LIMITED WARRANTY AND DISCLAIMER OF WARRANTY:** The Company warrants that the SOFTWARE, when properly used in accordance with the Documentation, will operate in substantial conformity with the description of the SOFTWARE set forth in the Documentation. The Company does not warrant that the SOFTWARE will meet your requirements or that the operation of the SOFTWARE will be uninterrupted or error-free. The Company warrants that the media on which the

SOFTWARE is delivered shall be free from defects in materials and workmanship under normal use for a period of thirty (30) days from the date of your purchase. Your only remedy and the Company's only obligation under these limited warranties is, at the Company's option, return of the warranted item for a refund of any amounts paid by you or replacement of the item. Any replacement of SOFTWARE or media under the warranties shall not extend the original warranty period. The limited warranty set forth above shall not apply to any SOFTWARE which the Company determines in good faith has been subject to misuse, neglect, improper installation, repair, alteration, or damage by you. EXCEPT FOR THE EXPRESSED WARRANTIES SET FORTH ABOVE, THE COMPANY DISCLAIMS ALL WARRANTIES, EXPRESS OR IMPLIED, INCLUDING WITHOUT LIMITATION, THE IMPLIED WARRANTIES OF MERCHANTABILITY AND FITNESS FOR A PARTICULAR PURPOSE. EXCEPT FOR THE EXPRESS WARRANTY SET FORTH ABOVE, THE COMPANY DOES NOT WARRANT, GUARANTEE, OR MAKE ANY REPRESENTATION REGARDING THE USE OR THE RESULTS OF THE USE OF THE SOFTWARE IN TERMS OF ITS CORRECTNESS, ACCURACY, RELIABILITY, CURRENTNESS, OR OTHERWISE.

IN NO EVENT, SHALL THE COMPANY OR ITS EMPLOYEES, AGENTS, SUPPLIERS, OR CONTRACTORS BE LIABLE FOR ANY INCIDENTAL, INDIRECT, SPECIAL, OR CONSEQUENTIAL DAMAGES ARISING OUT OF OR IN CONNECTION WITH THE LICENSE GRANTED UNDER THIS AGREEMENT, OR FOR LOSS OF USE, LOSS OF DATA, LOSS OF INCOME OR PROFIT, OR OTHER LOSSES, SUSTAINED AS A RESULT OF INJURY TO ANY PERSON, OR LOSS OF OR DAMAGE TO PROPERTY, OR CLAIMS OF THIRD PARTIES, EVEN IF THE COMPANY OR AN AUTHORIZED REPRESENTATIVE OF THE COMPANY HAS BEEN ADVISED OF THE POSSIBILITY OF SUCH DAMAGES. IN NO EVENT SHALL LIABILITY OF THE COMPANY FOR DAMAGES WITH RESPECT TO THE SOFTWARE EXCEED THE AMOUNTS ACTUALLY PAID BY YOU, IF ANY, FOR THE SOFTWARE.

SOME JURISDICTIONS DO NOT ALLOW THE LIMITATION OF IMPLIED WARRANTIES OR LIABILITY FOR INCIDENTAL, INDIRECT, SPECIAL, OR CONSEQUENTIAL DAMAGES, SO THE ABOVE LIMITATIONS MAY NOT ALWAYS APPLY. THE WARRANTIES IN THIS AGREEMENT GIVE YOU SPECIFIC LEGAL RIGHTS AND YOU MAY ALSO HAVE OTHER RIGHTS WHICH VARY IN ACCORDANCE WITH LOCAL LAW.

ACKNOWLEDGMENT

YOU ACKNOWLEDGE THAT YOU HAVE READ THIS AGREEMENT, UNDERSTAND IT, AND AGREE TO BE BOUND BY ITS TERMS AND CONDITIONS. YOU ALSO AGREE THAT THIS AGREEMENT IS THE COMPLETE AND EXCLUSIVE STATEMENT OF THE AGREEMENT BETWEEN YOU AND THE COMPANY AND SUPERSEDES ALL PROPOSALS OR PRIOR AGREEMENTS, ORAL, OR WRITTEN, AND ANY OTHER COMMUNICATIONS BETWEEN YOU AND THE COMPANY OR ANY REPRESENTATIVE OF THE COMPANY RELATING TO THE SUBJECT MATTER OF THIS AGREEMENT.

Should you have any questions concerning this Agreement or if you wish to contact the Company for any reason, please contact in writing at the address below or call the at the telephone number provided.

Robin Short
Prentice Hall PTR
One Lake Street
Upper Saddle River, New Jersey 07458

About LabVIEW

IN THIS CD-ROM YOU WILL FIND

The CD-ROM included with this book contains an evaluation version of LabVIEW for Windows 98/95/NT and PowerPC (MacOS). If you do not have the full version of LabVIEW already installed on your computer, you can use the evaluation version.

HOW TO INSTALL THE SOFTWARE

Install the LabVIEW evaluation software on your computer. To install this software, run the setup.exe program from the LabVIEW folder on the CD-ROM. Follow the instructions on the screen. If you already have the full version of LabVIEW installed, you do not need to install the evaluation version.

SYSTEM REQUIREMENTS

Mac OS (PowerPC only)	12 MB RAM, 16 MB recommended
	Mac OS 7.1.2 or later
	75 MB disk space for full installation
	CD ROM Drive
Windows NT	16 MB RAM, 32 MB recommended
	486 DX, Pentium recommended
	Windows NT 4.0
	75 MB disk space for full installation
	CD ROM Drive
Windows 95	16 MB RAM, 32 MB recommended
	486 DX, Pentium recommended
	75 MB disk space for full installation
	CD ROM Drive

IN ALL THE ABOVE CASES TO START THE INSTALLATION YOU MUST HAVE **100 MB** FREE ON YOUR HARD DRIVE EVEN THOUGH AFTER INSTALLATION IT MAY NOT REQUIRE 100 MB. USERS WHO WANT A COPY OF THE SENSORS, TRANSDUCERS AND LABVIEW LIBRARY VERSION 3.1 SHOULD GO TO "sensor.phys.dal.ca" FOR DOWNLOADING INSTRUCTIONS.

RESTRICTIONS ON THE LABVIEW EVALUATION VERSION

To run the evaluation version, launch the Labview.exe program from the folder in which you installed LabVIEW. Select the Exit to LabVIEW button in the lower right corner. This opens a window which gives you general information about LabVIEW and National Instruments. After reading this information, click on the OK button. This will then open the LabVIEW window. You can access all the features of the full version in the evaluation version with following restrictions

1. After you have installed the evaluation version, it will expire in 30 days. If you want to use the evaluation version after this time, you must reinstall the software.
2. A VI will execute for 5 minutes at any one time. After 5 minutes, you will have to run the VI again. In a single session, the total execution time is limited to 60 minutes. You can edit the block diagrams and front panels after this time-out. However, in order to run any VI, you will have to start LabVIEW again.
3. The evaluation version does not come with tools to build your own Code Interface Nodes (CINs) or with tools to build DLLs that can call back into LabVIEW.

FOR MORE INFORMATION ABOUT THE CD-ROM

For more information about the CD-ROM see pages xxvii and xxviii of the Preface.